Mathematics and Its Logics

In these essays Geoffrey Hellman presents a strong case for a healthy plural-ism in mathematics and its logics, supporting peaceful coexistence despite what appear to be contradictions between different systems, and positing different frameworks serving different legitimate purposes. The essays refine and extend Hellman's modal-structuralist account of mathematics, develop-ing a height-potentialist view of higher set theory which recognizes indefinite extendability of models and stages at which sets occur. In the first of three new essays written for this volume, Hellman shows how extendability can be deployed to derive the axiom of Infinity and that of Replacement, improving on earlier accounts; he also shows how extendability leads to attractive, novel resolutions of the set-theoretic paradoxes. Other essays explore advantages and limitations of restrictive systems – nominalist, predicativist, and con-structivist. Also included are two essays, with Solomon Feferman, on pre-dicative foundations of arithmetic.

GEOFFREY HELLMAN is Professor of Philosophy at the University of Minnesota, Twin Cities. His publications include *Mathematics without Numbers: Towards a Modal-Structural Interpretation* (Oxford, 1989), *Varieties of Continua: From Regions to Points and Back* (with Stewart Shapiro, Oxford, 2018), and *Mathematical Structuralism*, Cambridge Elements in Philosophy of Mathematics (with Stewart Shapiro, Cambridge, 2018).

Mathematics and Its Logics

Philosophical Essays

Geoffrey Hellman
University of Minnesota

CAMBRIDGE
UNIVERSITY PRESS

CAMBRIDGE
UNIVERSITY PRESS

University Printing House, Cambridge CB2 8BS, United Kingdom

One Liberty Plaza, 20th Floor, New York, NY 10006, USA

477 Williamstown Road, Port Melbourne, VIC 3207, Australia

314–321, 3rd Floor, Plot 3, Splendor Forum, Jasola District Centre,
New Delhi – 110025, India

79 Anson Road, #06–04/06, Singapore 079906

Cambridge University Press is part of the University of Cambridge.

It furthers the University's mission by disseminating knowledge in the pursuit of
education, learning, and research at the highest international levels of excellence.

www.cambridge.org
Information on this title: www.cambridge.org/9781108494182
DOI: 10.1017/9781108657419

© Geoffrey Hellman 2021

First published 2021

A catalog record for this publication is available from the British Library.

ISBN 978-1-108-49418-2 Hardback

Contents

Acknowledgements

Chapters 4, 10, and 14 are new and were written specifically for this volume. The remaining chapters originally appeared in the following locations. Permission to reprint these essays is gratefully acknowledged.

1 "Structuralism without structures," *Philosophia Mathematica* **4**(2) (1996): 100–123.

2 "What is categorical structuralism?," in van Benthem, J., Heinzmann, G., Rebuschi, M., and Visser, H., eds., *The Age of Alternative Logics: Assessing Philosophy of Logic and Mathematics Today* (Dordrecht: Springer, 2006), pp. 151–161.

3 "On the significance of the Burali-Forti paradox," *Analysis* **71**(4) (2011): 631–637.

5 "On nominalism," *Philosophy and Phenomenological Research LXII*(3) (2001): 691–705.

6 "Maoist mathematics? Critical study of John Burgess and Gideon Rosen, *A Subject with No Object: Strategies for Nominalist Interpretation of Mathematics* (Oxford, 1997)," *Philosophia Mathematica* **6** (1998): 334–345.

7 "Predicative foundations of arithmetic," *Journal of Philosophical Logic* **24** (1995): 1–17 (with Solomon Feferman).

8 "Challenges to predicative foundations of arithmetic," in Sher, G. and Tieszen, R., eds., *Between Logic and Intuition: Essays in Honor of Charles Parsons* (Cambridge: Cambridge University Press, 2000), pp. 317–338 (with Solomon Feferman).

9 "Predicativism as a philosophical position," *Revue Internationale de Philosophie* **58**(3) (2004): 295–312.

11 "Logical truth by linguistic convention," in Hahn, L. E. and Schilpp, P. A., eds., *The Philosophy of W. V. Quine* (La Salle, IL: Open Court, 1986), pp. 189–205.

12 "Never say 'never'! On the communication problem between intuitionism and classicism," *Philosophical Topics* **17**(2) (1989): 47–67.

13 "Constructive mathematics and quantum mechanics: unbounded operators and the spectral theorem," *Journal of Philosophical Logic* **22**(3) (1993): 221–248.

15 "Mathematical pluralism: the case of smooth infinitesimal analysis," *Journal of Philosophical Logic* **35** (2006): 621–651.

Introduction

Abstract mathematics, from its earliest times in ancient Greece right up to the present, has always presented a major challenge for philosophical understanding. On the one hand, mathematics is widely considered a paradigm of providing genuine knowledge, achieving a degree of certainty and security as great as or greater than knowledge in any other domain. A part of this, no doubt, is that it proceeds by means of deductive proofs, thereby inheriting the security of necessary truth preservation of deductive logical inference. But proofs have to start somewhere: ultimately there need to be axioms, and these are the starting points, not end points, of logical inference. But what then grounds or justifies axioms? The question becomes especially urgent when it is considered that the subject matter of pure mathematics, including its axioms, apparently consists of *abstracta* such as numbers, functions, classes, and relations, which are non-spatiotemporal and do not enter into causal interactions (with us or anything else). This is true even for Euclidean geometry, which originally was conceived as investigating properties of actual physical space and time, but nevertheless treats directly of dimensionless points, perfectly straight lines of 0 breadth, ideal perfect figures such as triangles and rectangles, etc., none of which exist in the material world.

In our own recent history, the logical empiricists, led by Rudolf Carnap and inspired by Gottlob Frege, proffered the doctrine of *analyticity*, that mathematical axioms are "analytic," but now in the sense of "guaranteed true solely in virtue of meanings of terms," or "true entirely by linguistic convention."[1] This was regarded as compatible with the obvious fact that theorems could be highly surprising and informative, as logical deductions can be highly complex and intricate, so that logical consequences of given axioms may appear quite unpredictable.

---

[1] Though Frege explicitly declared in the *Grundlagen der Arithmetik* that (in effect, second-order) arithmetic was "analytic," this was explained as "derivable from logic," whereas Frege's "logic" was nowhere said to be "true by linguistic convention." Indeed, not only was that logic committed to many infinitistic functions, it was also proved inconsistent due to implying Russell's paradox.

Now, as is well known, the doctrine of analyticity was severely critiqued by W. V. Quine (and others), due to the questionable scientific status of the concept of linguistic "meaning." But even granting – for the sake of argument – the scientific status of meaning concepts, there remains the problem that at least some of the axioms of going mathematics just seem not to qualify as "analytic" at all. Famously, the Euclidean parallels postulate (EPP) seems not to qualify, resisting attempts to derive it from anything more basic. And the case seemed to be sealed when, in the nineteenth century, it was discovered that the EPP could consistently be negated, giving rise to genuine non-Euclidean geometries, provably consistent relative to Euclidean geometry.

Now it may occur to the reader that one could reinterpret the EPP as claiming only what is true in a genuinely Euclidean space, thereby ensuring its analyticity, as a space would be "Euclidean" only if it satisfied the EPP. But now we have to accept the existence of a (possibly non-physical) Euclidean space, and what yields the analyticity of *that*? Indeed, that involves the existence of an infinite totality, and how can that be guaranteed true solely by the meanings of words?

Indeed, for another striking example of a non-analytic axiom, consider the Axiom of Infinity, that there exists an infinite set. Any attempt to derive this appears quite circular. For example, Dedekind thought he could derive it from reflecting on his capacity to entertain the thought of any thought that he could entertain. But this failed, not only for the reason that it is dubious that we can even understand enough iterates of "the thought of . . . the thought of my own ego (or whatever the initial object may be)," but because Dedekind needed to assume that "all objects that could be objects of [his] thought" form a "system" or set. But what guarantees this? Perhaps all subsets of objects of his thought could be objects of his thought, in which case Cantor's theorem (that there are always more subsets of a given set than members of it) would rule out that all objects of his thought form a set, as that set would have all its subsets as members, which is a contradiction!

The famous Axiom of Choice of set theory provides another example of an axiom whose "analyticity" seems impossible to secure.[2]

Such examples led to a view known as "if-thenism" or "deductivism" (espoused by Russell [1903], and later, at least in part, by Hempel [1945]), according to which mathematics need not assert its axioms and can confine itself to conditional claims of the form, "If these axioms, then this theorem." (A recent version of this is examined critically in some detail in the essay of Chapter 14.)

---

[2] The Axiom of Choice says that, given any set S of non-empty sets, s, there exists a "choice function" f on S, i.e. such that the value $f(s)$ of f is a member of s. Such a choice function on any infinite S "does an infinite amount of choosing at once" and is thus itself an infinitistic object.

Thus, even in the late twentieth century, the challenge regarding philosophical understanding presented by mathematics was far from being met. One overarching question addressed by a number of the essays of this collection is to what extent the rise of modern structuralism – in particular modal-structuralism (MS), begun by Putnam [1967] and further developed by me (Hellman [1989] and Hellman [1996], reproduced as Chapter 1 in this volume) – makes significant progress toward meeting this challenge.

The crucial starting point of modal-structuralism is to adopt the Dedekindian-Hilbertian view of mathematical axioms as *defining conditions* on types of mathematical structures of interest, rather than as asserted truths outright (as in the traditional Euclidean-Fregean view). The next step is to interpret ordinary mathematical statements S of a branch of mathematics as stating what would necessarily hold of structures of the appropriate type that there might be (logically speaking), structures characterized by the axioms of the branch of mathematics in question. Thus S is construed as a quantified modal conditional whose antecedent is the conjunction of the relevant axioms (relativized to an arbitrary given domain) and whose consequent is S (with its quantifiers restricted to the given domain). If that were all, then we would have a second-order logical version of if-thenism; and then there would be the problem that, if the axioms were inconsistent, then any such conditional would count as true or valid, regardless of the consequent – an intolerable situation. In order to block this "problem of vacuity," a further step is required, viz. to assert the (second-order logical) possibility of there being a structure fulfilling the relevant axioms (hence ruling out inconsistency), what we have called "the categorical component" of a modal-structural interpretation (the quantified conditionals constituting "the hypothetical component"). And it is just here that the logicist idea, that all mathematical truths are analytic, breaks down, for these modal existence postulates cannot be determined true solely in virtue of the meanings of the words involved. (Compare the ontological argument, which purports to "prove" the existence of a necessary, perfect being by counting existence as a perfection! True enough, *if* G is a necessary, perfect being, *then* G exists. But nothing guarantees that there actually *are* any necessary, perfect beings. Moral: you cannot *define* objects into existence!)

At the time that I was developing MS, I was unaware of the full potential of some crucial logical machinery already developed by Boolos [1985], that of plural quantification. Specifically, as shown by Burgess, Hazen, and Lewis [1991] (BHL) in their Appendix to Lewis [1991], the *combination* of plural quantification and atomic mereology (but neither separately) enables explicit general constructions of ordered-pairing, which achieves the expressive power of a full classical theory of relations (polyadic second-order logic). In the absence of this, Hellman [1989] got by with some rather ad hoc machinery to achieve the requisite expressive power to carry out MS interpretations. But the

BHL constructions afford a more general and smoother development, as explained in the first essay of this volume, "Structuralism without Structures," Chapter 1.

As the reader may well know, there are several other structural approaches to mathematics besides modal-structuralism. Indeed, set theory and category theory have each provided such from within mathematics itself. (For a systematic comparison of these along with Shapiro's *sui generis* structuralism, see Hellman and Shapiro [2019], also Hellman [2005].) Set-theoretic structuralism (STS), based on model theory, is probably the best known of all, and for much mathematics it does very well. There are, however, two main problems with it. First, it fails to treat set theory itself structurally, despite the fact that there are multiple, conflicting but perfectly legitimate set theories deserving of recognition. But second, and most important here, STS based on the Zermelo–Fraenkel axioms, with Choice (ZFC), as usually understood, is committed to a unique, maximal universe of "all sets," despite the fact that the very notion of "set" is indefinitely extensible, as are the notions of "ordinal" and "cardinal." That is, in the case of "set," by our understanding of sets as subject to certain operations, any totality of sets can be transcended by means of those operations, for instance the operation of forming singletons or forming powers (that is, forming the totality of all subtotalities). Thus, it is a central postulate of the modal-structural interpretation of set theory that any domain of sets (or, more neutrally, "set-like objects," objects obeying the axioms of ZFC) can be properly extended to a more comprehensive domain. We call this the Extendability Principle (EP). This applies also to category theory's (CT's) version of set theory, as explained in Chapter 2, and it rules defective commitment to a category of "all sets."

The essay of Chapter 2, "What is categorical structuralism?" assesses responses by Colin McLarty and Steve Awodey to my earlier critique of category theory's approach to mathematical structuralism.[3] There I had complained about the lack of assertory axioms governing existence of category-theoretic structures, and McLarty had countered that two axiom systems due to Lawvere met my concerns, axioms on the (sic) category of sets and axioms on the (sic) category of categories. On the other hand, Awodey had maintained that such axioms are unnecessary, that CT can get by with an entirely schematic approach to mathematical structures. The essay of Chapter 2 exposes a common problem with both of these approaches, viz. that both implicitly rely on some concept of *satisfaction* (of sentences by systems of objects), usually articulated via set theory, although second-order logic can be used

---

[3] References to McLarty and Awodey are given in Chapter 2.

instead. Thus CT's autonomy, as a foundational framework, from set theory or second-order logic has yet to be established by its proponents.[4]

Returning now to our theme of indefinite extensibility: in contrast to both set theory and category theory, as usually understood, the MS approach adopts a "height-potentialist" perspective, based on the EP framed modally. As Putnam [1967] forcefully put it, "Even God could not make a model of Zermelo set theory that it would be *mathematically* impossible to extend" (p. 310). But since MS allows modal quantification over arbitrary set-like objects – for example "for any set that there might be, there would also be its power set," etc. – what blocks commitment to a totality of "all possible sets or set-like objects"? Since MS eschews possible worlds or *possibilia* of any sort, the answer is that collections can only contain what would exist under given circumstances, not anything that merely *might* then have existed. Invoking "worlds" as heuristic only, we can say that sets or collections or set-like objects are "world-bound." It is impossible to form collections "across worlds." It literally makes no sense to speak of "the collection of all possible set-like objects."[5] Thus, in contrast to Zermelo's [1930] effort to articulate a height-potentialist view (which did not employ modal operators), MS naturally avoids commitment to "proper classes" or "ultimate infinities" in an absolute sense. The notion of "proper class" can only be *relative to a domain*: what qualifies as a proper class (hence not a member of anything) relative to domain D, functions as a *bona fide* set relative to any possible proper extension of D; and there can be no "union of all possible domains."[6]

The essays of Chapters 3 and 4 develop some important consequences of the height-potentialist view just sketched. Chapter 3 describes the MS resolutions of the set-theoretic paradoxes, concentrating on the so-called Burali-Forti paradox, of "the largest ordinal." There it is shown how, in a potentialist sense, MS can respect the desideratum that any well-order relation whatever can have an ordinal representing it. This is in contrast to standard resolutions of Burali-Forti based on a single fixed universe of "all sets." Chapter 4 then shows how a natural modal principle on the extendability of "stages" of sets on the well-known iterative conception of set leads to new derivations of the axioms of Infinity and Replacement, not available to Boolos' original [1971]

---

[4] Our analysis thus sustains the well-known earlier analysis by Feferman [1977], but focuses on the problem of articulating structuralism rather than foundations generally.

[5] Here we follow Kripke's [1980] "actualist" conception of the alethic modalities, as contrasted with Lewis' [1986] "possibilist" or "modal-realist" conception.

[6] While Zermelo [1930] did clearly state that the "set/proper class" distinction is relative to a domain, still that work is naturally formalized in axiomatic second-order logic, an axiom of which guarantees a class of all members of any domains, an ultimate proper class, contrary to Zermelo's (non-modal) Extendability Principle. In avoiding this "explosion," modality does essential conceptual work for the MS interpretation.

stage theory, or any other height-actualist theory recognizing a plurality of "all stages."

Thus far, we have seen that MS requires as assertory axioms statements affirming the possibility of structures of the appropriate type, geared to the mathematical axioms defining that type of structures. Thus, both Hilbertian-style and Fregean-style axioms are needed. But we have not yet said anything explicitly about the ontological commitments of mathematical theories interpreted according to MS. Take the most elementary case of arithmetic. Does the MS interpretation quantify over numbers as objects? No; all it requires is that there possibly be a progression; it makes no difference at all what objects make up the progression. All that matters is that, whatever the objects, they be arranged in the right way, as required by the Dedekind–Peano axioms. Thus, on interpretation, the predicate "__is a natural number" is *eliminated*. Similarly for the integers, the rationals, the reals, and the complexes. MS is, after all, "mathematics without numbers" (as explained in detail in Hellman [1989]). Now one of the virtues of the BHL machinery, deploying the logic of plurals combined with atomic mereology, is that it allows us to eliminate even *structures* as objects, in favor of speaking directly of enough objects – of whatever sort – interrelated in the right ways, as dictated by the proper mathematical axioms. Thus, as indicated in the title of Chapter 1, we have a "Structuralism without Structures." This thus raises the prospect that MS may be fully nominalistic, at least in the sense that abstract entities need not be recognized. To a surprising degree, this is correct. But, as will now be explained, it is not *entirely* correct.

Suppose we begin with a postulate asserting the possibility of a countable infinity of mereological atoms, say, satisfying the Dedekind–Peano axioms for natural numbers. Such objects can readily be conceived as part of space or space-time, for example non-overlapping space-time regions. They need not be abstract. Then applying mereology, we have all wholes of such atoms, also not abstract. This gives us a continuum of concrete objects, at the level of the classical real numbers, or classical second-order number theory, arguably enough to support virtually all of scientifically applicable mathematics. But we can go even further, using plural quantification over these real-number surrogates, yielding the equivalent of full, classical third-order number theory, again within the confines of nominalism. Furthermore, now consider that we could start off with a *continuum* of atoms instead of a countable infinity of such. Then applying mereology we obtain all wholes of such atoms, corresponding to all non-empty subsets of the atoms. Finally, with plural quantification over such wholes, we attain the level of third-order real analysis or fourth-order number theory, all within a nominalist framework. Thus, vast amounts of pure and applied mathematics (including e.g. differentiable geometry of Riemannian manifolds, measure theory, and much more) are nominalistically reducible.

(Indeed, we do not even need to invoke modality for this much, as actual space-time regions furnish us with enough objects.)

Such constructions thus serve to undermine the well-known indispensability arguments for the need to recognize mathematical *abstracta* in order to do justice to scientific applications of mathematics. In this sense, a nominalist ontology is enough to support virtually all such applications, depriving platonism of arguably its most powerful argument. Such considerations led us to reassess nominalism in the essay of Chapter 5. They are there coupled with an effort to reformulate nominalism as an epistemic thesis, rather than a strictly ontological one, viz. that there is no compelling evidence or reason for invoking mathematical *abstracta*, appearances to the contrary notwithstanding.

There is, however, a consideration that suggests that that conclusion may go too far. That is that the above constructions do not extend to abstract set theory, even Zermelo set theory, not to say ZFC. While the modal-structural interpretations of those set theories do eliminate the predicate 'is a set', much as 'is a number' is eliminated, still the postulate of the possibility of enough objects to form a model of those set theories goes well beyond the reaches of nominalist ontology, as described above. Not all such objects could be conceived as part of space-time, even a space-time of higher dimension. (Even postulating a continuum of dimensions would not take one far enough!) Yet, as the work of Harvey Friedman suggests, abstract set theory may well be required to solve problems at the level of sets of integers (see the essay, "On the Gödel–Friedman program," of Chapter 10). It is not inconceivable that such problems might even arise within physics. In that case, the Quine–Putnam indispensability argument could be restored. But for now that remains quite speculative. Thus, the thrust of "On nominalism" is that at present, except for achieving a realist understanding of higher set theory, a nominalist ontology qualifies as a default position, trading places with platonist ontologies that have dominated in the past.

Finally, on the topic of nominalism, the essay of Chapter 6, "Maoist mathematics?", which is a critical study of Burgess and Rosen's [1997] book, *A Subject with No Object: Strategies for Nominalist Interpretation of Mathematics*, defends nominalistic reconstruction programs against the charge of facing a dichotomy of either proposing (unjustifiably) to uncover the "real, deeper meaning" of mathematical theories, or (recklessly) advocating a revolutionary revision of mathematics as practiced. We argue that this is a false dichotomy, that the nominalist programs considered by Burgess and Rosen – those due to Field, Chihara, and Hellman – are proposed as neither "hermeneutic" nor "revolutionary," but rather serve as *rational reconstructions* designed to mitigate epistemological and metaphysical problems confronting platonistically construed mathematics.

This brings us to Part II, which consists mainly of essays on predicative mathematics, two of which were done jointly with Solomon Feferman. Now predicative foundations grew out of the work of Russell, viz. his ramified type theory, and writings of Poincaré and Weyl, and was designed to avoid so-called impredicative definitions or specifications of sets, that is specifications by formulas with quantifiers ranging over totalities which include the very set being introduced. For example, consider the classical least upper bound principle governing sets of real numbers. This says that any non-empty bounded set of reals has a least upper bound (or greatest lower one). Specification of such a bound involves a quantifier ranging over all (upper) bounds of the given set, which of course includes the least bound. Thus the principle is called "impredicative." Now classical set theorists have no problem with such specifications, as they regard the totalities of sets involved as objectively existing, independent of their being picked out by our languages. But those with constructivist inclinations find such specifications viciously circular (recall Russell's "vicious circle principle," essentially a ban on impredicative specifications of sets or other objects). Ultimately, this traces back to skepticism over the power-set operation, passing from an infinite set to the set of *all* of its subsets. Instead, the predicativist restricts this operation to taking the set of all *definable* subsets of the given (infinite) set, where the definitions lack quantifiers ranging over the very subset being specified.

As another paradigmatic example, consider the classical Fregean and Dedekindian specifications of the totality of natural numbers as the intersection of all classes containing the initial number (0 for Frege, 1 for Dedekind) and closed under the successor operation. Again, this reference to "all classes" includes the very class being introduced; hence the specification is impredicative.

The effect of this ban on impredicative specifications of sets is to avoid uncountably infinite sets, in an absolute sense, as formulas of countable languages are required to specify predicatively subsets of an infinite set. Instead, the predicativist recognizes a kind of *relative* uncountability of the real numbers, based on the negative conclusion of Cantor's diagonalization argument: given any putative enumeration of all the reals, the diagonal argument produces a real that differs from each real of the given enumeration. So far so good. But this is just interpreted to mean that more reals need to be recognized at any stage of construction. The continuum is thus viewed as an incompleteable, indefinitely extensible totality, something like what even some (but not all) classical set theorists recognize as true of the putative universe of "all sets or ordinals."

Two more features of predicative mathematics need to be mentioned here by way of background. The first is that classical logic is accepted, distinguishing predicative from constructive mathematics with its renunciation of the law of excluded middle (framed either classically or with intuitionistic logical connectives). Predicative mathematics is thus known as "semi-constructivist." Second, as usually presented, predicative mathematics begins by taking the natural numbers as given. Predicativity is understood as *relative to the natural numbers*. Thus, Poincaré's misgivings concerning logicism (which sought a logical foundation of arithmetic) are taken to heart: the natural-numbers system is regarded as more fundamental than the full battery of logicist machinery (which included second- and higher-order logical notions, not merely first-order ones).

There arises here, however, a nagging question: given that classical logicist foundations of arithmetic had to resort to impredicative definitions to obtain a natural-numbers structure, as reviewed above, would it not be better for predicative mathematics to begin without taking that structure for granted, but somehow to *derive* it (its existence)? After all, it is an infinite structure of a special type.

As had been pointed out by Dan Isaacson [1987], the framework of "weak second-order logic," with axioms quantifying over *finite sets* as well as individuals, does permit a characterization of a natural-numbers structure, unique up to isomorphism. This suggested to me that it should be possible for predicative mathematics to begin with an elementary theory of finite sets and build up a natural-numbers structure *predicatively relative to the notion "finite set."* Extensive correspondence with Solomon Feferman, leading proof theorist well known for developing predicative mathematics, then resulted eventually in the two papers reprinted here as Chapters 7 and 8. Notably, the theory of finite sets developed there is quite weak, lacking any axiom of finite-set induction, thereby avoiding the charge of circularity of our construction, which effectively derives mathematical induction governing natural numbers. Furthermore, our derivation of existence and unicity of a natural-numbers structure brings out an important difference between the notion of "finite set" as compared with "natural number," namely that finite sets are "self-standing" rather than inherently part of a structure, whereas the opposite is true of "natural number" (especially in Dedekind's sense of "finite ordinal"). In our view, it makes no sense to consider a natural number in isolation from a structure of at least a segment of natural numbers; whereas reference to finite sets makes sense apart from their belonging to a structure of, say, hereditarily finite sets (of some given individuals) ordered by set-inclusion. The upshot is that we provide a predicativist-logicist foundation of arithmetic, thereby meeting Poincaré's challenge.

This brings us to the essay of Chapter 9, "Predicativism as a Philosophical Position," where we assess the philosophical import of predicative foundations. Various limitative theses are examined and found wanting, mainly because their very assertion requires transcending the limits of predicativist mathematics. Instead, we find the main contributions of predicative foundations to be in the area of mathematical epistemology, through its detailed examination of "what rests on what" and "why a little bit goes a long way" (both of which phrases are titles of insightful papers of Feferman).[7] Unlike radical constructivism (examined in Part III), predicativism does not purport to set limits to classical mathematics, but rather seeks to show the sufficiency of predicative methods for the vast bulk of scientifically applicable mathematics. It thus poses a major challenge to the Gödel–Friedman program, which seeks to justify abstract set theory as needed to solve ordinary (or ordinary-appearing) mathematical problems.

The main point made in the last essay (Chapter 10) of Part II, "On the Gödel–Friedman program," is that Bayesian confirmation theory is relevant to meeting this challenge posed by predicative foundations. As explained in that essay, there is a major gap between statements of consistency of large cardinals and statements asserting directly the mathematical existence of such cardinals. Yet it is the former, not the latter, that recent work of Friedman demonstrates equivalent to certain low-level combinatorial statements very similar to statements that are provable without higher set theory (as in Friedman's Boolean Relation Theory (BRT)). As it stands, predicative mathematics is adequate for proving the statements of BRT that have been shown equivalent to statements of consistency of certain large cardinals. However, from the standpoint of ordinary mathematical practice, such consistency statements are arcane renderings of metamathematical content of precisely the kind that the Gödel–Friedman program seeks to improve upon. Our essay sketches how, in principle, a kind of inductive evidence can be gained to support the assertions of mathematical existence of the relevant large cardinals. The latter can thus emerge as the *best explanation* for a variety of independently justifiable consistency statements, in line with Gödel's ideas set out in his well-known [1947] paper, "What is Cantor's continuum problem?"

Turning to the essays of Part III, these focus on various logical systems used in different approaches to mathematics and its foundations. Now, to one who shares the popular misconception of mathematics as a cut-and-dried discipline of universally agreed upon results, it may come as something of a shock to learn that there are actually vastly different "schools" of mathematics favoring even different logics. But that is actually the case, as manifested in the divergence between mainstream classical mathematics based on (first- or higher-order) classical logic on the one hand, and, on the other, various versions of constructive mathematics based on intuitionistic logic, well known for renouncing certain classical logical laws, especially the law of excluded middle (that either p or not-p holds for any mathematical sentence, p) and proof of existence by

---

[7] See Feferman [1998].

reductio ad absurdum (deriving existence of something by proof that the assumption of inexistence leads to a contradiction).

Our perspective on such divergences counsels a healthy pluralism (in the tradition of Carnap), recognizing different legitimate goals of different systems of mathematics and logics. The choice of logic is thus largely a matter of selecting the best means to a given end. If one is concerned to track objective truth-values as in the development of idealized mathematical models of physical systems, then classical logic is a good choice. Its rules are designed to be necessarily truth preserving. But if one seeks to maintain constructivity in one's reasoning, ensuring, for example, that existence statements are backed up by in-principle available methods of construction of relevant objects, then intuitionistic logic is a good choice, as its rules are designed to preserve constructivity. Both sorts of goals are perfectly legitimate in their own right. As Carnap famously put it: "In logic, there are no morals," by which he meant that there are no prohibitions against pursuing diverse goals with different logics, so long as one is clear about what one is doing and how one is proceeding.[8]

The vast bulk of mathematics, historically and in modern times, can be formalized in a system of set theory, such as ZFC, or in category theory, both based on first-order classical logic (with equality as sole predicate constant). But what is the status, regarding meaning and justification, of first-order logic itself? The first essay of Part III, "Logical truth by linguistic convention" (Chapter 11), grapples with this question, via a dialogue between Carnap* and Quine*, broadly representing and developing views of the actual Carnap and Quine on the grounds of (first-order) logical truth. (The point of the *s was to allow myself freedom to modify and develop views of the actual Carnap and Quine without being tied to their own writings.) The upshot is to sustain the Carnapian thesis that (free) first-order logic is indeed *analytic* in the sense of being empty of content since true solely in virtue of the meanings of the logical vocabulary (the sentential connectives, the quantifiers, and the identity sign). On this view, classical first-order logic is a kind of scaffolding that codifies deductive reasoning between contentful premises and conclusions, but in itself asserts nothing and thus requires no justification. What does require justification is the claim, *of* a given classical logical truth (or if-then statement connecting premises with a conclusion) *that* it *is* a logical truth. As Carnap* insists, the use of logical rules to establish this type of claim is perfectly in order and does not lead to an infinite regress of justification, as Quine had argued it does

---

[8] See Carnap [1937], p. 52: "*In logic, there are no morals.* Everyone is at liberty to build up his own logic, i.e. his own form of language, as he wishes. All that is required of him is that, if he wishes to discuss it, he must state his methods clearly, and give syntactical rules instead of philosophical arguments." This was an expression of Carnap's "principle of tolerance," which we here endorse. (Intolerance of misuses of philosophy was, of course, seen as compatible with the principle!)

in his early, influential essay, "Truth by convention" [1936]. Notably, however, Carnap*, unlike the historical Carnap, does *not* claim that mathematics as a whole is analytic.

Now, although Carnap* retracts Carnap's claim that mathematics as a whole is analytic, he clearly does not concede to Quine* – or to Quine – that the very distinction between analytic and synthetic sentences is too unclear to do philosophical work. On the contrary, the case Carnap* mounts for the analytic status of first-order logic obviously depends on the viability of the distinction. Equally clearly, securing the analytic status of first-order logic is of philosophical importance, as it resolves the question of how to explain the *a priori* status of first-order logical knowledge, without invoking special faculties of intuition, for example. Furthermore, the notion of *meaning* of logical words plays a central role in the next essay, "Never say 'never'! On the communication problem between intuitionism and classicism" (Chapter 12), as the first main point there is that there really is no dispute between the two philosophies of mathematics over any logical laws, despite the fact that they are commonly said to disagree over *the* (sic) law of excluded middle (LEM), *the* (sic) law of double negation (DN), and *the* (sic) rule of *reductio ad absurdum* when used to infer the existence of a mathematical object (by deriving a contradiction from the assumption of inexistence). The main claim here is that – based on how leading intuitionists have explained their own usage – the intuitionistic logical terms differ markedly in meaning from their classical counterparts, even though they are often represented by the same symbols. Moreover, as explained in the essay, these differences are closely connected to radically different conceptions of mathematical propositions: the classicist regards them as objectively true or false of actual or possible mathematical objects and structures, whereas the intuitionist interprets them in terms, not of objective truth conditions, but rather via *proof conditions*, conditions of their constructive provability or refutability by an idealized human mathematician. In a sense, then, the differences between their perspectives run even deeper than disagreements over which logical laws to recognize. Indeed, once we appropriately subscript the two sets of logical particles (connectives and quantifiers), we see that they do not even disagree about logical laws at all. Both sides agree, for instance, that the *intuitionistic* version of LEM should not be adopted as a logical law at all (as it expresses that every mathematical sentence is either constructively provable or constructively refutable!). Furthermore, the *radical* constructivist, like Brouwer or Dummett, claims not even to *understand* the classical laws in the domain of pure mathematics, with their commitment to objective truth conditions.

The remainder of "Never say 'never'!" investigates consequences of the meanings of the intuitionistic logical terms regarding what can and cannot be expressed in intuitionistic logical languages. Surprisingly, perhaps, the very motivating statement that "We may never be able to prove or refute some mathematical propositions," turns out to be contradictory when written out in intuitionistic language! This, moreover, implies unfortunate limits on what can be expressed in domains of scientific applications of mathematics.

Ever since intuitionism came on the scene, it has been a significant question whether intuitionistic mathematics could meet the mathematical needs of the sciences. Early on, Hilbert expressed doubts, saying, according to anecdote, that denying the mathematician use of the law of excluded middle was like depriving the boxer of the use of his fists. This changed dramatically, however, in the 1960s with Errett Bishop's work,[9] which showed how a great deal of classical analysis could be reconstructed using intuitionistic logic (but otherwise no non-classical axioms such as Brouwer's governing choice sequences). In light of such work, it has not been easy to come up with examples of scientifically applicable mathematics that cannot be constructivized. Our essay of Chapter 13, however, provides a significant example. Adapting and applying an important theorem of Pour-El and Richards in recursive analysis, we argue that the *theory* of unbounded linear operators on Hilbert spaces of the sort used in quantum mechanics (QM) goes beyond the reaches of constructive mathematics. Such operators play a major role in QM, as they include position, linear momentum, and total energy, for example. While many individual applications of such operators are indeed constructive, the general theory of their mathematical behavior is not, as it involves quantification over such operators as non-constructive functions. But such theory is an important foundational component of quantum physics, and a mathematical framework adequate for the sciences needs to incorporate it. This, of course, is not to deny that mathematics built on intuitionistic logic may be of great mathematical interest. Indeed, an example of such, completely different from Brouwerian intuitionism, viz. smooth infinitesimal analysis, is the subject of the final essay of this volume.

As we have already seen above, the centrality of the axiomatic method in modern mathematics has naturally led some thinkers to propose if-thenism as a philosophy of mathematics as preferable to both objects-platonism and truth-values realism (generally "robust realism" in Maddy's [2011] phrase). One might refer to if-thenism as a kind of "easy-road" logicism, as the if-then statements at issue are supposed to be truths of pure logic (most commonly, classical logic.) And, if the logic is first-order (as it can be for any mathematics codifiable in ZFC), then, if Carnap* is right, mathematics on an if-thenist construal is indeed "analytic" in the Carnapian sense, after all. But a major challenge to if-thenism has been to provide a convincing account of the selection of axiom systems in the first place,

---

[9] Bishop [1967].

without embracing a robust-realist perspective, as arises, for instance, on a structuralist view (taking axioms as describing either actual or possible structures of mathematical interest). This challenge, however, arguably has been met in a new paper by Penelope Maddy,[10] which has led to the new essay of Chapter 14, "If 'if-then' then what?" Whether indeed Maddy's "Enhanced if-thenism" provides a viable philosophy of mathematics is critically examined in our essay, with the result that the view still faces major challenges despite the progress made over earlier versions.

It remains for us to provide some background to the final essay, "Mathematical pluralism: the case of smooth infinitesimal analysis." Until the method of limits was introduced into the calculus in the nineteenth century, an apparatus of infinitesimals was used to compute derivatives and integrals, although no satisfactory formalism was developed. As the reader may know, some raised sharp criticisms of the reliance on infinitesimals. Famously, Berkeley called them "ghosts of departed quantities," casting doubt on the coherence of the very idea of the infinitely small. And in modern times, Russell[11] voiced similar doubts and championed the method of limits as the definitive theory undergirding continuum mathematics, as reflected in modern textbooks and treatises. Remarkably, however, in the mid and late twentieth century, this landscape changed again rather dramatically, with the novel developments of Abraham Robinson's non-standard analysis (NSA), and then the more radical smooth infinitesimal analysis (SIA), each setting forth demonstrably consistent formalisms (relative to standard, limit-based analysis) governing algebraic treatments of infinitesimals without relying on the method of limits, thereby simplifying proofs in many cases.

Our final essay concentrates on SIA, as it poses special interpretative challenges. NSA employs fully classical logic applied to a non-standard model of classical analysis to obtain both points infinitely far from the standard reals along with infinitesimals (as reciprocals of the non-standard, infinitely distant reals). In contrast, SIA is forced to use intuitionistic logic in order to avoid outright contradictions. Thus, for example, it cannot even prove that nilsquares different from 0 exist, nor can it prove that they do not exist. (Thus the disjunction of those statements cannot be a theorem, violating the law of excluded middle.) But SIA's fundamental axiom (known as the Kock–Lawvere axiom), guaranteeing a microtangent vector to each point of a smooth curve, actually lying on the curve (and whose slope represents the derivative at that point of the function determining the curve), is not itself constructively interpretable. Thus, the relation between classical analysis and SIA is very different from that of classical analysis and intuitionistic analysis: some other justification than the need for constructive meanings of the logical

[10] "Enhanced if-thenism," Maddy [to appear].   [11] Russell [1903], Ch. 40.

connectives is required to understand SIA and its relation to classical analysis. Our essay grapples with this problem and proposes a surprising solution, appealing to the notion of "vague objects," contrary to a strand of philosophical literature claiming that such things are not even conceptually possible (a view voiced by Russell, Gareth Evans, and others, locating vagueness entirely in the behavior of language, not its subject matter).[12] We argue, to the contrary, that Evans' argument is flawed, and that SIA actually provides a coherent conception of "vague objects." Thus, not only must we recognize pluralism in mathematics based on the legitimate uses of different logics, but we must even recognize a plurality of mathematical frameworks based on intuitionistic logic itself, but for very different reasons.[13]

References

Bishop, E. [1967] *Foundations of Constructive Analysis* (New York: McGraw Hill).

Boolos, G. [1971] "The iterative conception of set," reprinted in *Logic, Logic, and Logic*, Jeffrey, R., ed. (Cambridge, MA: Harvard University Press, 1998), pp. 13–29.

Boolos, G. [1985] "Nominalist Platonism," *Philosophical Review* **94**: 327–344, reprinted in *Logic, Logic, and Logic*, Jeffrey, R., ed. (Cambridge, MA: Harvard University Press, 1998), pp. 73–87.

Boolos, G. [1998] "Must we believe in set theory?," reprinted in *Logic, Logic, and Logic*, Jeffrey, R., ed. (Cambridge, MA: Harvard University Press, 1998), pp. 120–132.

Burgess, J. and Rosen, G. [1997] *A Subject with No Object: Strategies for Nominalist Interpretation of Mathematics* (Oxford: Oxford University Press).

Burgess, J., Hazen, A. and Lewis, D. [1991] "Appendix," in Lewis, D., *Parts of Classes* (Oxford: Blackwell), pp. 121–149.

Carnap, R. [1937] *The Logical Syntax of Language* (London: Routledge & Kegan Paul).

Feferman, S. [1977] "Categorical foundations and foundations of category theory," in Butts, R. E. and Hintikka, J., eds., *Logic, Foundations of Mathematics and Computability Theory* (Dordrecht: Reidel), pp. 149–169.

Feferman, S. [1998] *In the Light of Logic* (New York: Oxford University Press).

Gödel, K. [1947] "What is Cantor's continuum problem?," reprinted in Benacerraf, P. and Putnam, H., eds., *Philosophy of Mathematics: Selected Readings*, 2nd edn. (Cambridge: Cambridge University Press,1983), pp. 470–485.

Hellman, G. [1989] *Mathematics without Numbers: Towards a Modal-Structural Interpretation* (Oxford: Oxford University Press).

Hellman, G. [1996] "Structuralism without structures," *Philosophia Mathematica* **4**(2): 100–123.

Hellman, G. [2003] "Does category theory provide a framework for mathematical structuralism?" *Philosophia Mathematica* **11**: 129–157.

---

[12] See references to the essay of Chapter 15.

[13] For further information on the relation of SIA to other theories of continua, see Hellman and Shapiro [2018], especially pp. 195–196.

16   Introduction

Hellman, G. [2005] "Structuralism," in Shapiro, S., ed., *Oxford Handbook of Philosophy of Mathematics and Logic* (Oxford: Oxford University Press), pp. 536–562.

Hellman, G. and Shapiro, S. [2018] *Varieties of Continua: From Regions to Points and Back* (Oxford: Oxford University Press).

Hellman, G. and Shapiro, S. [2019] *Mathematical Structuralism* (Cambridge: Cambridge University Press).

Hempel, C. G. [1945] "On the nature of mathematical truth," reprinted in Benacerraf, P. and Putnam, H., eds., *Philosophy of Mathematics: Selected Readings*, 2nd edn. (Cambridge: Cambridge University Press, 1983), pp. 377–393.

Isaacson, D. [1987] "Arithmetical truth and hidden higher-order concepts," reprinted in Hart, W. D., ed., *Oxford Readings in Philosophy of Mathematics* (Oxford: Oxford University Press, 1996), pp. 203–224.

Kripke, S. [1980] *Naming and Necessity* (Cambridge, MA: Harvard University Press).

Lewis, D. [1986] *On the Plurality of Worlds* (Oxford: Blackwell).

Lewis, D. [1991] *Parts of Classes* (Oxford: Blackwell).

Maddy, P. [2011] *Defending the Axioms: On the Philosophical Foundations of Set Theory* (Oxford: Oxford University Press).

Maddy, P. [to appear] "Enhanced if-thenism," in *A Plea for Natural Philosophy and Other Essays* (Oxford: Oxford University Press).

Putnam, H. [1967] "Mathematics without foundations," reprinted in Benacerraf, P. and Putnam, H., eds. *Philosophy of Mathematics: Selected Readings*, 2nd edn. (Cambridge: Cambridge University Press, 1983), pp. 295–311.

Quine, W.V. [1936] "Truth by convention," reprinted in Benacerraf, P. and Putnam, H., eds., *Philosophy of Mathematics: Selected Readings*, 2nd edn. (Cambridge: Cambridge University Press, 1983), pp. 329–354.

Russell, B. [1903] *The Principles of Mathematics* (New York: Norton).

Zermelo, E. [1930] "Über Grenzzahlen und Mengenbereiche: Neue Untersuchungen über die Grundlagen der Mengenlehre," *Fundamenta Mathematicae*, **16**: 29–47; translated as "On boundary numbers and domains of sets: new investigations in the foundations of set theory," in Ewald, W., ed., *From Kant to Hilbert: A Source Book in the Foundations of Mathematics*, Volume 2 (Oxford: Oxford University Press, 1996), pp. 1219–1233.

Part I

Structuralism, Extendability, and Nominalism

1   Structuralism without Structures

1.1   Introduction: Approaches to Structuralism

As with many "isms," "structuralism" is rooted in some intuitive views or theses which are capable of being explicated and developed in a variety of distinct and apparently conflicting ways. One such way, the modal-structuralist approach, was partially articulated in Hellman [1989] (hereinafter MWON). That account, however, was incomplete in certain important respects bearing on the overall structuralist enterprise. In particular, it left open how to treat generally some of the most important structures or spaces in mathematics, for example, metric spaces, topological spaces, differentiable manifolds, and so forth. This may have left the impression that such structures would have to be conceived as embedded in models of set theory, whose modal-structural interpretation depends on a rather bold conjecture, for example, the logical possibility of full models of the second-order ZF axioms. Furthermore, the presentation in MWON did not avail itself of certain technical machinery (developed by Boolos [1985] and Burgess, Hazen, and Lewis [1991]) which can be used to strengthen the program substantially. Indeed, these two aspects are closely interrelated; as will emerge, the machinery can be used to fill in the incompleteness so as to avoid dependence on models of set theory. The principal aim of this paper is to take the program forward by elaborating on these developments.

It will be helpful first, however, to remind ourselves of the main intuitive ideas underlying "structuralism" and to indicate at least roughly where in the landscape of alternative approaches the one pursued here resides.

One intuitive thesis (one I explicitly highlighted in MWON) is this:

Mathematics is the free exploration of structural possibilities, pursued by (more or less) rigorous deductive means.

Vague as this is, it already at least suggests the modern view of geometry, abstract algebra, number systems, and other "abstract spaces," in which we attempt to characterize the structures of interest by laying down "axioms" understood as "defining conditions," which we may be able to show succeed in their role by producing a proof of their categoricity, and then proceeding to explore their (interesting) consequences. This in turn reveals the importance of second-order logical notions in mathematical foundations, for, as is well known, first-order renditions of defining conditions will inevitably fail to characterize certain of the most central structures in all of mathematics, including the natural-number structure, the reals, the complexes, and initial segments of the cumulative set-theoretic hierarchy (cf. Shapiro [1991], also Mayberry [1994]).

A second intuitive principle, traceable to certain ideas of Dedekind [1888] and widely noted by philosophers and logicians, can be put thus:

> In mathematics, it is not particular objects which matter but rather certain "structural" properties and relations, both within and among relevant totalities (domains).

To this one may wish to add:

> The very identity of individual mathematical objects depends on such structural relations (i.e., on "relative positions" in structures).

This is illustrated by pointing out that it is nonsensical, for example, to postulate a single real number (as Field [1980], p. 31, entertained; cf. Shapiro [1997]); to be a real number is to be part of a complete, separable, ordered continuum. Particular constructions or definitions (e.g., as convergent rational sequences) may, in given contexts, allow one to recover such structure, and, by focusing on a particular construction, it may appear that one could sensibly postulate a single such item (say, sequence); but we can only regard this as postulating a real number after we have recovered the structure, and so Shapiro's point stands.

Note that, in stating this second intuitive thesis, we have been intentionally vague about its scope. Is it understood as saying that *all* mathematical reference to objects is to be interpreted structurally (whatever that means precisely), or does it say, more modestly, that salient cases are? What of notions such as "finite set" and "finite sequence" of given objects? In contrast to numbers, the identity of a finite set of objects A, for example, seems determined by its members without considering its relative position within the naturally associated structure, the totality of finite A – sets ordered by inclusion, itself a fairly complicated infinitistic object. (Cf. Parsons' [1990] related points concerning "quasi-concrete" mathematical objects.) And what of mathematical reference to structures themselves? Is there a regress involved in interpreting such reference structurally, and, if so, is it a vicious one (cf. Shapiro [1997])? An adequate structuralism should somehow account for these apparent differences among mathematical concepts. As different approaches may be expected to treat such matters differently, let us take the second thesis in its limited, modest sense, allowing for supplementation as a particular view may require.

So understood, there are at least four main approaches to structuralism that have been proposed and should be distinguished.

(i) The framework of Model Theory (MT), carried out in Set Theory (say, ZFC). Structures are understood as models (sets as domains, together with distinguished relations and possibly individuals), and one can also speak of isomorphism types as "structures" (at least one can if one is careful in either avoiding or admitting proper classes). (Of course, different choices of set theory yield different *explicanda*.) Equivalence of nominally distinct structures can be defined in terms of "definitional extension" and related notions. (Thus, for example, the full second-order natural-number structure with just successor distingushed is equivalent to that obtained by adding addition and multiplication.)

(ii) The framework of Category Theory (CT) (which itself can be axiomatized, as in Mac Lane [1986]). Structures are taken as the "objects" of a category, treated as simples or "points" by the axioms, and the "morphisms" between the "objects" typically preserve the characteristic "structural properties" of the branch of mathematics in question. Thus, for example, isometries preserve metric structure, homeomorphisms preserve topological structure, diffeomorphisms preserve differentiable manifold structure, etc. Categories themselves can be treated as "objects" in a category, and one can make sense of morphisms ("functors") preserving structural relations among the maps in the original categories. One even makes sense of morphisms ("natural transformations") of functors, giving rise to a functor category. (For an overview, see Mac Lane [1986], pp. 386–406. For a categorial recovery of number theory, see McLarty [1993].)

(iii) Rather than realizing structuralism within an overarching existing mathematical theory, one may pursue a *sui generis* approach, taking structures to be patterns or universals in their own right. (See e.g., Resnik [1981], Shapiro [1983, 1989, 1997]; for critical analysis, see Parsons [1990].) Different conceptions under this heading are possible, depending on the conception of universals (see Shapiro [1997]).

(iv) A modal-structural (ms) approach, as in MWON. Here literal quantification over structures and mappings among them is eliminated in favor of sentences with modal operators. (Hence the term "eliminative structuralism," see Parsons [1990], Shapiro [1997]. And hence the title of this paper, which I owe to Shapiro.) The framework is a modal second-order logic with a restricted (extensional) comprehension scheme (for details, see MWON, Ch. 1). Categorical axioms of logical possibility of various types of structures replace ordinary existence axioms of MT or CT, and typical mathematical theorems are represented as modal universal conditionals asserting what would necessarily hold in any structure of the appropriate type that there might be. It turns out that a great deal of ordinary mathematics may thus be represented nominalistically, without the language of classes at all, even under modality (see Hellman [1994] and below). Just how far this approach can be pushed is a somewhat open question, to be pursued further below. This brings us to the question already broached at the outset, whether this approach can do justice to "structuralism" without a detour through (modal-structurally interpreted) set theory or category theory. In the following sections, we will present some evidence in favor of a positive answer.

Now this is not the place to undertake a systematic comparison of these alternative approaches. There are, however, two related contrasts of immediate concern between (i) through (iii) on the one hand and (iv) on the other that require our attention. The first pertains to the trade-off between platonist ontology and modality. The first three approaches are framed in modal-free languages but they are entangled well above the neck (naturally) in Plato's beard. Sets, categories, or universals are just taken as part of reality, leading to perennial disputation as to the nature of such "things," how we can have knowledge of them or refer to them, etc., and (of course) whether or not such questions are somehow misguided in the first place. Modal structuralism avoids commitment to such *abstracta*, at least in its initial stages (in treating, say, the number systems, prior to reconstructing set theory itself), and raises the prospect that a (modal) nominalistic framework may suffice to represent the bulk of ordinary mathematics.[1] (This depends on treating the second-order variables of the ms language nominalistically, but in ordinary contexts this can be done (see Hellman [1994] and below).) The price of course is taking a logical modality as primitive, raising questions of evidence and epistemic

---

[1] The phrase "ordinary mathematics" is not a precise one, but we intend it more broadly than do Friedman–Simpson et al. in the program of reverse mathematics, where explicitly excluded are "those branches of mathematics which ... make essential use of the concepts and methods of abstract set theory," such as "abstract functional analysis, general topology, or uncountable algebra" (Brown and Simpson [1986], p. 123). We do mean to exclude set theory and category theory themselves, but not the three fields just listed, nor the theory of non-separable Banach and Hilbert spaces, which by implication are also excluded by the Friedman–Simpson usage. The latter is motivated primarily by the question, "What portions of ordinary mathematics can be carried out, in which interesting subsystems of classical analysis (PA^2)?" and for this purpose, mathematical questions which cannot even be asked (even via suitable coding) in the language of PA^2 are sensibly excluded from "ordinary mathematics." Since we are under no such constraint, however, we can afford to be more liberal, counting as ordinary virtually any subfield short of those devoted to the grand foundational schemes. In our usage, certain "concepts and methods of abstract set theory" can be deployed to some extent without commitment to abstract sets. But this can be spelled out without a precise use of the phrase "ordinary mathematics."

access not unlike those raised by platonist ontologies. This trade-off is a subject of ongoing discussion, and will not be resolved here. We would point out, however, that assessing the trade-off depends on a better understanding of the alternatives themselves, including the ms approach. In particular, *just what modal-existence postulates are required to implement structuralism?* The more modest they are, the better the prospects for the ms approach. The results of our reflections below will bear directly on this. As will emerge, only rather modest modal-existence postulates are required; for much of mathematics, only countably many atoms need be postulated (as logically possible); for much more, including a great many topological structures and manifolds, uncountably many atoms are needed, but this need not transcend the scope of nominalism.

The second contrast between (i)–(iii) and (iv) concerns the wealth of mathematical structures incorporated within the respective framework. In the cases of (i)–(iii), the extent of richness is literally endless. While any particular set theory has its limitations, there are still boundless riches as regards the structures and spaces of ordinary mathematics. (In particular, there are no limits on cardinality or on type.) Category theory, especially with its large categories, is *prima facie* even more generous. And, presumably, *sui generis* universals are, as Quine might say, "free for the thinking up." Not that these frameworks avoid honest toil; nor that they are larcenous; they merely rely on the bountifulness of reality as they conceive it.

The case of the ms approach is more complex; indeed, we should distinguish two sub-approaches: (a) first develop a modal-structural interpretation of set theory (or of category theory), and then simply translate the MT (or CT) treatment of structures of interest accordingly; (b) seek a direct ms interpretation of theory of any such structures, avoiding set-theoretic commitments to whatever extent possible. From the perspective of ontology, it is (b) that is of greater interest. Moreover it confronts the challenge of describing interrelations of different types of structures, something that both MT and CT are set up to handle. If approach (b) were to be successful, structuralism would then stand independently of set theory rather than being just a chapter in it, even as interpreted; and it would represent a rather remarkable extension of nominalistic methods. It is this approach that we shall now continue to pursue.

(Traditionally, the problem with nominalism in mathematics has been not so much that Occam's razor has been dulled by Plato's beard, but rather that it has managed to remove the beard only by severing the head at the neck. Modal structuralism, as it has been extended (in Hellman [1994] and below), manages a fairly clean shave while leaving the brain quite intact.)

1.2   Extending the Reach of Nominalism to Third-Order Arithmetic and Third-Order Analysis

The plan of this section is as follows. First we shall review the modal-structural frameworks for arithmetic and for real analysis developed in MWON, taking advantage of certain improvements since developed. These improvements consist principally in (1) the combined use (due to Burgess, Hazen, and Lewis [1991], henceforth BHL) of plural quantification (Boolos [1985]) and mereology to *define* nominalistic ordered pairing in a general way (as opposed to adopting a new primitive pairing relation, as was suggested in MWON); and (2) the development of predicative foundations for arithmetic (in Feferman and Hellman [1995], reproduced here as Chapter 7) which enables modal-structuralism to get started, at least, in a manner compatible with predicativist principles. Having reviewed this, we will then indicate how the machinery just referred to under (1) can be used to extend the reach of nominalism one level beyond each of the core systems of arithmetic and real analysis described in MWON – essentially how to pass from PA^2 to PA^3 and from RA^2 to RA^3. Some of the benefits of these extensions regarding structuralism will then be explained in the Section 1.3.

Beginning with the standard Peano–Dedekind axioms for the natural numbers, PA^2, involving just successor, $'$, and the second-order statement of mathematical induction, we treat an arbitrary sentence S of first- or second-order arithmetic (in which any function constants have been eliminated by means of definitions in terms of $'$) as elliptical for the modal conditional

$$\Box \forall X \forall f [\wedge PA^2 \rightarrow S]^X ('/f),$$

in which a unary function variable f replaces $'$ throughout and the superscript X indicates relativization of all quantifiers to the domain X. This is a direct, modal, second-order statement to the effect that "S holds in any model of PA^2 there might be." (Note that use of model-theoretic *satisfaction* is avoided.) This was called the "hypothetical component" of the modal-structural interpretation (MSI) of arithmetic. In order that this provide a faithful representation of classical arithmetic, it is also necessary to add a "categorical component," a statement that such structures (ω-sequences or $\mathbb{N}$-structures) are logically possible:

$$\Diamond \exists X \exists f [\wedge PA^2]^X ('/f).$$ (Poss $\mathbb{N}$)

This is the characteristic modal-existence (mathematical existence) claim underlying classical arithmetic (or "classical analysis," logicians' term for PA^2). It distinguishes the modal-structural approach from "deductivism" and from "if-thenism." All sentences of the original mathematical language (for PA^2) are regarded as truth-determinate regardless of their formal provability or refutability. Furthermore, various arguments show that the translation scheme respects classical truth-values. A key step is the recovery of Dedekind's categoricity proof, that any pair of models of PA^2 are isomorphic. This can be carried out within modal second-order logic, using just the ordinary second-order (extensional) comprehension scheme (with ordinary universal quantifiers in the prefix, not boxed ones) and basic quantified modal logic (although S-5 is the preferred background). Here the reasoning is straightforward mathematical reasoning under the assumption that a pair of PA^2 structures is given. Appeal to intensions – relations across possible worlds, as it were – can be avoided if we assume an "accumulation principle" to the effect that if it is possible there is an ω-sequence with (PA^2-definable) property P (which, by quantifier relativization, involves only items internal to the given sequence) and it is possible there is another ω-sequence with such property Q (internal to its sequence), then the conjunction of these existential statements is also possible, i.e., these two sorts of ω-sequences occur in the same world, so to speak. (For further details, see MWON, Ch. 1; also Hellman [1990].)

Note that talk of "possible worlds" is heuristic only; the modal operators are primitive in the framework and are not required to be given a set-theoretical semantics. Note further that the accumulation principle derives its plausibility from the combination of two considerations: first, it is only logical possibility that is at issue, and second, the mathematical properties labelled "P" and "Q" are entirely "internal" to their respective sequences, as relativization to the respective domains of any quantifiers they may contain insures. The essential point is that anything internal to a given structure cannot conflict with anything internal to another, so that structures satisfying the respective conditions are logically compossible. Thus, there is no requirement that the structures involved be of the same general type. One could be an ω-sequence and another could be an ordered continuum, or whatever. Moreover, the principle can be generalized in the obvious way to cover any finite number of structures. These generalizations are important for this approach to structuralism, since, as in set theory or category theory, we often wish to speak of relations among a variety of structures. Finally, note that the formulas above quantify over structures by quantifying directly over their domains and distinguished relations or functions; it is not necessary to ascend a further level in type as is commonly done in model theory. From the rest of the formula, it can always be made clear which relations or functions are defined on which domains.

So far we have used some of the ordinary language of mathematics – the language of "domains" and "functions" – to eliminate reference to "numbers" as special objects. Arithmetic is not about special objects; it is rather about a special type of structure. In accordance with part of Dedekind's conception, it investigates facts which hold of any "simply infinite system," but we have explicitly used modal operators both to get away from commitments to any special instantiation of the structure-type and to achieve an open-ended generality appropriate to mathematics.[2] On this conception, mathematics investigates a certain category of necessary *truths*, not confined to what happens to exist. But it does not have to postulate a special realm of necessary *existents* in the process. Only one level of *abstracts* has been invoked, corresponding to the second-order variables. And, it should be noted, the second-order comprehension scheme does *prima facie* commit us to actual classes and relations of whatever actual first-order objects we recognize among the *relata* of the relations (or relation variables) of our language. Significantly, there is no iteration of collecting, so this is not Plato's full beard, to be sure; but it does seem more than just a six o'clock shadow! So how do we get a clean shave?

Well, it turns out, there are many ways to shave a beard, at least at the stage of second-order arithmetic. One way is that of *predicative foundations*, in the tradition of Poincaré, Weyl, and Feferman et al. The central idea here is to restrict comprehension axioms to *definable* classes (and relations), where this is spelled out in terms of formulas of mathematical language whose quantifiers range over already defined or specified objects. (For details on various options, including systems of variable type, see writings of Feferman, e.g., [1964, 1968, 1977, 1988].) Typically one begins by taking the natural numbers for granted and considering, first, those sets of natural numbers definable by arithmetic formulas (with quantifiers only over natural numbers) – the first-order sets – and then sets of natural numbers definable by formulas with quantifiers over numbers and first-order sets, and so on. (How far this may be iterated is a delicate matter.) This can qualify as nominalistic – relative to the natural numbers – in that one can eliminate reference to sets and relations in favor of the semantic notion of *satisfaction* of formulas by natural numbers, or by other nominalistically acceptable objects, for example, predicates themselves (cf., e.g., Chihara [1973], Burgess [1983]).

The problem with this as a nominalization program, however, is that the natural-number structure has been taken as given. And this has appeared unavoidable, for there is *prima facie* reliance on *impredicative* class existence principles in the classical constructions of the natural-number structure, for example, the Dedekind–Frege–Russell definition as (essentially) the intersection of all inductive classes containing 1 (or 0). Such principles are standardly used also to prove the existence of an isomorphism between any structures satisfying the PA^2 axioms; and it is well known that the second-order statement of induction is necessary for this result. Contrary to these appearances,

---

[2] As Tait [1986] and Parsons [1990] have pointed out, it would be a mistake to attribute to Dedekind himself an eliminativist structuralist position (cf. MWON, Ch. 1). In the text, however, we are referring to Dedekind's conception of "the science of arithmetic" as investigating what holds in any simply infinite system. This, surely, is the starting point of any eliminativist approach.

however, the natural-number structure can itself be constructed predicatively, beginning with the notion of *finite set* governed by axioms that are intuitively evident or of a stipulative character, as carried out by Feferman and Hellman [1995]. Of particular interest here are the facts that mathematical induction is itself derivable from within an elementary theory of finite sets and classes (EFSC),[3] and that Dedekind's categoricity proof ("*unicity* of the natural number structure") is also recoverable (cf. Feferman and Hellman [1995]).

This still leaves us with the non-nominalist notion of "finite set" governed by the EFSC axioms. But even this vestige of platonist commitment (5 o'clock shadow?) can be eliminated. One can first postulate the logical possibility of an infinitude of atoms (atomic individuals, governed by the axioms of atomic mereology; cf. Goodman [1977], also MWON, Ch. 1), and then interpret "finite set" as "finite sum (or whole, or fusion) of atoms." To express "infinitude of atoms" one can use the device of plural quantification and postulate the following:

There are (possibly) some individuals one of which is an atom and each one of which fused with a unique atom not overlapping that individual is also one of them.   (Ax ∞)

With this postulate, one has the essentials of a mereological model of the EFSC axioms: first-order variables can be taken to range over arbitrary individuals (atoms and fusions of atoms); finite set variables range over finite fusions of atoms; class variables range over arbitrary fusions of atoms. If a null individual is admitted (as a convenience), the axioms are satisfied as they stand; otherwise the comprehension axioms can be complicated slightly to avoid the null

---

[3] The system EFSC is formulated in a three-sorted language with individual variables, variables for finite sets of individuals, and variables for classes of individuals. Formulas with no bound class variables are called WS formulas (for "weak second-order"). A pairing operation-symbol is primitive as is ∈ relating individuals to finite sets and classes. The logic is classical (with equality in the first sort). The axioms of EFSC are, in words, as follows:

(WS-CA)	Weak second-order comprehension: existence of classes as extensions of WS-formulas;
(Sep)	Separation for finite sets: existence of a finite set as the intersection of any given finite set and the extension of a WS formula;
(Empty)	Existence of the empty finite set;
(Adjunction)	Existence of a finite set obtained by adjoining any single individual to a given finite set;
(Pairing I)	"Pairs are distinct just in the case either first or second members are";
(Pairing II)	Existence of urelements under pairing.

The system EFSC* is obtained from EFSC by adding the axiom,

| (Card) | "Any finite set is Dedekind-finite." |

In Feferman and Hellman [1995], the existence of structures satisfying mathematical induction is derived in EFSC*; however, as Peter Aczel has pointed out, this can already be proved in EFSC.

individual. (EFSC also takes a pairing function as primitive, governed by two axioms: P-I, the standard identity condition, $(x, y) = (u, v)$ iff $x = u$ and $y = v$, and P-II, existence of an urelement under pairing. It turns out that both these are satisfied on, say, the Burgess construction of nominalistic pairing in BHL [1991]. So we can invoke this in interpreting pairing in EFSC.) Now, within such a model of EFSC there is a mereological model of the PA^2 axioms. (This follows from the construction of an $\mathbb{N}$-structure in EFSC.) Moreover, as the class variables are taken to range over arbitrary individuals (fusions of atoms), we even have a *full* second-order PA^2 model in the classical sense, in which arbitrary sets of numbers correspond to arbitrary fusions of the individuals serving as numbers. The predicativist may stop short of this, confining oneself to "definable fusions" of atoms in specifying the range of the class variables (cf. Hellman [1994], sec. 3). But the essential point here is that both predicativist and full second-order arithmetic are interpreted nominalistically.

From here, one could continue on with predicativist analysis, constructing countable analogues of the classical continuum, made up of "definable" or "specifiable" reals, which support rather rich portions of functional analysis and related subjects (cf., e.g., Feferman [1988]). As Feferman has pointed out, within various systems of predicative analysis, one can even prove the unicity of the real number structure, as seen from within that system: impredicativity is avoided because one requires, not the full classical Principle of Continuity (least upper bound axiom), but only the sequential form, "Every non-empty bounded sequence of reals has a least upper bound." (For details, see Hellman [1994], sec. 2.)

Continuing along the predicativist route, one can introduce symbolism for reasoning about (specifiable) classes (and relations and functions) of reals, classes (etc.) of classes of reals, and so on through the finite types (see, e.g., Feferman's system W [1988, 1992]). At each level, one is considering not the full classical totalities, of ever higher uncountable cardinality, but subtotalities of objects predicatively specifiable in mathematical language.[4] Thus, the ranges of the quantifiers at each level are really countable, although from within the predicativist system they may be described as "uncountable." (The predicativist can carry out the reasoning of Cantor's diagonal argument, but, implicitly, only *specifiable* enumerations are considered.)

To what extent can predicativism carry out a structuralist program for mathematics? This is a large question which cannot be fully answered here. But the following points towards an answer may be offered. First, one must be more precise about what it means to "carry out a structuralist program." Presumably this includes these things: (i) characterizing the types of structures

---

[4] For ways of making this precise in connection with unramified systems such as W, see Hellman [1994], n. 2.

or spaces that arise in the various branches of mathematics (or at least "ordinary mathematics" as we have used that term above); (ii) describing the main types of relationships among these structures, including the various morphisms within and among them (isomorphisms, homomorphisms, embeddings of various sorts, etc.); (iii) recovering the important theorems concerning the various structures and relations among them, including existence theorems.

Judged by these standards, predicativism gets mixed reviews. It does surprisingly well, for example, in recovering theories of various types of metric spaces central to scientifically applicable mathematics. Although the concept of Lebesgue outer measure is not predicatively available, theories of measurable sets and measurable functions can be developed (cf. Feferman [1977], sec. 3.2.5), and then one can obtain the L_p spaces and carry out a structuralist treatment of Banach and Hilbert spaces. On the other hand, clearly there are important limitations. (1) Various objects of importance in the classical (set-theoretic) treatments are simply not available, for example, outer measures, as just indicated, and general descriptions of major types of spaces are *prima facie* impredicative, for example, topological spaces with families of open sets closed under *arbitrary* unions. (2) Proofs of key theorems, even if predicatively statable, may require impredicative constructions essentially. A known example is Friedman's finite form of Kruskal's Theorem on embeddability of finite trees (see e.g., Smorynski [1982]). Even if this example does not pertain to scientifically applicable mathematics, it surely pertains to significant mathematical structures. (3) Even in the cases in which predicative proofs of key theorems are possible, for example, the unicity of the real-number structure, in reality we know – as does the predicativist – that the structures to which the result pertains are countable, hence only small parts of the structures classically conceived. Even if the predicativist makes no direct sense of the latter phrase, one can pass to more encompassing, predicatively graspable totalities, essentially by enriching a given language with predicatively intelligible semantic machinery for defining new objects, for example, real numbers (cf. Hellman [1994], sec. 2). If one is a skeptic about uncountable totalities generally, then presumably one is willing to pay the price of this language-relativity of much of mathematics. If, however, one follows the classicist in taking seriously the absoluteness of uncountability – for example, if one treats totalities such as "all sets of natural numbers" or "all fusions of countably many atoms" as having a definite and maximal sense – then one will regard the predicativist substitutes as falling far short of the genuine articles.

This much should be clear: if the objection to the uncountable is motivated by nominalist concerns – the desire to avoid commitment to classes or universals, etc. – then it is misplaced. A fusion of atoms is just as "concrete" – just as much not a class or a universal, etc. – as the atoms themselves. In the language of types, both are of type 0. It does not matter how many there are.

If we are given countably infinitely many atoms – by definition pairwise discrete – then we may speak of arbitrary fusions of them without nominalistic qualms. Then to go on to say how many of such fusions there are requires further reasoning, to be sure; but in fact there is no problem in carrying out Cantor's diagonal argument nominalistically to convince oneself that there are uncountably many. And if in the course of this reasoning, one meant to consider "any possible enumeration" – not merely any that could be specified in some privileged symbolism as the predicativist intends – then the conclusion has its absolute force.

This leads us to relax the "definitionist" stance of predicativism in pursuit of a nominalist structuralism. In our nominalist Σ-comprehension scheme,

$$\exists x \Phi(x) \;\rightarrow\; \exists u \forall y [y \circ u \;\leftrightarrow\; \exists z (\Phi(z) \,\&\, z \circ y)] \tag{CΣ}$$

(in which $\circ$ is "overlaps" or "contains a common part with"), we allow the predicate Φ (lacking free u) to contain quantifiers over *arbitrary* individuals, whether or not specifiable by any particular symbolic means. An immediate consequence is that, once we have postulated (Ax ∞) – guaranteeing an ω-sequence of atoms (Poss $\mathbb{N}$) – we already have embedded within such a sequence enough subsequences to serve as arbitrary real numbers. Standard arithmetization procedures can be used to introduce negative integers, rationals, and then reals (either as Cauchy sequences of rationals or as Dedekind cuts). (By the device of numerical pairing, that is pairing via the atoms of the postulated ω-sequence, one can remain within that structure; reals are then just certain fusions of atoms.) The following important facts should be noted.

(1) The full classical Continuity Principle (lub principle for arbitrary non-empty bounded sets of reals) is derivable along logicist lines (using (CΣ)), without exceeding the bounds of nominalism. (The usual set-theoretical arguments are available making use of plural quantifiers to get the effect of quantification over sets and functions of reals.)

(2) The categoricity of real analysis (RA) is also derivable within this nomin-alist framework, in the sense that any two concrete $\mathbb{R}$-structures (i.e., with reals built up from concrete ω-sequences in logicist fashion as just alluded to, together with the usual ordering $<$ on reals) are isomorphic. (The proof of this requires even less than (1), viz. Sequential Completeness rather than the full lub principle suffices. For a visualizable nominalistic construction, see Hellman [1994].) Furthermore, since all fusions of atoms count as individuals, regardless of specifiability by formulas, this categoricity proof has the absolute significance of the standard set-theoretical one.

Thus, (Poss $\mathbb{N}$) – hence (Ax ∞) – suffices for a nominalist structuralist treatment of full classical analysis (PA^2). In particular, the criteria (i)–(iii) above are met with respect to PA^2-structures. But, as the alert reader may have noticed, we have enough machinery at our disposal to ascend one more level, to third-order number theory (PA^3). Plural quantifiers achieve the effect of quantification over sets of reals, as just described; and the device of BHL pairing reduces polyadic quantification at this level (over relations of reals) to monadic. Still we have only had to hypothesize a countable infinity of atoms.

There may be the concern that plural quantifiers (at a given level, say pluralities of reals) do not really get around classes (of reals); that a sentence such as, "Any reals that are all less than or equal to some real are all less than or equal to a least such," really concerns classes of reals as values of a hidden variable. (No first-order conditions determine the same class of models.) I agree with Boolos [1985] and others, however, that we do have an independent grasp of plural quantifiers and their accompanying constructions, and that we can use them to formulate many truths at any given level that would have to be regarded as false on an ontology that repudiates classes of objects at that level. (The case based on this and related points has been made very effectively by Lewis [1991], sec. 3.2.) The school teacher who says, "I've got some boys in my class who congregate only with each other," should not be ascribed on that basis a commitment to classes other than school classes. The EPA official who says, "Some cars are tied with one another in being the most polluting vehicles on the road," need not be committed to classes of vehicles other than the usual predicative ones (two-door sedan, station wagon, etc.). And the nominalist who entertains just some atoms and their fusions can even go on to say things like, "Some of those fusions can be matched up in a one-one manner with the atoms whereas not all of them can," without implying anything about any objects of higher type than the fusions themselves. (In virtue of BHL pairing, even the talk of one-one correspondences is innocent.)

Thus the strength of full, classical third-order number theory – equivalently, second-order real analysis – is attained within a nominalist modal-structural system without postulating more than a countable infinity of atoms. This is already quite a rich framework for carrying out structuralism. But if we are prepared to entertain at least the logical possibility of a continuum of atoms, we can attain one full level more, that of *fourth-order number theory*, equivalently *third-order real analysis*. Suppose we postulate the possibility of a complete, separable ordered continuum ($\mathbb{R}$-structure for short) of atoms, which we may write

$$\Diamond \exists X \, \exists f [\wedge RA^2]^X (< / f),$$ (Poss $\mathbb{R}$)

in which RA^2 denotes the axioms for such a structure (with first-order quantifiers stipulated to range over atoms), including the second-order statement of Continuity, and where a relation variable f replaces the ordering relation constant $<$ throughout; then our second-order variables already range over arbitrary fusions of these, at the level of sets of reals. Functions and relations of reals are reducible to sets via pairing. Quantifying plurally over sets of reals then gives the effect of quantifying (singularly) over sets of sets of reals and, via BHL pairing, over functions and relations of sets of reals (hence also of functions and relations of reals). This is mathematically at the level of RA^3 or PA^4, a very rich framework indeed. (Note that, since fusions of (atoms serving as) reals are entertained as *bona fide* objects, it is pluralities of them that we plurally quantify over to achieve the next level. Even if plural quantification is already invoked lower down, to introduce pairs, triples, etc. of reals so as to reduce relations of reals to sets, we get the effect of plurally quantifying over relations of reals when we plurally quantify over sets of reals. We are not "plurally quantifying over pluralities" except in this innocent sense.)

In order to characterize $\mathbb{R}$-structures, it is necessary that we use the second-order statement of the Continuity Principle (lub axiom); it is one of the defining conditions built into the modal-existence postulate (Poss $\mathbb{R}$). Does this mean that we have given up the advantages of a logicist-style *derivation* of Continuity from more elementary, general principles? Well, yes – but also no! Yes, in that the principle (or a geometric equivalent) is essential in asserting the possibility of an $\mathbb{R}$-structure *whose first-order objects are atoms*. But no, in that we have already derived Continuity in logicist fashion (but nominalistically) above as it governs $\mathbb{R}$-structures built from fusions of atoms of an $\mathbb{N}$-structure. If reals are taken as constructed objects (e.g., Dedekind sections, etc.), then we already have (Poss $\mathbb{R}$) based on (Poss $\mathbb{N}$), as already described. (This point was not made in MWON, for, not utilizing the resources of plural quantifiers, we did not have the means to state Continuity in full generality, nominalistically, given just (Poss $\mathbb{N}$). A major advantage of the present approach is that, now, we do have the means to speak of arbitrary pluralities of reals, given just (Poss $\mathbb{N}$).) We give none of this up when we *add* the postulate of an *atomic* $\mathbb{R}$-structure. But of course we are adding something substantial. In terms of the familiar aphorism, there are two cakes: we are having one while eating the other.

Now the worry arises that in entertaining anything so idealized and remote from experience as spatiotemporal or geometric points (typically thought of as realizing the RA axioms) we have exceeded the bounds of nominalism. Surely a case can be made that "points of space-time" – especially "unoccupied points" – are in some sense "abstract," even if invoked by standard formulations of physical theory. Field [1980], who based his nominalization program on the acceptability of space-time points and regions, sought to answer such objections, in part by appealing to substantivalist interpretations of space-time physics which do seem to accord a kind of causal role to (even unoccupied) space-time (e.g., as a field source with clearly physical effects). And certainly

this neo-Newtonian perspective is not without its adherents. But it is important
to realize that the modal-structural axiom (Poss ℝ) does not ultimately depend
on a substantivalist view of space-time. For, just as the logical modality frees us
from having to posit actual countable infinities, so it frees us from any particular view of actual space, time, or space-time. It suffices, for example, that
Newtonians who posited a continuous luminiferous medium were entertaining
a logically coherent possibility. Similarly one can coherently imagine a perfect
fluid filling, say, a bounded open region and satisfying whatever causal conditions one likes. All that matters is that it be conceived as made up of parts
satisfying the RA^2 axioms; it need not involve "unoccupied space"; it need not
be useful to any explanatory purpose; it need not be (period). Furthermore,
supposing that such possibilities are "world-independent" – that they have the
absolute status generally accorded logical possibilities – frees mathematics
based on (Poss ℝ) of any contingency. (In the modal system S-5, appropriate to
logical possibility, we have Nec Poss ℝ once we have Poss ℝ.) There is no
"lucky accident" involved in the truth of modal-structuralist mathematics.

Since the comparison with Field's program has come up, it is appropriate to
emphasize that, for Field also, there is no "lucky accident" involved in the truth
of mathematics, for (except in a vacuous sense) Field's program is instrumentalist, not recognizing mathematical truth at all, except in vacuously assigning
"false" to all existential statements and "true" to anything equivalent to the
negation of one, since "there are no numbers"! This is as great a contrast with
modal-structuralism – nominalist or not – as there could be.

Yet perhaps the contrast is at bottom illusory. For, in seeking to recover
nominalistically various arguments that certain mathematical-physical theories
are semantically conservative with respect to their nominalized counterparts,
Field [1980, 1992] introduced logical modality as a means of bypassing
reference to models. Along the way he seems committed to affirming such
things we would write as,

$$\Diamond \wedge PA^2,$$ (FC)

that the conjunction of the second-order Peano–Dedekind axioms is logically
possible (FC is for "Field's commitment"). Now this is not quite the same as
our (Poss ℕ), which involves domain and function variables, but (Poss ℕ)
follows in second-order logic from (FC). Mathematically speaking, they are not
essentially different. (Philosophically they are. As explained in MWON, Ch. 1,
problems arise in interpreting the function constant of (FC) under the modal
operator, problems that (Poss ℕ) avoids. And explicit reference to domains is
natural in a structuralist interpretation.) But then *everything needed for a
modal-structuralist treatment of mathematics in PA^2 – or even PA^3, if plural
quantifiers are added – is available. So Field's system is not really*

instrumentalist after all. It embraces the core of the ms interpretation but then does not go on to utilize it. In this sense, much genuine mathematics is still present.[5]

1.3   Realizing Structuralism

Without attempting anything like a survey of mathematical structures, let us indicate by salient examples how the above described frameworks suffice for a great deal of the structuralist enterprise. Let us refer to those frameworks as PA^3 (or RA^2) and PA^4 (or RA^3), understanding of course that it is the nominalistic modal-structural version of these systems that is meant, that is, the relevant modal-existence postulate, for example (Ax ∞), with the background of S-5 modal logic, mereology with the comprehension scheme (CΣ) for modal-free formulas, plural quantifiers, and the assumptions needed for BHL pairing (as in BHL [1991]). (The second-order logical notation of the modal-existence postulates may be retained, understanding the monadic second-order variables to range over arbitrary fusions of atoms and the polyadic second-order variables to range over fusions of n-tuples of individuals based, say, on b-pairing of BHL.) It should also be understood that, in treating many abstract structures, what we really require of these frameworks is not the full structure of an $\mathbb{N}$-structure or an $\mathbb{R}$-structure, but merely a domain with sufficiently many atoms, for example, a countable infinity or continuum many. Then one needs the specific functions, relations, or other items characteristic of the type of abstract structure in question, for example, a metric, a topology, a chart-system,

---

[5] Concerning Field's commitment (FC), let Ax stand for the conjunction of axioms of a finitely axiomatized mathematical theory T (such as PA^2), and let S be a nominalistically formulated statement (of applied mathematics); then Field's modal formulation of the conservativeness of T is

$$\text{If } \Diamond S, \text{ then } \Diamond(Ax \ \& \ S), \qquad\qquad\qquad (\text{Conserv } T)$$

where Field's $\Diamond$ operator is said to mean "it is not logically false that" (cf. Field [1992], p. 112). Initially, Field focuses on first-order theories (for which (Conserv T) may be a *schema* with finite conjunctions of axioms), but later he considers some second-order cases as "not completely without appeal." Moreover, since the "complete logic of Goodmanian sums" is invoked (in Field [1980]) in connection with representation theorems for $\mathbb{R}^4$ – i.e., even arbitrary fusions of space-time points are recognized – there should certainly be no objection to (Conserv PA^2). (Indeed, as we have seen, Ax in this case is nominalistically interpretable without anything so strong as space-time substantivalism. Field's reservations in connection with second-order principles seem to have primarily to do with transfinite set theory and not with arithmetic or analysis; cf. Field [1992], p. 119.) But of course there are some S such that Field accepts $\Diamond S$, and then $\Diamond Ax$ follows.

It may well be that Field's informal understanding of the logical modality differs in some respects from that expressed in MWON, but his core modal logic, like that of MWON, is a version of S-5, and it seems clear that a great deal of modal mathematics (of PA^2 and even RA^2) can be carried out within Field's system.

etc. Thus when we appeal to PA^2 (or PA^3), say, we may only be appealing to the modal-existence of a countable infinity of atoms; and when we appeal to RA^2 (or RA^3), we may only be appealing to the modal-existence of continuum-many atoms. Alternatively, we may only wish to appeal to the modal-existence of uncountably many atoms, leaving open the cardinality. It should be clear how to formulate this using the available machinery, mereology, plural quantifiers, and BHL pairing. We need merely translate the following statement:

$\diamond \exists X$ ['X is a fusion of infinitely many atoms & a proper part Y of X is a fusion of countably many atoms (which we can say directly by saying Y is in one-one correspondence with any infinite part of Y) & X is not in one-one correspondence with Y'].

If we wished to specify a fusion of $\aleph_1$ atoms, we could simply add to this that X is in one-one correspondence with any infinite part which is not in one-one correspondence with Y. Similarly we could specify a fusion of $\aleph_2$ atoms, and so on. (Writing such things out in primitive notation with all the plural quantifiers needed for BHL pairing would not be pleasant, but that is why we like quotes. For more on the expressibility of cardinality, see Shapiro [1991].)

Let us begin with the important example of *metric spaces*. Abstractly considered, these are pairs consisting of a domain X and a function d (positive definite metric) from $X \times X$ into $\mathbb{R}$ satisfying the familiar first-order conditions:

$$d(x, x) = 0, \ x \neq y \rightarrow d(x, y) > 0, \ d(x, y) = d(y, x), \text{and}$$

$$d(x, y) + d(y, z) \geq d(x, z) \qquad \text{(triangle inequality).}$$

Bijective mappings ϕ between two such spaces, (X, d) and (X', d'), preserving metrical relations are called *isometries*. (That is, $d(x, y) = d'(\phi(x), \phi(y))$.) Even if we require that the domains be uncountable, all this is describable at the level of PA^3: the domains are at the level of sets of reals, metrics are at the level of sets of ordered triples of reals, hence sets of reals via pairing, and isometries are at the level of sets of ordered pairs of reals, hence sets of reals. Thus we can quantify over structures of this type and isometries (and similar relations) among them without positing more than a countable infinity of atoms.

Of course, many metric spaces of importance carry additional structure, often embedded in the elements of the spaces themselves. For example, Banach spaces are normed linear spaces whose elements are often real- or complex-valued functions; they become metric spaces under the definition $d(f, g) = \|f - g\|$. Hilbert spaces have the additional structure of an inner product, $(\ ,\ )$, which gives rise to a norm via $\|f\|^2 = (f, f)$. Such spaces are called *separable* if they include a countable dense set of vectors (where density means that any f in the space can be approximated arbitrarily closely in the metric by elements of the countable subset). The vectors of a separable Hilbert space are codable as real numbers, and inner products are then at the level of sets of reals (via pairing). Linear operators are also at this level, as are subspaces. The latter, however, can be coded as reals since a subspace is spanned by countably many vectors. The same is true for a great many operators, for example, the continuous (= bounded) ones. Norm-preserving maps, *unitary transformations*, are also at the level of sets of reals. Thus, the metrical structure of separable Hilbert spaces is describable in the PA^3 framework, again not exceeding the postulation of countably many atoms.[6] Significantly, the classical demonstration that the axioms for separable Hilbert spaces (of infinite dimension) are categorical carries over intact. (The standard examples ℓ^2 and L^2 are available.) Indeed, even measure-theoretic structure can be captured at this level. Note, finally, that – in virtue of the unrestricted second-order (extensional) comprehension principles of our background logic – theorems of the standard classical theories of these structures translate into theorems of the modal-structuralist framework, as in the cases of arithmetic and analysis (cf. MWON, Ch. 1). This preservation of theorems holds regarding all the structures we shall consider, so we need not keep repeating the point.

For some theorems involving multiple structures, instances of accumulation principles, described above, may be used. It should be mentioned, however, that in many cases, appeal to such principles can be bypassed and one can reproduce constructions and theorems regarding multiple structures – indeed even infinite classes of structures – by considering certain parts of a single possible universe of infinitely many atoms, each such part endowed with relevant mathematical structure. The example of product spaces in topology is a good illustration and will be described briefly below. (Whether the class of structures can be uncountably infinite depends on the case.)

Let us consider now some structures from *measure theory*, which has posed a challenge to various constructive programs. (We follow Halmos [1974].) Central are *measure spaces*, triples (X, S, μ) where X is a domain of points (e.g., real numbers), S is a class of (measurable) subsets of X (a σ-ring) whose union is X, and μ is an extended real-valued, non-negative, countably additive set function on S assigning 0 to the empty set. Now, if S is the σ-ring generated by a collection of cardinality of $\mathbb{R}$ (e.g., S is the collection of Borel sets of reals, generated by the bounded left semi-closed intervals), it has cardinality of $\mathbb{R}$ also. Then the members of S can be coded as reals so that μ is of the type of a function from $\mathbb{R}$ to $\mathbb{R}$, at the level of PA^3, and so within the framework of ms arithmetic.

---

[6] Indeed, if the vectors of the dense set are codable as natural numbers, as they are in standard examples such as the ℓ^p spaces, then the metrical structure of the separable Hilbert space is describable in PA^2, and indeed much of the theory can be developed in very weak subsystems (see Brown and Simpson [1986]). However, the completion of the countable dense substructure does not formally exist in PA^2; a general structuralist treatment naturally will distinguish the uncountable completions from their countable codes, and this requires ascent past PA^2.

Beyond this, however, one quickly encounters measure-theoretic structures that require resources beyond PA^3. Let X be the real line, $\mathbb{R}^1$, S the class of Borel sets, and μ the Lebesgue measure on S. Let $\bar{S}$ be the result of adding to S all sets of the form $E \cup N$ where $E \in S$ and N is a subset of a member of S of measure 0, and let $\bar{\mu}$ be the completion of μ on $\bar{S}$, i.e. $\bar{\mu}(E \triangle N) = \mu(E)$, E and N as just described, where $\triangle$ is symmetric difference. $\bar{\mu}$ is the complete Lebesgue measure, and the sets of $\bar{S}$ are the Lebesgue measurable sets. Now, since there are uncountable (Borel) sets of (μ) measure 0 (of cardinality of the continuum, in fact, by Cantor's middle third construction), and since $E \in S$, $\mu(E) = 0$, and $F \subset E$ together imply $F \in \bar{S}$, the cardinality of $\bar{S}$ must be that of the power set of $\mathbb{R}$ (by Cantor's cardinality theorem) (cf. Halmos [1974], p. 65 (5), (6)). Thus $\bar{S}$ and $\bar{\mu}$ are essentially at the level of classes of sets of reals, i.e., at RA^3, one step beyond PA^3. But RA^3 suffices; this structure is describable within (nominalistic) ms analysis, and its modal-existence assured.

Similar remarks apply to the theory of outer measure. The relevant structures are extensions of a measure space, (X, S, μ), of the form $(X, H(S), \mu^*)$, where $H(S)$, the *hereditary σ-ring generated by S*, is the smallest σ-ring containing the sets of S and closed with respect to (arbitrary) subsets of such sets, and μ^* is the outer measure defined on $H(S)$ by

$$\mu^*(E) = \inf\{\mu(F) : E \subset F \in S\}.$$

Here again both the class of (outer) measurable sets and the (outer) measure are at the level of classes of sets of reals, hence at RA^3 (PA^4). Once again, the resources of this framework are needed but suffice for this type of structures.

Turning now briefly to topological spaces, the story is similar. In general, these are structures of the form $(X, \mathcal{O})$, where X is a set and $\mathcal{O}$ is a class of *open subsets* of X such that $\mathcal{O}$ contains the empty set and X and is closed under finite intersections and *arbitrary* unions. This leaves open a vast array of possibilities. At one extreme, the *trivial* topology contains just the empty set and X, and at the other extreme the *discrete* topology contains every subset of X. In between lie the most familiar topologies of the real line, the plane, etc. (the usual topologies of $\mathbb{R}^n$), which are *separable*, i.e., have a countable base (e.g., the open n-spheres of rational radii). This means that for every point x in an open set U there is a basic open set B such that $x \in B \subset U$. Thus, every open set in such a topology can be represented as a countable union of basic open sets. In the case of $\mathbb{R}^n$, this means that every open set can be represented by a real number, so that $\mathcal{O}$ itself can be represented as a set of reals. Thus, such topological spaces can be described in PA^3, modal-structural arithmetic. (Similarly for the various spaces encountered in point-set topology.) However, this representation relies on metrical information which is generally not available. A long story here can be shortened considerably by noting that the discrete topology (on X) is a worst case, concerning cardinality (which is the chief guide in determining type level). Thus if X has cardinality of the continuum, any topology on X has cardinality no greater than the power set of the continuum, i.e., $\mathcal{O}$ is at or below the level of a class of sets of reals, at RA^3. And morphisms between such spaces (especially homeomorphisms) are generally capturable at RA^2. Thus, a very rich variety of topological spaces is certainly describable within the nominalist ms framework. Even if higher set theory (even Morse–Kelley) is needed for a completely general theory of topological spaces (free of any cardinality restrictions whatever), clearly a great deal of the subject can be developed on the basis of our two modal-existence postulates above, that is without officially countenancing sets or classes at all. (Note here that the predicativist alternative does not extend nearly so far, as the very notion of closure of a topology under arbitrary unions is not available. This serves further to highlight the contrast between nominalism and even liberal varieties of constructivism.)

Let us end this cursory tour by mentioning three examples of topological spaces of some complexity, as well as importance, that can be treated within our framework. One example is that of n-dimensional manifolds, of great importance in space-time physics. As was already described in MWON, Ch. 3, the general theory of such structures does exceed the reach of RA^2, but only by one level. The effect of adding plural quantifiers is that we now have RA^3 at our disposal, and this does suffice to capture maximal systems of charts, needed to describe manifold structure. It may seem somewhat remarkable that, for example, so abstract and extensive a work as O'Neill's [1983] *Semi-Riemannian Geometry* can be translated virtually entirely without loss into a nominalistic framework, but that does seem to be the case.

Topological spaces often arise from set-theoretic constructions out of classes of functions, and one might expect that set theory is inevitably encountered. Our two final examples serve as an antidote. Consider first the construction of *product spaces* of a given family $\mathcal{F}$ of topological spaces. (We follow Kelley [1955].) Suppose that $\mathcal{F}$ is countably infinite and that each of the spaces, $(X_i, \mathcal{O}_i)$ ($i \in \mathbb{N}$) has a domain X_i at the level of a set of reals. The domain of the product space $\Pi(\mathcal{F})$ consists of the Cartesian product $\Pi\{X_i\}$ of the X_i together with the smallest topology for which inverses of projections onto open sets in the coordinate spaces form a subbase. That is, the domain consists of all functions f from $\mathbb{N}$ to the union of the X_i such that $f(i) \in X_i$, i.e., $\Pi\{X_i\} = \{f : f(i) \in X_i\}$. Since each f is (codable as) a real, this Cartesian product is at the level of a set of reals, i.e., at RA^2. The *product topology* on this domain is motivated by the requirement that the projections P_i onto the factor spaces be continuous, i.e., that for U open in $\mathcal{O}_i$, $P_i^{-1}[U]$ be open. The sets of this form are stipulated to be a subbase for the product topology, i.e., finite intersections of such sets form a base. These are of the form $V = \{f : f(i) \in U_i \text{ for } i \in F\}$, F a finite index set (e.g., a subset of $\mathbb{N}$), U_i an open set in $\mathcal{O}_i$. Such V are at the level of sets of reals, and so the standard set-theoretic construction of the least collection of these closed under arbitrary unions is available in RA^3. Note that for this construction, it is not really necessary to recognize the family $\mathcal{F}$ as an object; it suffices if one is given the collection (or whole) of the domains X_i together with the assumption that for each of these there is a collection (plurality) $\mathcal{O}_i$ of open sets of X_i; and this can be said in RA^3 on our assumption that the X_i are at the level of sets of reals. Note further that this construction in RA^3 depends on the countability of the family of given spaces. If it is uncountable, functions in the Cartesian product will be at the level of sets of reals, not reals, and the product topology itself will be a collection at RA^4. Obviously, fully general topology transcends RA^3. The same limitation does not arise, however, for products of, say, metric spaces, for which the structure is capturable at the level of relations on the points.

Finally, consider the notion of a *sheaf*, which arises as follows. One begins with holomorphic (complex differentiable) functions f, g, etc. on neighborhoods U, V, etc., of a point c in the complex plane. f and g are said to be *germ equivalent at c* just in case f and g agree on some open neighborhood W of c such that $W \subset U \cap V$. The equivalence class $[f]_c$ of functions under this relation is called the *germ of f at c*. One then forms the class A_c of germs at c and then introduces $A =^{\mathrm{df}} \cup_{c \in C} A_c$, called *the sheaf of germs* of holomorphic functions on $\mathbb{C}$. This becomes a topological space upon taking as basic open neighborhoods of $[f]_c$ the class of all germs of f at points of the domain U of f, that is $\{[f]_c : c \in U\}$. (Neighboring germs come from the same holomorphic function.) One then considers morphisms such as the natural projection p of A onto $\mathbb{C}$ sending each germ $[f]_c$ to c, which is a local homeomorphism. Also one has a continuous function $F : A \rightarrow \mathbb{C}$ such that $F([f]_c) = f(c)$ which can be used to represent all holomorphic functions by means of cross-sections (see Mac Lane [1986], pp. 352 f). Now it turns out that the holomorphic functions can be coded as reals and, moreover, that germs can also be so represented via power series in $z - c$ convergent in some open circle about c. Thus the sheaf A becomes identifiable as a set of reals, i.e., at RA^2, as do the basic open neighborhoods. The topology $\mathcal{O}$ for this space is itself then at the next level, at our familiar RA^3. Thus, even this somewhat elaborate and abstract set-theoretic construction is within the reach of the ms framework. Finally, it should be noted, all the diagrams of category theory involving these structures and the various arrows between them can be described as well, as they represent relations among finite tuples of structures (illustrating propositions which may, of course, involve universally or existentially quantified variables ranging over structures or morphisms).

It should be clear that it is not being claimed that set theory is "never needed," whatever that might mean specifically, or that mathematics "ought to" restrict itself to what can be nominalistically described. Rather the point has been simply to illustrate the far-reaching scope of the PA^3 and RA^3 frameworks in the interests of class-free structuralism, to give some idea of how rich a structuralism one may actually have without yet embracing anything so strong as general model theory or general category theory. All these frameworks have their points and their places, and our task has been to understand better just what these are.

There is a sort of corollary worth noting, however, regarding scientific indispensability arguments. As Feferman has already emphasized, in connection with predicativist mathematics, it is remarkable how much of scientifically applicable mathematics can be captured within predicativist systems, and this tends to undercut Quinean arguments for set theory based on indispensability for scientific applications (see, e.g., Feferman [1992]). This holds *a fortiori* for nominalist systems as above, for these reach much further than predicativist systems, as we have already indicated. While some impredicative constructions do arise at the outer limits of applicable mathematics, it would be a real challenge to find anything in the sciences requiring mathematical power beyond the RA^3 framework. Thus, indispensability arguments should be seen in a new light, not as justifying set theory *per se*, but rather as helping (to some degree) to justify key mathematical existence assumptions such as (modal nominalistic) axioms of infinity, including not just (Ax ∞), or (Poss $\mathbb{N}$), and (Poss $\mathbb{R}$), but also the unrestricted comprehension scheme (CΣ) and related principles, such as the full comprehension principle of second-order logic (cf. Hellman [1999]). Surprisingly perhaps, if classes are "genuinely needed," that is probably not because of scientific applications but rather because of needs from within mathematics proper, for instance, because they allow the greatest freedom and ease of construction of anything yet devised. But whether and to what extent we have or can have evidence for the truth or possibility of models of powerful set-theoretic axioms remains unresolved.

Acknowledgement: I am grateful to Stewart Shapiro for helpful comments on an earlier draft. I am also grateful to the National Science Foundation (USA) for support, through Award No. SBER93-10667, of work on which this material is based.

References

Boolos, G. [1985] "Nominalist Platonism," *Philosophical Review* **94**: 327–344.

Brown, D. K. and Simpson, S. G. [1986] "Which set-existence axioms are needed to prove the separable Hahn-Banach theorem?," *Annals of Pure and Applied Logic* **31**: 123–144.

Burgess, J. [1983] "Why I am not a nominalist," *Notre Dame Journal of Formal Logic* **24**(1): 93–105.

Burgess, J., Hazen, A., and Lewis, D. [1991] "Appendix," in Lewis, D., *Parts of Classes* (Oxford: Blackwell), pp. 121–149.

Chihara C. [1973] *Ontology and the Vicious Circle Principle* (Ithaca, NY: Cornell University Press).

Dedekind, R. [1888] "The nature and meaning of numbers," reprinted in Beman, W. W., ed., *Essays on the Theory of Numbers* (New York: Dover, 1963), pp. 31–115, translated from the German original, *Was sind und was sollen die Zahlen?* (Brunswick: Vieweg, 1888).

Feferman, S. [1964] "Systems of predicative analysis," *Journal of Symbolic Logic* **29**: 1–30.

Feferman, S. [1968] "Autonomous transfinite progressions and the extent of predicative mathematics," in *Logic, Methodology, and Philosophy of Science III* (Amsterdam: North Holland), pp. 121–135.

Feferman, S. [1977] "Theories of finite type related to mathematical practice," in Barwise, J., ed., *Handbook of Mathematical Logic* (Amsterdam: North Holland), pp. 913–971.

Feferman, S. [1988] "Weyl vindicated: *Das Kontinuum* 70 years later," in *Temi e Prospettive della Logica e della Filosofia della Scienza Contemporanee* (Bologna: CLUEB), pp. 59–93.

Feferman, S. [1992] "Why a little bit goes a long way: logical foundations of scientifically applicable mathematics," in Hull, D., Forbes, M., and Okruhlik, K., eds., *Philosophy of Science Association 1992*, Volume 2 (East Lansing, MI: Philosophy of Science Association), pp. 442–455.

Feferman, S. and Hellman, G. [1995] "Predicative foundations of arithmetic," *Journal of Philosophical Logic* **24**: 1–17.

Field, H. [1980] *Science without Numbers* (Princeton, NJ: Princeton University Press).

Field, H. [1992] "A nominalist proof of the conservativeness of set theory," *Journal of Philosophical Logic* **21**: 111–123.

Goodman, N. [1977] *The Structure of Appearance*, 3rd edn. (Dordrecht: Reidel).

Halmos, P. R. [1974] *Measure Theory* (New York: Springer-Verlag).

Hellman, G. [1989] *Mathematics without Numbers: Towards a Modal-Structural Interpretation* (Oxford: Oxford University Press).

Hellman, G. [1990] "Modal-structural mathematics," in Irvine, A. D., ed., *Physicalism in Mathematics* (Dordrecht: Kluwer), pp. 307–330.

Hellman, G. [1994] "Real analysis without classes," *Philosophia Mathematica* **2**(3): 228–250.

Hellman, G. [1999] "Some ins and outs of indispensability," in Cantini, A., et al., eds., *Logic and Foundations of Mathematics* (Dordrecht: Kluwer), pp. 25–39.

Kelley, J. L. [1955] *General Topology* (New York: Van Nostrand).

Lewis, D. [1991] *Parts of Classes* (Oxford: Blackwell).

Mac Lane, S. [1986] *Mathematics: Form and function* (New York: Springer-Verlag).

Mayberry, J. [1994] "What is required of a foundation for mathematics," *Philosophia Mathematica* **2**(1): 16–35.

McLarty, C. [1993] "Numbers can be just what they have to," *Noûs* **27**: 487–498.

O'Neill, B. [1983] *Semi-Riemannian Geometry with Applications to General Relativity* (Orlando, FL: Academic Press).

Parsons, C. [1990] "The structuralist view of mathematical objects," *Synthese* **84**: 303–346.

Resnik, M. [1981] "Mathematics as a science of patterns: ontology and reference," *Noûs* **15**: 529–550.

Shapiro, S. [1983] "Mathematics and reality," *Philosophy of Science* **50**: 522–548.

Shapiro, S. [1989] "Structure and ontology," *Philosophical Topics* **17**: 145–171.

Shapiro, S. [1991] *Foundations without Foundationalism: A Case for Second-Order Logic* (Oxford: Oxford University Press).

Shapiro, S. [1997] *Philosophy of Mathematics: Structure and Ontology* (New York: Oxford University Press).

Smorynski, C. [1982] "The varieties of arboreal experience," *Mathematical Intelligencer* **4**: 182–189.

Tait, W. W. [1986] "Critical notice: Charles Parsons," *Mathematics in Philosophy, Philosophy of Science* **53**: 588–606.

2   What Is Categorical Structuralism?

In a recent paper, Hellman [2003], we examined to what extent category theory (CT) provides an autonomous framework for mathematical structuralism. The upshot of that investigation was that, as it stands, while CT provides many valuable insights into mathematical structure – specific structures and structure in general – it does not sufficiently address certain key questions of logic and ontology that, in our view, any structuralist framework needs to address. On the positive side, however, a *theory of large domains* was sketched as a way of supplying answers to those key questions, answers intended to be friendly to CT both in demonstrating its autonomy vis-à-vis set theory and in preserving its "arrows only" methods of describing and interrelating structures and the insights that those methods provide. The "large domains," hypothesized as logico-mathematical possibilities, are intended as suitably rich background universes of discourse relative to which both category-and-topos theory and set theory can be developed side by side, without either emerging as "prior to" the other. Although those domains, as described, resemble natural models of set theory (on an iterative conception) or toposes suitably enriched with an equivalent of the Replacement Axiom, they are defined without set-membership as a primitive, and also without 'function' or 'category' or 'functor' as primitives; all that is required is a combination of 'part/whole' and plural quantification (in effect, the resources of monadic second-order logic). This background logic (including suitable comprehension axioms for wholes and "pluralities") suffices; and ontological commitments are limited to claims of the possibility of indefinitely large domains, any one extendable to a more encompassing one, without end.

Two interesting responses to this have already emerged on behalf of CT proponents, one by Colin McLarty [2004] and the other by Steve Awodey [2004]. Here we take the opportunity to come to terms with these and to assess their bearing on our original assessment and proposal. We will begin with a brief review of the main critical points of Hellman [2003]; then we will take up the responses of McLarty and Awodey in turn; and finally, we will try to draw appropriate conclusions.

2.1   "Category Theory" and Structuralist Frameworks

The first point to stress is that the very term "category theory" is ambiguous, and the ambiguity follows closely on the heels of another, more basic one, that of "axiom" itself. On the one hand, axioms traditionally are conceived as basic truths simpliciter, as in the traditional conception of Euclidean geometry, or the axioms of arithmetic, or the axioms of, say, Zermelo–Fraenkel set theory. Call this the "Fregean conception" of axioms. In the geometric case, primitive terms such as 'point', 'line', 'plane', 'coincident', 'between', 'congruent' are taken as determinate in meaning, so that axioms employing them have a determinate truth-value. For number theory, 'successor', 'plus', 'times', 'zero', etc. have definite meanings leading to true axioms (say the Dedekind–Peano axioms); and for set theory, of course, 'membership' is taken as understood, and the axioms framed in its terms true (or true of the real world of sets). In contrast, there are algebraic-structural axioms for groups, modules, rings, fields, etc., where now they are not even assertions, but rather *defining conditions* on types of structures of interest. The primitive terms are not thought of as already determinate in meaning but only as schematically playing certain roles as required by the "axioms." Call this the "Hilbertian conception." Any objects whatever interrelated in the ways required by the defining axioms constitute a structure of the relevant type, say, a group, or a module, . . ., or a category. In the latter case, primitive terms such as 'object', 'morphism', 'domain', 'codomain', and 'composition' are not definite in meaning, but acquire meaning only in the context of a particular interpretation which satisfies the axioms.

Thus, "morphisms" need not be functions, and (so) "composition" need not be the usual composition of functions, etc.[1] A large part of the debate between Frege and Hilbert on foundations turned on their respective, very different understandings of "axioms" along precisely these lines.

It is worth noting, incidentally, that Dedekind [1888] presented his "axioms" for arithmetic explicitly as defining conditions, i.e. axioms in the Hilbertian sense, as part of his definition of a "*simply infinite system*," and not as assertions.

One may, oversimplifying a bit, say that the tendency of modern mathematics toward a structuralist conception has been marked by the rise and proliferation of Hilbertian axiom systems (practically necessitated by the rise of non-Euclidean geometries), with relegation of Fregean axioms to a set-theoretic background, usually only mentioned in passing in introductory remarks. Category theory surely has contributed to this trend; we now even have explorations of "Zermelo–Fraenkel algebras" [Joyal and Moerdijk, 1995].

---

[1] Thus, "morphisms" may be realized as homotopy classes of maps (between topological spaces), as formal deductions of formulas in a logical system, as directed line segments in a diagram, etc.

This ambiguity over "axioms" is, of course, passed on to "theories" of algebraic structures, as in "group theory," "field theory," ..., and, indeed, "category theory" and (with some qualifications to be discussed below) "topos theory" as well. On the one hand, there is the first-order theory (definition) of groups, or of categories or toposes; but, on the other, there is a body of substantive theorizing *about* such structures, which, while constantly appealing to the first-order definitional axioms, is intended as *assertory*, and takes place in an informal background whose primitive notions and assumptions usually require logical analysis and reconstruction to be identified. Standard practice refers (in passing) to a background set theory, as it is well known that that suffices for most purposes. But of course that cannot serve in the context of "categorical foundations" where autonomy from set theory is the name of the game.

So what is the background theory? It is not clear. And so we find ourselves uncertain when it comes to comparing categorical structuralism with other frameworks that have been proposed. Here are five fundamental questions that we would submit any such framework should address:

(1) What is the background logic? Is it classical? Is it modal? Is it higher-order logic? If so, what is the status of relations as objects?

(2) What are the extra-logical primitives and what axioms – presumably assertory – govern them? Are 'collection', 'operation', 'category', 'functor', for example, on the list? Especially, what axioms of mathematical existence are assumed?

(3) Is indefinite extendability of mathematical structures recognized or is there commitment to absolutely maximal structures, for example of absolutely all sets, all groups, etc.?

(4) Are structures eliminated as objects, and, if not, what is their nature?

(5) What account, if any, is given of our reference and epistemic access to structures?

In the case of set-theoretic structuralism, it is fairly clear how to answer at least (1)–(4); similarly, in the case of Shapiro's [1997] *ante rem* structuralism, and he takes a stab at (5) as well. In the case of modal-structuralism, answers to (1)–(4) are also forthcoming. (As an eliminativist structuralism, questions of reference are replaced with questions of knowledge of possibilities, related to, but even more difficult than, questions of knowledge of consistency, for in central cases we are interested in *standard* structures.)[2] But when it comes to categorical structuralism, it is not clear what to say even with regard to (1)–(4). At most, bearing on (3), one finds widespread opposition to the view that a fixed background of sets is the privileged arena of mathematics.[3]

---

[2] For detailed comparisons of these varieties of structuralism, see Hellman [2001].

[3] For further details, see Hellman [2003].

2.2   McLarty's "Fregean" response

In a nutshell, McLarty claims that, while the algebraic-structuralist reading of CT axioms and general topos axioms is correct, nevertheless specific axioms for certain particular categories and toposes are intended as assertory. In particular, he singles out ETCS, the elementary theory of the category of sets, CCAF, the category of categories as a foundation, and synthetic differential geometry (SDG) as a theory of the category of smooth spaces. The axioms of these systems are not to be read merely as defining types of structures but rather as assertions, true of existing parts of mathematical reality, much as the axioms of ZFC are normally understood. Indeed, in the case of ETCS, this could be understood as describing the very same subject matter as ZFC, although with the characteristic arrows machinery rather than a primitive set-membership relation.

It appears to me that CCAF, or, better, McLarty's [1991] own approach to axiomatizing a category of categories, is actually the most promising in relation to the above questions. Let us return to consider this below. First, let us take up the other two examples, ETCS and SDG.

Now, I would not wish to deny that ETCS provides an important part of a structuralist analysis of sets. Through its "arrows only" formulations and generalizations, it abstracts from a fixed set-membership relation and analyzes sets in their functional roles, "up to isomorphism," which is all that really matters for mathematics. What remains problematic, however, regarding McLarty's reading of ETCS (which he attributes to Mac Lane), is its apparent commitment to a fixed, presumably maximal, real-world universe of sets, "*the* category of sets." This just strikes me as a convenient fiction. First, there is the question of multiplicity of conceptions of sets, for example non-well-founded as well as well-founded, possibly choice-less as well as with choice, with or without Replacement, the various large cardinal extensions, and so forth. Presumably, all of these conceptions are mathematically legitimate, and it would be arbitrary to treat just one as ontologically privileged. But even if suitable qualifications of the "intended universe" are added to the (meta) description, the problem of indefinite extendability still looms. Whatever domain of sets we recognize can be transcended by the very operations that set theory seeks to codify, collecting, collecting everything "already collected," passing to collections of subcollections, iterating along available ordinals, etc. (This, incidentally, is entirely in accord with Mac Lane's expressed views [Mac Lane, 1986] on the open-endedness of mathematics.) Set-theoretic structural-ism can be faulted precisely for failing to apply to set theory itself, especially in regard to the very multiplicity of universes of sets that it naturally engenders. Categorical structuralism promises to do better, but it is hard put to keep that promise if it falls back on a maximal universe of sets or, more generally, on an absolute notion of "large category."

When it comes to a realist interpretation of SDG, the problems are quite different but equally challenging. This is a non-classical theory of continua which can be developed independently of category theory, known as "smooth infinitesimal analysis" (SIA). (Topos theory has proved useful in providing models of SIA, but the essential analytic ideas do not depend on the topos machinery.) This is a theory intended as an alternative to classical, "puncti-form" analysis; it introduces nilsquare (and nilpotent) infinitesimals, while at the same time limiting the class of functions of reals to smooth (C^∞) ones. A central axiom, the Kock–Lawvere (KL) axiom, stipulates that any function on the infinitesimals about 0 obeys the equation of a straight line. (The axiom is also called the Principle of Microaffineness.)

This actually implies the restriction to continuous functions. And the constant slope of the "linelet" given by the axiom serves to define the derivative of a function. (The "linelet" can be translated and rotated, but not "bent.") To accommodate nilsquares, certain restrictions apply to the classical ordered field axioms for the reals: nilsquares do not have multiplicative inverses, nor are they ordered with respect to one another or with respect to 0. Indeed, one proves that "not every x is either = 0 or not = 0." Not only does SIA refrain from using the Law of Excluded Middle (LEM), it derives results that are formally inconsistent with it (similar in this respect to Brouwerian intuitionism but contrasting with Bishop constructivism, which, with LEM added, gives back classical analysis). But SIA, consisting of the (restricted) ordered field axioms, the KL axiom, and a certain "constancy principle," suffices for a remarkable development of calculus in which limit computations are replaced with fairly straightforward algebra, placing on an alternative, consistent and rigorous footing early pre-limit geometric methods in analysis and mechanics (cf. Bell [1998] for a nice survey of such results).

Why is there a problem with thinking of this theory as an objective description of continuous functions or phenomena? After all, the charge by Russell and fellow classicists that infinitesimals lead to inconsistencies, while true of some naïve, informal practice, is demonstrably not true here, at least relative to the consistency of classical analysis. One simply must renounce LEM and, as already said, tolerate things like negations of generalizations of it, as just noted.

The difficulty comes when we attempt to explain why LEM fails, even though (under the realistic hypothesis we are entertaining) there really are nilsquares making up the "glue" of actual continua, the points of classical analysis now regarded as a useful but fictitious formal artifact of analytic methods. If there really are such "things," does not logical identity apply to them just as to everything else, regardless of our abilities to discriminate them from one another or from 0? The situation is really very different from that posed by intuitionistic analysis. There constructive meanings of the logical operators, disjunction, negation, the conditional, both existential and universal quantifiers, obviously do not sustain the formal LEM or related classically valid principles, for example quantifier conversions such as "not for every x $\varphi(x)$" to "there exists x such that not $\varphi(x)$." Apparent conflicts with classical analysis are only apparent due to these radically different meanings. But none of this is applicable in interpreting SIA, for constructive meanings do not seem appropriate to the subject. Nilsquares, for example, are not constructed at all; indeed, their existence cannot be asserted any more than it can be denied, on pain of contradiction. Rather one must settle for the double negation of existence. While this in itself might seem compatible with a constructive reading, the KL axiom itself seems, if anything, more dubious on such a reading. (What method do we have for finding the slopes of the linelets from a given constructive function on the nilsquares? Indeed, in what sense are we ever presented with such a function?) Lawvere and others have taken failure of LEM for nilsquares to express "non-discreteness," perhaps in analogy with familiar kinds of vagueness. But, once we are speaking of objects at all, however invisible or intangible, how can the predicate '$\_ = 0$' itself be vague? Though from a setting he never contemplated, one harks back to Quine: "No entity without identity!"

Indeed, with the tendency to speak of $\neq$ as "distinguishable" [Bell, 1998], it is natural to seek an interpretation of '=' in SIA as an equivalence relation broader than true identity, and this suggests trying to recover SIA in a *classical* interpretation. Such an interpretation has actually been carried out in detail [Giordano, 2001]. Certain differences emerge: the class of functions treated is narrower than all continuous ones (a Lipschitz condition is invoked), but the KL axiom and much of the theory are recovered on a fully classical basis. Whether proponents of SIA and SDG will plead "change of subject" remains to be seen.

Turning to "category of categories," efforts towards axiomatization are at least grabbing the bull by the horns, laying down explicit assertory axioms on the mathematical existence of categories and providing a unified framework for a large body of informal work on categories and toposes (hence mathematics, generally). Three questions demand our attention. (1) What concepts are presupposed in such an axiomatization? (2) Are these such as to sustain the autonomy of CT vis-à-vis set theory or related background, or do they reveal a (possibly hidden) dependence thereon? (3) What is the scope of such a (meta) theory, in particular, what are the prospects for self-applicability and the idea of "*the* category of (absolutely) *all* categories"?

On (1) and (2), it is clear that these axioms (as in McLarty [1991]) are not employing the CT primitives ('object', 'morphism', 'domain', 'codomain', 'composition') schematically, as in the algebraic defining conditions, but with intended meanings presumably supporting at least plausible truth of the axioms. The objects are *categories*, the morphisms are *functors between categories*, etc. Commenting on this, Bell and I recently wrote:

Primitives such as 'category' and 'functor' must be taken as having definite, understood meanings, yet they are in practice treated algebraically or structurally, which leads one to consider interpretations of such axiom systems, i.e. their semantics. But such semantics, as of first-order theories generally, rests on the set concept: a model of a first-order theory is, after all, a set. The foundational status of first-order axiomatizations of the [better: a] metacategory of categories is thus still somewhat unclear. [Hellman and Bell, 2006]

In other words, when we speak of the "objects" and "arrows" of a metacategory of categories as *categories* and *functors*, respectively, what we really mean is "structures (or at least "interrelated things") satisfying the algebraic axioms of CT," i.e. we are using "satisfaction" which is normally understood set-theoretically. That is not to say that there are no alternative ways of understanding "satisfaction"; second-order logic or a surrogate such as the combination of mereology and (monadic) plural quantification of modal-structuralism would also suffice. But clearly there is some dependence on a background that explicates *satisfaction* of sentences by structures, and this background is not "category theory" itself, either as a schematic system of definitions or as a substantive theory of a metacategory of categories. But this need for a background theory explicating *satisfaction* was precisely the conclusion we came to in Hellman [2003], reinforcing the well-known critique of Feferman [1977], which exposed a reliance on general notions of "collection" and "operation." It was precisely to demonstrate that this in itself does not leave CT structuralism dependent on a background set theory that I proffered a membership-free theory of large domains as an alternative. Although the reaction, "Thanks, but no thanks!" frankly did not entirely surprise me, it will also not be surprising if a perception of dependence on a background set theory persists.

As to the third question of scope, I think it is salutary that McLarty calls his system "a (meta) category of categories," rather than "*the* category of categories," which flies in the face of general extendability. No structuralist framework should pretend to "all-embracing completeness," in Zermelo's [1930] apt phrase. And we certainly had better avoid such things as "the category of exactly the non-self-applicable categories"! But it is, I believe, an open question just what instances of impredicative separation should be allowed.

To conclude this section, it seems a fair assessment to say that, while axioms for a (meta) category of categories do make some progress toward providing answers to some of our five questions put to the various versions of mathematical structuralism, we are left still well short of satisfactory, full answers, even to the first four.

2.3   Awodey's "Hilbertian" Response

In contrast to the foregoing, this response takes as its point of departure an "anti-foundationalist" stance: mathematics should not be seen as based on a fixed universe of special objects, the elements of domains of structures, the *relata* of structural relations, as on the set-theoretic view. Instead, mathematics has a *schematic* character, which seems to mean two things: any theorem includes hypothetical conditions, which govern just what aspects of structure are relevant; and, in any case, the particular nature of individual objects is irrelevant. Moreover, whereas modal-structuralism tries to get at this by open-ended modal quantification (which is not to be interpreted as ordinary quantification over a fixed background domain of *possibilia*, as Awodey seems to recognize), category theory itself provides a more direct expression, standing on its own without need of any further (assertory) background principles.[4] The central, general but flexible primitive notion is "morphism," capable of grounding talk of relations, operations, etc. (A summary list of relevant "arrows only" categorical concepts illustrates this.) A "top-down" metaphor, as opposed to "bottom-up," is used; it seems that it is sufficient simply to describe whatever ambient background structure we deem relevant to the mathematical purpose at hand, without needing to worry about any absolute claims of existence. (Clearly, this is reminiscent of Hilbert's view that sought to eliminate metaphysics from mathematics by, in effect, replacing absolute claims of existence with a combination of proofs from formal axioms, as defining conditions, together with a proof of formal consistency of the relevant axioms, although presumably, since we are in a post-Gödelian era, the latter demand is omitted.)

I see a dilemma in understanding all this. Either mathematics is adequately understood as just a complex network of deductive and conceptual interconnections, or it is not. On the first horn, what we are really presented with is a kind of formalism, in which theorems in conditional form, together with definitions, are all there is to mathematics, that is, we just give up on the notion of mathematical truth as anything beyond deductive logical validity. In this case, we really need not worry about primitives with meanings supporting basic axioms in the Fregean sense. Questions of truth (beyond first-order logical truth of conditionals) would arise only in certain limited cases, typically in applications of mathematics where we might be in a position to assert that the antecedent conditions are indeed fulfilled (e.g. that certain finite structures, say, are actually instantiated, or that even certain infinite ones are, say space or

---

[4] It should be clear that modal-structuralism, although it does provide such background principles, is also quite explicitly "non-foundationalist" in Awodey's sense. To avoid confusion, I have preferred to speak of "structuralism frameworks" rather than "foundations," but I would certainly plead guilty to "foundational concerns," much in the spirit of Shapiro [1991]. Clearly, this is reflected in the questions (1)–(5) we have been putting to the various versions of structuralism.

time or space-time as continua). Whether this is intended and is viable, after all (i.e. after all the criticisms that have been levelled against deductivism), remain to be determined.[5]

On the other horn of the dilemma, we take seriously the idea that 'morphism' is a primitive with definite, if multifaceted, meaning, giving genuine content to mathematics beyond mere inferential relations. But what is that meaning? And what is the content beyond inferential relations? Awodey himself [1996] has stressed the algebraic-structural character of the CT axioms, and, unlike McLarty, he does not appeal to any special topos axioms as assertions, nor does he appeal to a special category of categories. It seems clear, then, that the notion of morphism – which, unlike the notion of 'part/whole' employed in modal-structuralism and also exhibiting a kind of "schematic character," is a mathematical term of art, not a familiar one in ordinary English – depends on the context, viz. *on the category or categories presupposed or in which one is working*. As indicated at the outset, arrows (i.e. morphisms) need not be ordinary functions. They need only satisfy the conditions on "arrows" of the CT axioms or extensions thereof. Surely, this is what should be said in explaining "what morphisms are." Moreover, functors do more than ordinary functions, and they are centrally involved in "the usual language and methods of category theory" (Awodey [2004], p. 62). But then what we really have as primitives are "*satisfaction of axioms*" and "*functor between categories*"; i.e. we are presupposing "*category*" as a primitive as well. But this brings us right up against the same problem that confronted the previous view, namely that we are falling back on *prima facie* set-theoretic notions after all. The main difference seems to be that, whereas on the McLarty view we were at least being given axioms asserting the mathematical existence of various categories, here we are not even being given that. In any case, the CT "arrows only" explications of "relations," "operations," and so forth, are of no avail until we first understand "morphism" (i.e. "arrow"), "functor" and "category," i.e. until we already understand *satisfaction* or equivalent (second- or higher-order) notions. It would be plainly circular to appeal to "morphisms" to explain this!

2.4   Conclusion

The contrast between Fregean and Hilbertian axioms seems to present us with a stark choice. But really, unless we go back to formalism, mathematics requires both. For all the axiom systems of ordinary mathematics, for number theory, analysis, algebra, pure geometry, topology, and surely much of category

---

[5] I have been taking as a ground rule for articulating structuralism that it should not collapse to formalism or deductivism. If CT structuralism is playing by different rules, that certainly should be made explicit.

and topos theory, i.e. for all commonly studied structures and spaces, not only is the Hilbertian conception appropriate, it is part and parcel of standard modern practice. But when we step back and contemplate fundamental and foundational issues – when we ask questions about what principles govern the mathematical existence of structures generally, or when we consider the closely related "unfinished business" of Hilbert's own program (as Shapiro puts it), the place of metamathematics, questions of absolute and relative formal consistency, questions of (informal) higher-order "consistency" or "coherence," relative interpretability, independence, etc. – then we are in the realm of outright claims, not mere hypotheticals as to what holds or would hold in any "structures" satisfying putative algebraic (meta) axioms of metamathematics. Rather we are seeking assertory axioms in the Fregean sense.[6] Thus, in connection with category theory, the advice of Berra [1998], "When you come to a fork in the road, take it!"[7] is quite apt, and the theory of large domains I sketched in Hellman [2003] was one way of taking the advice (and the fork!). The alternative responses considered here, categories of categories (Fregean) or category theory as schematic mathematics (Hilbertian), lead us straight back to *prima facie* set-theoretic notions, only slightly beneath the surface, and so do not sustain category theory as providing an autonomous structuralist framework adequate to the needs of both mathematics and metamathematics.

Acknowledgement: Support of this work by the National Science Foundation (USA, Award SES-0349804) is gratefully acknowledged.

References

Awodey, S. [1996] "Structure in mathematics and logic: a categorical perspective," *Philosophia Mathematica* **4**: 209–237.

Awodey, S. [2004] "An answer to Hellman's question: 'Does category theory provide a framework for mathematical structuralism?'," *Philosophia Mathematica* **12**: 54–64.

Bell, J. L. [1998] *A Primer of Infinitesimal Analysis* (Cambridge: Cambridge University Press).

Berra, Y. [1998] *The Yogi Book: "I Really Didn't Say Everything I Said!"* (New York: Workman Publishing).

Dedekind, R. [1888] "The nature and meaning of numbers," reprinted in Beman, W. W., ed., *Essays on the Theory of Numbers* (New York: Dover, 1963), pp. 31–115, translated from the German original, *Was sind und was sollen die Zahlen?* (Brunswick: Vieweg, 1888).

---

[6] This is highlighted in Shapiro [2005]. As he reminds us, Hilbert clearly regarded the claims of proof theory as "contentful," closely related to statements of number theory (insofar, "mathematical" as well as "metamathematical"), and not conditional with respect to implicitly defined (meta) structures of metatheory.

[7] One road leads to Göttingen, the other to Jena.

Feferman, S. [1977] "Categorical foundations and foundations of category theory," in Butts, R. E. and Hintikka, J., eds., *Logic, Foundations of Mathematics, and Computability Theory* (Dordrecht: Reidel), pp. 149–169.

Giordano, P. [2001] "Nilpotent infinitesimals and synthetic differential geometry in classical logic," in Berger, U., Osswald, H., and Schuster, P., eds., *Reuniting the Antipodes–Constructive and Nonstandard Views of the Continuum* (Dordrecht: Kluwer), pp. 75–92.

Hellman, G. [2001] "Three varieties of mathematical structuralism," *Philosophia Mathematica* 9: 184–211.

Hellman, G. [2003] "Does category theory provide a framework for mathematical structuralism?" *Philosophia Mathematica* 11: 129–157.

Hellman, G. and Bell, J. L. [2006] "Pluralism and the foundations of mathematics," in Bell, J. L., ed., *Minnesota Studies in the Philosophy of Science Series*, Volume 19 (Minneapolis, MN: University of Minnesota Press), pp. 64–79.

Joyal, A. and Moerdijk, I. [1995] *Algebraic Set Theory* (Cambridge: Cambridge University Press).

Mac Lane, S. [1986] *Mathematics: Form and Function* (New York: Springer-Verlag).

McLarty, C. [1991] "Axiomatizing a category of categories," *Journal of Symbolic Logic* 56: 1243–1260.

McLarty, C. [2004] "Exploring categorical structuralism," *Philosophia Mathematica* 12: 37–53.

Shapiro, S. [1991] *Foundations without Foundationalism: A Case for Second-Order Logic* (Oxford: Oxford University Press).

Shapiro, S. [1997] *Philosophy of Mathematics: Structure and Ontology* (New York: Oxford University Press).

Shapiro, S. [2005] "Categories, structures, and the Frege–Hilbert controversy: The status of metamathematics," *Philosophia Mathematica* 13: 61–77.

Zermelo, E. [1930] "Über Grenzzahlen und Mengenbereiche: Neue Untersuchungen über die Grundlagen der Mengenlehre," *Fundamenta Mathematicae*, 16: 29–47; translated as "On boundary numbers and domains of sets: new investigations in the foundations of set theory," in Ewald, W., ed., *From Kant to Hilbert: A Source Book in the Foundations of Mathematics*, Volume 2 (Oxford: Oxford University Press, 1996), pp. 1219–1233.

3   On the Significance of the Burali-Forti Paradox

3.1   Background

Often the Burali-Forti paradox is referred to as the paradox of "the largest ordinal," which goes as follows: Let Ω be the class of all (von Neumann, say) ordinals. Then, since Ω represents the order-type of the well-ordering $<$ on ordinals (i.e. $\in$ restricted to the ordinals), Ω itself qualifies as an ordinal. But then it has a successor, $\Omega + 1$, which is an ordinal and so must occur as a member of Ω, by definition of the latter as the class of *all* ordinals. But then we have that $\Omega + 1 < \Omega < \Omega + 1$, a contradiction. (Or, more directly, $\Omega \in \Omega$ whence $\Omega < \Omega$, contradiction.)[1]

How this is handled depends of course on how one formalizes set theory. On the standard set theory of contemporary mathematics, the first-order system, ZFC, the "paradox" is blocked from the start as it cannot even be presented there, as the language of ZFC lacks any way of designating proper classes generally, hence no means of introducing Ω. Thus, ZFC resolves the Burali-Forti paradox in the same way that it resolves the paradoxes of Russell ("set of all and only non-self-membered sets") and Cantor ("set of all sets"), i.e. by inferring that there simply cannot exist a set which contains all ordinals and hence dominates them (and itself) as itself an ordinal.

What happens if, instead, set theory is formalized as a two-sorted theory à la NBG (von Neumann, Bernays, Gödel), with explicit machinery for referring to proper classes and distinguishing them from sets? Then Ω can be introduced by legitimate definition, but that refers to the (proper) class of all ordinals which are sets. So, even though there is also recognized the proper class of all ordered pairs constituting the ordering relation, $<$, holding among set-ordinals, with Ω obviously representing its order-type, it is not even well formed in NBG to write $\Omega < \Omega$ since that means $\Omega \in \Omega$, violating the restriction that proper-class

---

[1] For some interesting history of the Burali-Forti paradox, see Menzel [1984], Moore and Garciadiego [1981], Ferreirós [2007].

terms cannot occur in the first position of the ϵ-relation. Once again, the reasoning of the Burali-Forti paradox cannot get off the ground.

The treatment of the Burali-Forti paradox in second-order ZFC is exactly analogous to that of two-sorted NBG.

What happens, however, if instead of a two-sorted first-order or standard second-order formalization of set theory, we choose to work with a version formalized with the aid of plural quantification? As we shall see, then matters are considerably more interesting, and, as will be argued, the implications deeper and more far-reaching.

3.2   Burali-Forti in Set Theory with Plural Quantification

Let us begin by laying down some desiderata on a theory of ordinals. The first two are of a mathematical character, the third more logico-philosophical.

(1) It should be demonstrable that ordinals satisfy transfinite induction.
(2) (A generalization of) the theorem of Hartogs should hold, namely that for any given ordinals, there exists a least strict cardinal upper bound in the sense of having greater cardinality than that of any of the given ordinals.
(3) We should also strive to meet the condition that any well-ordering, as a relation, should be represented by a unique ordinal, in the sense that the pairs of the given well-order relation should be in one–one order-preserving correspondence with the pairs of ordinals strictly less than the representing ordinal (where the ordering of ordinals is that used in defining them).

Let us write $\gamma \, \eta \, \alpha\alpha$ as abbreviating 'γ is among the ordinals $\alpha\alpha$', where $\alpha\alpha$ is a plural variable ranging over ordinals; also, let us write $\alpha\alpha_<$ as a plural variable ranging over ordered pairs (β, γ) of ordinals, β, γ, among the $\alpha\alpha$ such that $\beta < \gamma$. Then our desideratum (3) says that we should have, for any given well-order relation R, construed as a plurality zz of pairs, $\langle x, y \rangle$, that there is a unique ordinal λ representing R in the sense that there exists a bijective mapping f from relata x, y of R to exactly the ordinals $\alpha\alpha < \lambda$ satisfying

$$\langle x, y \rangle \eta \, zz \;\; \leftrightarrow \;\; \langle f(x), f(y) \rangle \eta \, \alpha\alpha_<.$$

Here, we assume that ordered pairing is available, either from the background set theory or from a substitute such as a modal-structural reconstruction (utilizing say the methods of pairing of Burgess, Hazen, and Lewis [1991]). Furthermore, note that this method of describing relations is general enough to apply both to relations as sets and to relations as classes. But the advantage of plural variables is that they serve to replace variables ranging over objects of any given type. (Talk of "pluralities" is strictly a *façon de parler*, not a reference to any special objects.) Thus, we have the means of expressing a theory of relations literally without treating relations as special objects distinct from the relata and pairs or n-tuples thereof.

Now let us observe an elementary fact about ordinals in their natural well-ordering, $<$. Let us assume that the ordinals begin, as do the von Neumann ordinals, with the null set representing the null well-ordering.

Furthermore, assume that ordinals $\beta\beta$ are downward closed, i.e. if $\gamma < \delta$ and δ η $\beta\beta$ then γ η $\beta\beta$. Then we have the following basic fact about ordinals as a good representation of well-order relations (e.g. the von Neumann ordinals).

Proposition: *If the $\beta\beta$ are represented by ordinal λ, then λ is the least strict upper bound of the $\beta\beta$, so that it is not the case that λ η $\beta\beta$.*

Proof: By transfinite induction: the cases of the null well-ordering and of successor ordinals are trivial. For limits, λ, by hypothesis λ is the order-type of all its predecessors (under $<$); so if λ η $\beta\beta$, the order-type of the $\beta\beta$ has the predecessors of λ as a proper initial segment, call them $\alpha\alpha$, but then the $\beta\beta$ cannot be order-isomorphic to the $\alpha\alpha$, since any bijective order-preserving map on the $\alpha\alpha$ to an extension thereof must be the identity map there. This contradicts the assumption that λ represents the well-ordering of the $\beta\beta$. (That λ must be *least* strict upper bound on the $\beta\beta$ follows similarly.)

But now we have the following corollary:

Corollary: *If it makes sense to refer to "absolutely all ordinals," then it is impossible to satisfy desideratum (3) above.*

For then obviously all the ordinals – call them $\omega\omega$ – are well-ordered by $<$ but if an ordinal Ω represents this well-order relation, it would have to occur as one of the $\omega\omega$, which by the Proposition is impossible. This is just the Burali-Forti "paradox" in this setting.[2]

What has emerged, however, is the point that we have a *choice*: *either* we stick with the above instance of "absolute generality" and give up on desideratum (3), *or* we seek to enforce the latter but deny the antecedent of the Corollary, that it makes sense to refer to "absolutely all ordinals," or "absolutely all well-order relations." Standard set-theoretic presentations of Burali-Forti obscure this choice because they build in type distinctions which then can be exploited in the ways reviewed above of blocking contradiction. But the formulation via plurals in effect collapses type distinctions, allowing full

---

[2] Note that the proof of the Proposition does not invoke desideratum (2). The alert reader will have noticed that that desideratum also leads to problems closely related to Burali-Forti. As usually stated, Hartogs' theorem governs arbitrary *sets* of ordinals, but the generalization in the language of plurals applies to "arbitrary collections," as it were. So again, if it makes sense to refer to "absolutely all ordinals," then this would present an exception to Hartogs' theorem, as there could then be no initial ordinal, cardinally – hence ordinally – greater than *all* ordinals! ("Initial ordinal" just means "ordinal cardinally greater than any earlier ordinal.")

generality in quantification over well-orderings. This opens up the choice just indicated, and the question then comes to the fore, how, without absolute generality, can we implement desideratum (3)?

A key idea goes back to Zermelo in his great [1930] paper: quantification over ordinals (and sets generally) makes sense only relative to a hypothesized model of given set-theoretic axioms, in the case of Zermelo's focus, the ZF^2C axioms (the superscript indicating that the axioms of Replacement and Separation were understood as second-order statements rather than first-order schemata, essential in Zermelo's proofs of the quasi-categoricity of ZF^2C (see the next note). We cannot speak of "all ordinals" in an absolute sense, only of "all ordinals of model $\mathfrak{M}$." As is clear from the Proposition above, the well-ordering of all such ordinals will not be represented by any ordinal of $\mathfrak{M}$, but it *will* be represented by an ordinal of any proper extension of $\mathfrak{M}$, and Zermelo postulated that any model of his axioms has a proper extension (indeed, a proper end-extension, see the next note). In general, within any model, the well-order relations will outstrip the ordinals available to represent them, but ordinals of proper extensions then become available for representing them. Of course, extensions will give rise to further well-order relations not representable by any ordinal of the least model in which they occur (as second-order objects or as in the range of plural variables), but they in turn will get represented in further extensions, and so on. This then appears to be an attractive way to accommodate desideratum (3) without falling prey to the Burali-Forti paradox.

But is it? Zermelo did not formalize the axioms of his [1930] paper, but if he had, he would have confronted a serious problem: had he stated the general comprehension principle for classes (and relations) usually taken as axiomatic in modern formalisms for second-order logic,

$$\forall x_1, \ldots, \forall x_n [(R(x_1, \ldots, x_n) \leftrightarrow \Phi(x_1, \ldots, x_n)]$$

(where Φ is a formula lacking free occurrences of R), he would have seen that indeed the Burali-Forti paradox re-emerges on taking Ω as the formula "x is an ordinal in domain of some model, $\mathfrak{M}$," for this "union of all ordinals of domains of models" would be well-ordered by a relation having no ordinal as representative, violating desideratum (3) after all.[3]

---

[3] These conditions are best understood as only 'up to isomorphism'. A major achievement of Zermelo [1930] was to present a general proof of the quasi-categoricity of the ZF^2C axioms, viz. that given any two models (without urelements), one is an end-extension of a model isomorphic to the other.

It should also be noted that Hilary Putnam [1967] independently formulated an extendability principle very much in the spirit of Zermelo's (although it applied to models of Zermelo set theory as well as to those of ZFC). Also, Putnam was the first to formulate it as explicitly a *modal* principle.

3.3   Solution

One way to preserve the core of Zermelo's idea of indefinite extendability while meeting desideratum (3) is to introduce a logico-mathematical modality, in terms of which the extendability principle (EP) takes the following form:

$$\Box \, \forall \, \mathfrak{M} \, \Diamond \, \exists \, \mathfrak{M}' \, [\mathfrak{M} \prec_e \mathfrak{M}'],$$

where the variables range over standard models of ZF^2C, and $\prec_e$ indicates the converse of "end-extension." With the modal operators available, we have a new degree of freedom, and we can now adopt natural restrictions on the appropriate comprehension principles of second-order or plurals logic, namely that (1) only modal-free formulas are taken to define classes, and (2) classes or pluralities are guaranteed only to contain or encompass items existing within a given "world"[4] (and similarly for relations), i.e. classes of objects drawn from the universes of different worlds are not recognized. Thus, for example, we do *not* write

$$\exists \, R \Box \, \forall x_1, \ldots, \Box \, \forall x_n [(R(x_1, \ldots, x_n) \leftrightarrow \Phi(x_1, \ldots, x_n)],$$

or even this preceded by '$\Diamond$', let alone '$\Box$'; and we stipulate that the formula Φ lack modal operators.[5] As a result, we cannot form, for example "the proper class of all possible ordinals," i.e. "all ordinals of any possible model," nor can we even speak plurally of "all possible ordinals," etc. Although of course we can speak of the class (or plurality) of all ordinals of a *given* hypothetical model, or all well-order relations of such, these totalities are subject to the EP and lead to no paradoxes or contradictions.

Are these restrictions on second-order comprehension really natural or are they ad hoc? We maintain the former. Indeed, they merely express a widely held actualist view of existence and modality. Put in terms of classes or collections, it makes no sense to speak of actual collections of merely possible but non-actual things. At most we can speak of collections that *would exist* if there *were* such things. Similarly with plural formations: there are, in fact, only those items (or *n*-tuples of items) which actually exist; if there were or had been others, then there would be or would have been those items along with any (subpluralities) of them you like.[6] How, though, does this block the cogency of speaking, not individually but plurally or

---

[4] Officially, there is no need to quantify over worlds. Everything needed for pure and applied mathematics can be expressed directly via the modal operators.

[5] To get the effect of assertions of relations of structures across worlds, e.g. that two such are isomorphic, we can get by with additional assumptions of compossibility of models satisfying the relevant conditions. For further details, see Hellman ([1989]), Ch. 1.

[6] Indeed, these expressions of actualism involving plural constructions sound even more tautologous than those involving classes. In effect, we get instances of plurals comprehension such as "There are exactly the things satisfying condition Φ that exist," and, in counterfactual circumstances C, "there would be exactly those things satisfying Φ that would then exist"!

collectively, of "any ordinals of any possible model"? The key here is to recognize that the following inference is invalid (without additional assumptions): to infer from

There might have been the Φs and there might have been the Ψs

to

Therefore, there might have been the Φs and the Ψs.

This depends on the compossibility of the Φs being entertained and the Ψs, an additional assumption. In the modal-structural framework for set theory, where Φ is, say, "ordinal of standard ZF^2C model of height κ," and Ψ is "ordinal of ZF^2C model of height $\kappa' > \kappa$," the EP does in fact guarantee their compossibility, and likewise for any finite generalization.[7] And transfinite generalizations can be licensed by stronger extendability principles, for example ensuring common proper extensions of given α-sequences of models, where α is an ordinal of a given model. But nothing permits generalizing to the compossibility of all possible models, and indeed that would lead to contradiction with the EP. Now those in favor of an ontology with *possibilia* may remain unsatisfied with this resolution. The development of modal-structural interpretations, however, shows that their extravagances are just that, at least as far as recovering pure and applied mathematics is concerned.

In sum, such a modal-structural account of ordinals and well-orderings in the context of set theory provides a natural way of meeting desideratum (3) while blocking the Burali-Forti paradox and all the other set-theoretic paradoxes as well. In Zermelo's terms, we must recognize limits on the tendency toward "all-embracing completeness," which seeks to recognize "ultimate infinities," for example "absolutely all ordinals," "absolutely all well-orderings," etc. Instead, we favor the opposing tendency of "creative progress," expressed in principles such as the EP. And then all the ordinals we ever could require become available for all the legitimate mathematical work they could be asked to perform.

Acknowledgement: I am grateful to an anonymous referee and to Roy Cook and other members of the Foundations Interest Group of the Center for Philosophy of Science at the University of Minnesota for helpful comments on an earlier draft.

---

[7] Note, however, that this is not guaranteed for arbitrary choices of Φ and Ψ. For example, if Φ is "pairs coding a bijective order-preserving map from all ordinals to all accessible ordinals," and Ψ is "pairs coding a bijective order-preserving map from all ordinals to all ordinals some of which are inaccessible," clearly there is no way for both $\exists\alpha a[\Phi(\alpha a)]$ and $\exists\alpha a[\Psi(\alpha a)]$ to be satisfied in a single (ZF^2C) model.

References

Burgess, J., Hazen, A., and Lewis, D. [1991] "Appendix," in Lewis, D., *Parts of Classes* (Oxford: Blackwell), pp. 121–149.

Ferreirós, J. [2007] The early development of set theory. In *The Stanford Encyclopedia of Philosophy.* http://plato.stanford.edu/entries/settheory-early/, last accessed September 2010.

Hellman, G. [1989] *Mathematics without Numbers: Towards a Modal-Structural Interpretation* (Oxford: Oxford University Press).

Menzel, C. [1984] "Cantor and the Burali-Forti paradox," *The Monist* **67**: 92–107.

Moore, G. H. and Garciadiego, A. [1981] "Burali-Forti's paradox: a reappraisal of its origins," *Historia Mathematica* **8**: 319–350.

Putnam, H. [1967] "Mathematics without foundations," reprinted in *Mathematics, Matter, and Method: Philosophical Papers*, Volume I (Cambridge: Cambridge University Press, 1975), pp. 43–59.

Zermelo, E. [1930] "Über Grenzzahlen und Mengenbereiche: Neue Untersuchungen über die Grundlagen der Mengenlehre," *Fundamenta Mathematicae*, **16**, 29–47; translated as "On boundary numbers and domains of sets: new investigations in the foundations of set theory," in Ewald, W., ed., *From Kant to Hilbert: A Source Book in the Foundations of Mathematics*, Volume 2 (Oxford: Oxford University Press, 1996), pp. 1219–1233.

4   Extending the Iterative Conception of Set: A Height-Potentialist Perspective

After reviewing shortcomings of Boolos' presentation of the "Iterative conception of set," we formulate simple axioms on "stages" of set formation, using modal logic and the logic of plurals, that imply both the Axiom of Infinity and the Axiom of Replacement. (Two routes to these results are presented, the second less open to charges of circularity than the first.) We then present two other advantages of the Height-Potentialist framework, pertaining to motivating the smallest large cardinals (strongly inaccessible and Mahlo), and to furnishing attractive resolutions of the set-theoretic paradoxes. Our routes to these results are not available from within the Height-Actualist framework.

4.1   Introduction

As is widely known, especially through the work of Boolos [1971] and [1989],[1] the iterative conception of sets motivates the axioms of Zermelo set theory (at least its axioms less Extensionality and Choice) via a scheme whereby sets are "formed" at (ordinally indexed) stages: at the 0th stage, we have the null set, and at successor stages are formed all sets of sets formed at all earlier stages. What about at limits? Boolos simply assumed that there is a least limit stage, ω, at which are formed all sets of sets formed at earlier stages, and then that there are always successor stages of any stages. How high up do the stages extend? This is left quite indefinite. But enough is specified so as to be able to derive, from natural axioms on stages, most of the axioms of Zermelo set theory, Z, including Foundation, Infinity, Power Sets, and Separation (*Aussonderung*); however, Extensionality and Choice have to be "put in by hand" separately. Note further that Infinity only follows from the *stipulation* that there is an ωth stage. *There is no derivation that there exists an ωth stage, so there is no real derivation of the Axiom of Infinity.*

---

[1] See also Shoenfield [1967], pp. 238 ff., who also attempts to motivate the Axiom (Scheme) of Replacement.

Do Boolos' axioms on stages and "set formation" allow derivation of the Axiom of Replacement (which, with the axioms of Z, yields the system ZF, Zermelo–Fraenkel set theory)? Decidedly not. Indeed, Boolos [2000] came (implicitly) to cast doubt that Replacement is even "true," let alone motivated by axioms on stages. But since the axioms of ZFC (Zermelo–Fraenkel plus Choice, which by definition includes Replacement) form the core system of modern set theory, this is a very bad situation for the iterative conception. Can one not do better?

Furthermore, modern set theory has developed the whole vast subject of large cardinals, cardinals so large that their existence cannot be proved in ZFC if it is consistent. Moreover, work in modern set theory and proof theory has shown that at least the consistency of large cardinals is quite useful, even necessary, in deriving good mathematical theorems concerning sets of real numbers and even sets of rationals and integers.[2] Is there any way in which the iterative conception can be extended so as to derive at least some of the smallest of the large cardinals?

So far, then, we have two ways in which the iterative conception as developed by Boolos falls short: first, it fails to derive the Axioms of Infinity and Replacement; and second, it fails to derive the existence of even the smallest of the large cardinals (known as "inaccessible cardinals"). The core results of this paper will show that, provided one works within a suitably rich *height-potentialist* framework, it is fairly easy to carry out derivations of both the Axiom of Infinity and the Axiom of Replacement, at least modally interpreted, and of the possible existence of strongly inaccessible and Mahlo cardinals. Moreover, it will become clear that these derivations are unavailable to the "height-actualist": modal postulates are indispensable to the derivations we will present, but unacceptable to the height-actualist, as we shall see.

After developing this much by way of demonstrating the power and utility of the height-potentialist framework (call it HP), we will close by reflecting on how HP delivers especially attractive resolutions of the set-theoretic paradoxes (known as Burali-Forti, Russell, and Cantor), again not achievable within the height-actualist (HA) framework.

4.2   Deriving Infinity and Replacement: First Route

What we are about to sketch is how, on a height-potentialist iterative conception of set theory, using modal logic (S-4.3, say, guaranteeing linearity along the accessibility relation) and logic of plural quantifiers, we can formulate three natural axioms on stages, the first two of which imply the set-theoretic Axiom of Infinity – that there exists an infinite set – and the first and third of which

---

[2] See Moschovakis [2009] and Friedman [2018a, 2018b].

imply the instances of first-order Replacement, and the second-order axiom as well. Notably, these arguments are not available on a height-actualist conception (Boolos'?), which we find interesting in its own right. Also, we find interesting the fact that – as will emerge momentarily – each of the axioms follows from the combination of a common axiom on indefinite extendability of stages along with a further appropriate compossibility axiom. Our overall moral is that Infinity and Replacement are equally easy to obtain on a height-potentialist iterative conception of sets.[3]

We use s, t, etc. as singular variables for stages, and ss, tt, etc. as plural variables for stages, and $<$ (or its inverse, $>$) for the well-ordering relation on stages (or their ordinal indices),[4] i.e. $t > s$ means "stage t comes after stage s," and likewise for plural variables, i.e. $tt > ss$ means that the stages tt all come after all the stages ss. Also, below we will write $x@s$ to mean that set x is formed at stage s. Then the first of the new axioms states that for any possible stages ss possibly there are further stages tt such that $tt > ss$. Formally, we have the following:

Axiom 2.1:[5]

$$\Box\, \forall ss\ \Diamond\, \exists\, tt\ (tt > ss), \qquad\qquad\qquad \text{(Poss Ext)}$$

"Possible Extension of Stages" or "Extendability of Stages."

This says that the possible stages (or possibilities of new stages) "go on and on" without any limit whatsoever. NB: This is not available to the height-actualist who recognizes the (putative) totality of "all the stages," at least as a plurality, *whether or not* a universe of all stages and sets is recognized. So both first-order and second-order actualist set theories are distinguished from potentialist set theories. So far then, so good.

Let us consider now the Axiom of Infinity. This says that there exists an infinite set, and this goes well beyond there existing infinitely many sets. Our first axiom above guarantees the latter but not the former. It guarantees that there could be infinitely many stages, hence infinitely many sets, but not that there is (possibly) a stage with an infinity of preceding stages. We cannot derive the existence or possibility of a limit stage from Poss Ext alone.

Nor can we derive merely that infinitely many stages are compossible from Poss Ext. That much we must simply assume as a (well-motivated) axiom on

---

[3] Recall that Boolos treated Infinity as somehow part and parcel of the iterative conception – he simply assumed the existence of transfinite stages of set-formation – whereas he expressed doubt that Replacement could be justified at all.

[4] Boolos did not assume that the stages are well-ordered, but derived that later from further assumptions. See Boolos [1989], pp. 88–89. We regard well-foundedness of an iterative hierarchy as a desideratum and so assume it at the outset.

[5] It would suffice to write $\Box\, \forall ss\ \Diamond\exists\, t\ (t > ss)$, but the version with plural variable tt follows from this by transitivity of $>$ and of the accessibility relation. Of course, such tt would still be "world-bound" or "limited."

stages, for example, letting $s \prec ss$ stand for "s is among the stages ss," and letting 0 be the initial stage (at which the null set occurs), we postulate the following.

Axiom 2.2:

$$\Diamond \exists\, ss\ (0 \prec ss \wedge \forall s(s \prec ss \rightarrow \exists t(t \prec ss \wedge t > s)))) \qquad\qquad \text{(Poss } \infty\text{)}$$

(Of course, we could stipulate that t is the least stage dominating s, but that is not necessary for present purposes.)

But now, armed with Poss ∞, we easily obtain an infinite limit stage by applying Poss Ext: for example, if the ss and the tt are minimal with their required properties, the least stage among the $tt > ss$, where the ss are given by Poss ∞, behaves as a limit stage, viz. the ωth stage, which we may stipulate to contain all the hereditarily finite sets (equal to the union of all the finite stages); then with one more application of Poss Ext, we have many infinite sets.[6] ∎

Now consider the Replacement Axiom Scheme (or the second-order statement), and let φ define a unary function f sending x to y, and assume that f is defined on a set u. By hypothesis, for each $x \in u$, $f(x) = y$ exists as a set first formed, say, at stage s_y. We want to be able to conclude that the range of f on u, $f[u]$, is also a set. Now, in order to apply Poss Ext, we need that all the stages s_y of values of f constitute a plurality, i.e. are compossible.[7] Poss Ext itself does not guarantee this, so we need to assume it as a further axiom.

Axiom 2.3:

$$\Diamond \exists\, ss\ \forall r(r \prec ss \leftrightarrow \exists y(y = f(x) \text{ for some } x \in u \wedge y@r))$$

$$\text{(Poss Stages of Values)}$$

But now we can infer that the values y of f on u form a set, $f[u]$, since this follows directly from Poss Ext, letting ss be the (plurality of) stages s_y at which the values y of f restricted to u are formed, as guaranteed by Poss Stages of Values, since $f[u]$ is formed at the least stage $t > ss$, guaranteed by Poss Ext, the assumption that the stages are well-ordered by $<$, and the stipulation that later (successor) stages contain *all* sets formable from sets at earlier stages as elements.[8] ∎

---

[6] Here we are stipulating that stages behave just like the *ranks* in Z and ZF set theory, where, at limit stages occur exactly all the sets occurring at earlier stages. So, at stage ω there occur all the hereditarily finite sets (the union of all the finite stages), but not yet any infinite sets, which are formed (in abundance) at the next stage, $\omega + 1$. In his original paper, Boolos [1971] stipulated that at stage ω are formed all sets of sets formed at earlier stages, which already includes the infinite sets of hereditarily finite sets. It really does not matter which definition is adopted.

[7] We are of course presupposing, as does Øystein Linnebo [2013], that pluralities are world-bound (although officially we do not quantify over "worlds"): the items quantified over plurally are assumed to be compossible. Exactly the same items are values of the plural variable under all modal and counterfactual uses.

[8] This derivation of Replacement is comparable to that of Linnebo [2013], pp. 221–222, although Linnebo's derivation does not include a formal treatment of stages.

Thus we see that, in contradistinction to Boolos' view, the justification of Replacement is just as simple and straightforward on an iterative conception as that of Infinity. Both axioms follow from Poss Ext together with one natural further assumption of compossibility of relevant stages. As should be clear, however, this requires a height-potentialist modal-logical formalism; it is not available to the height-actualist, who cannot accept the crucial axiom, Poss Ext.

There is, however, a criticism that can be raised pertaining to this derivation of Replacement, namely that the second axiom used in that derivation (Poss Stages of Values), is intuitively "too close" to the statement of Replacement itself, that any function from sets to sets should preserve set-hood of the domain of that function, i.e. set-hood of the forward image of the function (restricted to a given set-domain). For Poss Stages of Values just replaces "set-hood" with "compossibility".[9] Of course, compossibility is a precondition of set-hood; but it would be nice to formulate an alternative to Poss Stages of Values that is intuitively not so close to the language of Replacement, that "set-hood" of domains be preserved by functions. Section 4.3 attempts to provide such an alternative.

A similar criticism can be levelled against the above derivation of Infinity: the second axiom on stages, Poss ∞, is simply too close for comfort to the conclusion sought. Once you have an infinity of compossible things ("an infinite plurality"), not surprisingly you generate a set of those things at the next stage, guaranteed to exist by Poss Ext (along with the assumption of well-ordering of stages). Can one do better? Our second route promises an affirmative answer.

4.3   Deriving Infinity and Replacement: A Second Route

Here is a second way to derive the Axioms of Infinity and Replacement from axioms on stages and sets formed at stages, using modal logic (specifically S-4.3, as above) and logic of plurals. The argument is available on a *height-potentialist* interpretation of set *cum* stage theory, but not on the (more common) height-actualist view (recognizing a fixed universe of absolutely *all* sets and stages).

First, we will use an equivalent of the first axiom (Poss Ext) of Section 4.2, which stated that any plurality *ss* of stages possibly has further stages, *tt*, dominating all of the *ss*. That consequence is simply the following.

---

[9] In the derivation of Linnebo [2013], the near neighbor of Replacement used to derive it formalizes the statement that "extensional definiteness is a matter of size" (p. 221), where a plurality's or a set's being "extensionally definite" is defined (intuitively) as "having the same items or elements in any modal or counterfactual context, i.e. in any world in which the plurality or set occurs."

Axiom 3.1:

$$\Box\forall ss\ \Diamond\exists\ t\ (t > ss) \qquad\qquad\qquad\qquad \text{(Poss Ext)}$$

"Possible Extension of Stages" or "Extendability of Stages."

Let us now take up deriving Replacement, either the first-order scheme or the second-order axiom. Again we use singular variables s, t, etc. for stages (which could as well be understood as ordinals used to index stages), and plural variables, ss, tt, etc. for stages; and we use x, y etc. as singular variables for sets, and xx, yy, etc. as plural variables for sets. As above, we write $x@s$ to mean "x is formed or first occurs at s," and we write $s \prec ss$ to mean "s is among or one of the ss," etc.; and we use $>$ as (the converse of) the well-ordering of stages.

We need just one crucial definition, viz. of the notion, "Predicate P of pluralities of stages is *indefinitely extensible*" (abbreviated IE) iff

$$\Diamond\exists ss\ P(ss) \wedge \Box\forall ss\ (P(ss)\rightarrow\Diamond\exists t(t > ss \wedge P(ss \cup \{t\})))$$

(Indef Ext (Stages))

(Here we have borrowed set-theoretic notation in the last clause to designate the plurality of the ss augmented with one new stage, t.) A similar but somewhat more complex definition applies to predicates of pluralities of sets.[10]

Now the new axiom we propose that implies Replacement is stated as follows.

Axiom 3.2: *Let φ define (or be) a unary function from sets to sets and let D be a set (of sets) on which φ is defined. Then the predicate Val is not IE, where Val (ss) holds of all possible stages $s \prec ss$ at which a value of $\varphi\restriction_D$ is formed, where $\varphi\restriction_D$ denotes the restriction of the function φ to set-domain D.*

This easily implies Replacement. **Proof:** Since Val is not indefinitely extensible, then possibly there are ss at which are formed all the values of the function $\varphi\restriction_D$. Then, by Axiom 3.1, possibly there is a stage $t > $ any of the ss. So, by definition of the stages, at t are formed all sets of sets formed at the stages ss, including the range of $\varphi\restriction_D$, that is, this range (potentially) occurs as a set. ∎

Remark: First, note that Axiom 3.2 is interdeducible with the assumption of our earlier derivation that the values of $\varphi\restriction_D$ and their stages are all compossible.

---

[10] That definition states that a predicate P of pluralities of sets is IE just in case:
$\Diamond\exists xxP(xx) \wedge \Box\ \forall xx\ (P(xx)\rightarrow\Diamond\exists y\ \exists s(y@s \wedge \forall t(s > t \wedge$ "t is any stage at which
any of the xx is formed" $\wedge P(xx \cup \{y\}))))$ (Indef Ext (Sets))

Next, however, consider that the intuitive meanings of these assumptions are by no means the same. In particular, the present Axiom 3.2 seems especially compelling, as it does not sound like an axiom of replacement: there is nothing in our common, classical understanding of "function defined on a set" or "values of such" that implies that it is always possible to continue to add values indefinitely. Indeed, once the set domain of arguments of the given function has been exhausted, there is no possibility of "adding more values" to the range. This is very unlike the cases of "possible stage" or "possible set" itself, where operations on sets or stages are readily available to transcend any putative limit.

Not surprisingly, an analogous argument can be given for the Axiom of Infinity, that there exists an infinite set. Let N be a predicate that holds of the plurality of exactly those stages at each of which is formed a finite von Neumann ordinal (serving as a natural number). Then our axiom takes the following form.

Axiom 3.3: *The predicate N is not IE.*

Let ss be the plurality of all the stages at which a finite von Neumann ordinal is formed. (Such a plurality must exist by the definition of "not IE.") But then, by Axiom 3.1, there is possibly a stage beyond the ss; and by the assumption that the stages are well-ordered, there is a least such stage, which we may designate as stage ω. At this stage are formed all sets of sets occurring at earlier stages, including the sets of all finite von Neumann ordinals, in particular the set of all such, which is of course an infinite set. Or, if we prefer stages to follow the set-theoretic construction of *ranks*, one could stipulate that at limit stages occur just all the sets occurring at earlier stages. In that case, to form the set of all finite von Neumann ordinals, we must ascend one more stage, which exists (or is possible) by our Axiom 3.1, Poss Ext. Then, at stage $\omega + 1$ are formed all subsets of the sets occurring at stage ω.　■

Remark: Of course, Axiom 3.3, while evident to the classicist and also, it should be remarked, to the predicativist, is not acceptable to the finitist or constructivist mathematician. The latter accept only the potentially infinite, regarding even the actual infinity of *all* the natural numbers as unjustified at best, or even "unintelligible" (as implied in Dummett [1977], e.g. at p. 11). On the other hand, it is less evident that the finitist or constructivist would not accept Axioms 3.1 and 3.2, as those are framed without commitment to infinite totalities. Indeed, Replacement holds in the realm of the finite, although of course, for the classicist, it functions in the context of the Axioms of Infinity and Powersets to generate many large infinite sets.

Also, it should be emphasized that the above arguments are not available to the height-actualist. As in the case of our earlier derivation via axioms on compossibility of stages, the height-actualist does not accept Axioms 2.1 or 3.1, Poss Ext. But also, the language of the height-actualist lacks modal operators, so all our axioms and the crucial definition of "indefinitely extensible," are simply not expressible within the non-modal, height-actualist language. In short, the language of modalities and plurals is tailored to the expressive needs of the height-potentialist, but is not expressible in the height-actualist framework.

Conclusion: Once again, we see that natural axioms on stages lead to an iterative conception of sets that readily justifies both Infinity and Replacement. And, as in the case of our earlier proposal, the framework of the above axioms on stages shows that Replacement is no more difficult to justify on an iterative conception than Infinity, again contrary to Boolos' stance.

4.4   Small Large Cardinals from Below

Contemporary set theory has successfully pursued the subject of large cardinals, exploring concepts and implications of cardinals so large as to dwarf the so-called "small large cardinals" considered here. The latter include the so-called strongly inaccessible cardinals and the Mahlo cardinals, both of which come in multiple "orders," for example hyper-inaccessible, hyper-hyper-inaccessible, etc., which, in comparison with the large, large cardinals (large enough to be incompatible with Gödel's Axiom of Constructibility, abbreviated $V = L$), are such baby steps beyond the iterative conception as we have extended it thus far in this essay that the question naturally arises whether that conception could be further extended so as to include them.[11] As we aim to show in this section, the answer is affirmative, which we take to be another piece of good news for the height-potentialist iterative conception.

The very natural idea of how to extend the iterative conception so as to capture strong inaccessibles imports into the height-potentialist framework

---

[11] For an excellent introduction to the subject of large cardinals, see Drake [1974]. See also Kanamori [2003]. We note here that the term, "small large cardinals" applies to a richer universe than what we can treat here, extending at least through the so-called "indescribable cardinals." These are still "small" in the sense of being compatible with $V = L$, yet we know of no way to generate them from below comparably directly as in the cases of inaccessible and Mahlo cardinals. But even these baby steps are steps in the right direction and deserve to be articulated on behalf of the height-potentialist iterative conception. Notably, as "baby" as these steps are, they cannot be taken from within the height-actualist framework, i.e. they cannot be so readily derived from more elementary principles. Of course, they are still well-motivated and can be, and of course are, postulated.

a key result of Zermelo's [1930] paper, namely that the ordinal height (least upper bound of the ordinal ranks) of a standard model[12] of the second-order ZFC axioms is a strongly inaccessible cardinal, in the sense of a regular strong limit cardinal.[13] Once given such a model, provided there is the possibility of stages beyond those of the sets of the model, we reach the ordinal height of the model as a *bona fide* set ordinal or cardinal, viz. a strongly inaccessible cardinal.

Thus, we need to formalize an axiom expressing that possibly there exists a model of the ZF^2C axioms. Using our modal language with plurals, we can express that possibly there are stages and sets formed at those stages such that ZF^2C holds when relativized to those sets. In symbols:

$$\Diamond \exists ss \; \exists xx \; \forall x \; (x \prec xx \; \rightarrow \; \exists s \; (s \prec ss \; \wedge \; x@s \; \wedge \; (\wedge ZF^2 C)^{xx}))$$

(Poss ZF2C)

where the last conjunct indicates that all the quantifiers in the conjunction of the ZF^2C axioms are relativized to the xx.[14]

But now we are almost home, for we can invoke Poss Ext (Axiom 3.1) to infer that there possibly is a (least) stage s beyond all the ss provided by Poss ZF^2C; and at s is formed an ordinal κ marking the height of the ZF^2C "model," more precisely the l.u.b., κ, of the ordinals marking the stages ss; then by Zermelo's theorem, we infer the possible existence of a strongly inaccessible cardinal, viz. κ.[15]   ∎

A closely analogous story can be told leading to (strongly) Mahlo cardinals, by adding an axiom (labeled F by Drake) to the system ZF^2C expressing that every normal function[16] from ordinals to ordinals has a strongly inaccessible fixed point. Then the height of any standard model of ZF^2C + Axiom F is "strongly

---

[12] We call a model of second-order ZFC (ZF^2C) "standard" just in case it is well-founded and its power-sets are full, containing all subsets of the given set (The model is maximally "wide."). Of course no model of ZF^2C can be proved to exist provided that theory is consistent, as a consequence of Gödel's [1931] second incompleteness theorem.

[13] A cardinal κ is a "limit" cardinal just in case $\kappa \neq \lambda + 1$, for any cardinal λ. A cardinal κ is a "strong limit" just in case, for any $\lambda < \kappa$, also $2^\lambda < \kappa$. Finally, a cardinal κ is called "regular" just in case it is not the limit of a sequence of lesser ordinals of length $< \kappa$ itself. (This is also described as being equal to its own "co-finality.")

[14] Note that the direct use of logic of plurals (alternatively second-order logic) to express relativization of the quantifiers in the axioms obviates the need to invoke metalinguistic *satisfaction*. If we were aiming to *eliminate* set-membership via generalization on binary relations in the interests of a modal-structural interpretation, avoiding the need for *satisfaction* would be crucial, although this is not our aim here, where we have been freely quantifying over sets.

[15] For a sketch of how to derive hyper-inaccessibles, see Hellman [1989], p. 84. This pattern generalizes.

[16] A function f from ordinals to ordinals is called "normal" just in case f is increasing and continuous at limits, where the latter means that, for limit ordinals $\lambda, f(\lambda) = \cup_{\xi < \lambda} f(\xi)$.

Mahlo."[17] All that is needed is to translate this into the notation of the logic of plurals (to get the full effect of second-order quantification, on which this result, like the previous one on inaccessibles, relies), and add the appropriate modal operators on the plan of Poss ZF^2C. Then, by again invoking Axiom 3.1, Poss Ext, we infer that the height of the possible model of ZF^2C + Axiom F occurs as a set-ordinal (cardinal) at the next stage, and that it must be strongly Mahlo. ∎

In closing this section, let us ask whether results such as these are attainable on the height-actualist conception. In a sense, the answer is yes; but the usual story proceeds by way of *reflection principles*, in particular, second-order Reflection (known as Bernays reflection).[18] The big difference, however, is that this is "from above": one postulates that a certain largeness condition holds of the whole universe of sets, V, and then infers that the restriction of that condition to some segment V_α also holds. Now Reflection has its own motivation, viz. as a kind of largeness or richness condition on the (putative) universe of all sets: this is said to be "indescribable" in just the sense of satisfying Reflection. Still, especially since this is a second-order principle, one has to buy into a fixed, maximal, inextendable universe of sets, an immediate consequence of second-order logical comprehension. The advantage of the height-potentialist route outlined above is that it proceeds genuinely "from below," without invoking any absolutely maximal universe, while at the same time enjoying the expressive power of second-order logic, via the logic of plurals. Modal logic here plays a crucial role in blocking the modal equivalent of proper classes, since there is no reason to think it possible to collect "all possible sets or ordinals" – indeed, to the contrary, that is impossible precisely because of the Extendability Principle (or, in terms of stages, Poss Ext). Instead, the set/proper class distinction is relativized to a given model, where it renders harmless second-order or plurals comprehension.

4.5   Resolving Paradoxes

This, in our opinion, is one of the height-potentialist's strongest suits. For further details the reader may consult Hellman [2011] on resolving the Burali-Forti paradox, reproduced here as Chapter 3, which we take up first. As explained there, standard resolutions for working set theory, first- or second-order, under a height-actualist interpretation, while quite adequate for practicing mathematics, all share the common feature of failing to provide ordinal representatives of proper-class size well-order relations, despite the fact that the whole point of

---

[17] See Drake [1974], Ch. 4, Ex. 3.7 (3).

[18] First-order Reflection, as is well known, is a scheme, provable in ZF, saying that if a condition φ is true, then so is the restriction of φ to a segment V_α of the cumulative hierarchy. This also implies the Axiom of Infinity and the Replacement scheme; but to obtain many (small) large cardinals, higher-order reflection is needed and suffices. For a good presentation, see Tait (2005), Ch. 6.

introducing ordinal numbers is to codify reasoning about well-orderings. In contrast, this *desideratum* can be met from within a height-potentialist framework, with axioms such as Poss Ext on stages or analogous ones on (say, von Neumann) ordinals themselves. The core idea is this: relative to any given possible model M of, say, ZFC, any well-order relation forming a set relative to M, will be represented by a set-ordinal of M (in the precise sense of being order-isomorphic with the well-order relation). But M does not provide ordinal representatives of the proper-class size well-orderings of sets in M. However, in possible proper extensions M' of M, those proper-class size well-orderings occur as *bona fide* sets, and receive ordinal representatives in M'. This back-and-forth process then continues indefinitely. This of course depends on an Extendability Principle applicable to models (of ZFC, first- or second-order):

$$\Box \, \forall M \, \Diamond \exists M' \, (M \prec_e M') \tag{EP}$$

where $M \prec_e M'$ stands for "M' is a proper end-extension of M" (proper extension in which no new elements occur in sets of the lesser model M). From EP it also follows that there can be no "modal proper class" of "all possible sets or ordinals": sets and ordinals are "world-bound"; transworld collections are simply not recognized, based on the very intuitive idea that it is not possible to collect at a world any objects that merely might then have existed but do not occur in that world (just as, in the actual world, you cannot collect any merely possible objects). Thus, in the potentialist sense, any possible well-order relation possibly has an ordinal representative, satisfying our *desideratum*.

Closely analogous resolutions can also be provided for the Russell and Cantor paradoxes. Consider Russell: although relative to any ZFC model M there is at best a proper class of exactly the non-self-membered sets of M, relative to any proper extension, M' of M, such a class occurs as a *bona fide* set. Of course new collections of non-self-membered sets of M' occur only as proper classes of M', but those in turn occur as *bona fide* sets in proper extensions M'' of M'. And this continues without conceivable end, again in light of EP. The net effect is to sustain the full generality of set theory as a theory of collections.

The story with the Cantor paradox of the set of all sets is exactly analogous to the resolution of the Russell paradox.

4.6   Concluding Reflection

Clearly the three bodies of results of adopting the height-potentialist framework reviewed here belong squarely within the foundations and philosophy of mathematics. But, none of the results of this essay contributes directly to present-day working set theory. Indeed, from the perspective of the latter, none of these results appears particularly compelling in favor of adopting a height-potentialist framework. Modern set theory has no qualms whatever concerning the consistency or legitimacy of the Axioms of Infinity and Replacement.[19] Nor has it any need for new derivations of small large cardinals, whose existence is amply supplied by hosts of large, large cardinals of genuine interest to current set-theoretic practice. And to be sure, current set theory is happy enough with its own defusing of the paradoxes, getting by perfectly well without, for example, ordinal representatives of proper-class size well-orderings. For modern set theorists (or Maddy's "thin realist"[20]) to come to regard the height-potentialist framework as really worth the price of putting up with its logical complexity (modal logic + logic of plurals), it may well be necessary for that framework to lead to some genuinely new mathematics, for example results helpful in settling the Continuum Hypothesis, or – more modestly, perhaps – the invention or discovery of a new kind of large, large cardinal involved in some genuinely new and interesting theorems. (Not being a cutting-edge set theorist, the author of this paper could not provide anything like such advances.)

Be this as it may, the most we can have done here is to call attention to some of the foundational advantages of the height-potentialist's framework, including helping persuade skeptics of even ZFC (as, ironically enough, George Boolos became by the time of his 2000 paper, "Must we believe in set theory?"), let alone critics of classical mathematics, of the naturalness of ZF^2C; and further to suggest to leading set theorists that pursuit of their mathematics along the lines of the height-potentialist framework of this paper may indeed yet prove its worth, even from their own perspective.

Acknowledgement: The author is grateful for helpful correspondence on precursors of this paper with C. Anthony Anderson, Philip Ehrlich, Joel Hamkins, Akihiro Kanamori, Peter Koellner, Øystein Linnebo, and Penelope Maddy.

References

Boolos, G. [1971] "The iterative conception of set," reprinted in *Logic, Logic, and Logic*, Jeffrey, R., ed. (Cambridge, MA: Harvard University Press, 1998), pp. 13–29.

Boolos, G. [1989] "Iteration again," reprinted in *Logic, Logic, and Logic*, Jeffrey, R., ed. (Cambridge, MA: Harvard University Press, 1998), pp. 88–104.

Boolos, G. [2000] "Must we believe in set theory?," in Sher, G. and Tieszen, R., eds., *Between Logic and Intuition: Essays in Honor of Charles Parsons* (Cambridge: Cambridge University Press), pp. 257–268.

Drake, F. R. [1974] *Set Theory: An Introduction to Large Cardinals* (Amsterdam: North Holland).

---

[19] For a mathematical assessment of the Axiom of Replacement, see Kanamori [2012].
[20] See Maddy [2011], Ch. 3.

Dummett, M. [1977] *Elements of Intuitionism* (Oxford: Oxford University Press).

Friedman, H. [2018a] "Tangible mathematical incompleteness of ZFC," #108 at http://u.osu.edu/friedman.8/foundational-adventures/downloadable-manuscripts/.

Friedman, H. [2018b] "Concrete mathematical incompleteness: basic emulation theory," in Cook, R. T. and Hellman, G., eds., *Hilary Putnam on Logic and Mathematics* (Cham: Springer), pp. 179–234.

Gödel, K. [1931] "On formally undecidable propositions of *Principia Mathematica* and related systems," reprinted in van Heijenoort, J., ed., *From Frege to Gödel: A Source Book in Mathematical Logic, 1879–1931* (Cambridge, MA: Harvard University Press, 1967), pp. 596–616.

Hellman, G. [1989] *Mathematics without Numbers: Towards a Modal-Structural Interpretation* (Oxford: Oxford University Press).

Hellman, G. [2011] "On the significance of the Burali-Forti paradox," *Analysis* **71**(4): 631–637.

Kanamori, A. [2003] *The Higher Infinite* (Springer-Verlag).

Kanamori, A. [2012] "In praise of replacement," *Bulletin of Symbolic Logic* **18**(1): 45–90.

Linnebo, Ø. [2013] "The potential hierarchy of sets," *Review of Symbolic Logic* **6**(2): 205–228.

Maddy, P. [2011] *Defending the Axioms: On the Philosophical Foundations of Set Theory* (Oxford: Oxford University Press).

Moschovakis, Y. [2009] *Descriptive Set Theory*, 2nd edn (Providence, RI: American Mathematical Society).

Shoenfield, J. R. [1967] *Mathematical Logic* (Reading, MA: Addison-Wesley).

Tait, W. W. [2005] "Constructing cardinals from below," in *The Provenance of Pure Reason* (Oxford: Oxford University Press), Ch. 6.

Zermelo, E. [1930] "Über Grenzzahlen und Mengenbereiche: Neue Untersuchungen über die Grundlagen der Mengenlehre," *Fundamenta Mathematicae*, **16**: 29–47; translated as "On boundary numbers and domains of sets: new investigations in the foundations of set theory," in Ewald, W., ed., *From Kant to Hilbert: A Source Book in the Foundations of Mathematics*, Volume 2 (Oxford: Oxford University Press, 1996), pp. 1219–1233.

5   On Nominalism

5.1   Goodman's Nominalism and Set Theory

Probably there is no position in Goodman's corpus that has generated greater perplexity and criticism than Goodman's "nominalism." As is abundantly clear from Goodman's writings,[1] it is not "abstract entities" generally that he questions – indeed, he takes sensory qualia as "basic" in his Carnap-inspired constructional system in *Structure* [Goodman, 1977] – but rather just those *abstracta* that are so crystal clear in their identity conditions, so fundamental to our thought, so prevalent and seemingly unavoidable in our discourse and theorizing that they have come to form the generally accepted framework for the most time-honored, exact, sophisticated, refined, central, and secure branch of human knowledge yet devised, *mathematics* itself! Of all the *abstracta* to question, why *sets*? Of course, Goodman gave his "reasons," the unintelligibility of "generating" an infinitude of "constructed objects" automatically from any given object or objects. But critics have been quick to point out that set theory is intended not as a theory of what can be "generated" or "constructed" from given objects in any *literal* sense but rather as a theory of a certain realm of objects independently existing in their own right. "Construction" is a metaphor.

Now there are, in the foundations of mathematics, various conceptions of sets as "constructed objects," namely as "definable sets." As Goodman himself frequently cautioned, however, we must ask, "'Definable' *from what and by what means*?" A restricted interpretation leads to *predicative* set theory, usually beginning with a countably infinite set, namely the natural numbers (taken as unproblematic on this view, inspired by Poincaré and Weyl, and perfected by Feferman and his coworkers[2]), then recognizing what results from iterating definability over earlier (countable) totalities using formulas with quantifiers restricted to such, the iteration taken along predicatively acceptable countable

---

[1] See Goodman [1977], Ch. 2, and Goodman [1972].
[2] See, e.g., Feferman [1998], especially the essays in Part V.

ordinals less than a certain countable limit. Thus, predicative set theory is even more constrained than Goodmanian nominalism in this sense: beginning with a countable totality of objects, only countably many sets based on them ever get recognized. In contrast, Goodmanian nominalism recognizes *all parts* of any given (whole of) individuals, so, if countably many pairwise non-overlapping individuals ("pseudo-atoms") are somehow given, uncountably many fusions are also recognized. (It is not difficult to reformulate Cantor's diagonal argument in nominalistic terms.) *Definability* of fusions in any specified language is *not* required. Thus, predicativism and nominalism do agree in not respecting the Zermelo–Fraenkel (ZF) axiom of *power sets*, but their lack of respect is manifested differently. Nominalism allows in effect just one application of passing to "all subsets (except for the null set)" of a given set (i.e. whole of atoms or pseudo-atoms), finite or infinite; predicativism allows this only if the given set is finite. Nevertheless, predicativism does transcend Goodman's strictures, not necessarily in the starting point of assuming a countable infinity (for Goodman has no principled objection to infinite wholes *per se*), but in the appeal – implicit or explicit – to *possibility*, to the notion of what *can* or *could* be defined in an "in-principle" sense. Indeed, as we shall see below, this points to a way in which a "modal nominalist" can recover a great deal of set theory. Here, at least, the metaphor of "construction" is taken rather seriously, and, as we shall remark, a *semantic elimination* of predicative sets can even be effected.

The other main notion of "constructible" in set theory is Gödel's notion of the constructible hierarchy, which is just like the predicative conception but with one crucial difference: rather than limiting iteration of definability to predicatively acceptable countable levels, transfinite iteration along *arbitrary* ordinals is allowed. For purposes of investigating Gödel's constructible hierarchy as an inner model of set theory, its levels go on and on as guaranteed by the Zermelo–Fraenkel axioms. This is even less amenable to a Goodmanian interpretation, for it requires an objective reading of strong axioms of set theory, or, at any rate, an objective understanding of "transfinite ordinal," and so is quite at odds with Goodman's construction metaphor.

Thus, we have the curious situation that *predicative* set theory, which takes the construction metaphor seriously, is ontologically even more conservative than Goodmanian nominalism in the domain of the infinite. And, in the case of *impredicative* set theory, even Gödel's "constructible universe" treats ordinals objectively and so evades Goodman's objection.

Indeed, I do not think Goodman ever gave a genuine *argument* against sets, much less set theory; at best he gave a *criterion* for distinguishing certain mathematically weak systems, in a particular sense, from stronger ones and announced that he would limit himself to the weak ones.

5.2   Nominalism by Default

In the middle of the twentieth century, it was common to regard platonism (regarding mathematics) as a "default" position in the sense that no non-platonist position was available as a serious alternative preserving mathematics as a science. As is well known, Goodman and Quine [1947] managed to provide a nominalization of merely an actualized, finite portion of the syntax of (suitably logicized) mathematics. Later, Quine, admitting defeat, acquiesced in platonism in so far as he regarded platonistic assumptions as indispensable for the empirical sciences, while Goodman went about work on other projects.

Since that time, however, a lot has happened. Several nominalistic or quasi-nominalistic approaches to mathematics have been developed, sometimes incorporating surprising technical advances not available to Goodman et al.[3] Combined with other results in logic demonstrating the remarkable power of weak mathematical systems – especially the work of Friedman, Simpson et al. in "reverse mathematics"[4] and the work of Feferman et al. in predicative analysis – the indispensability argument on behalf of platonism developed by Quine and Putnam has been seriously weakened, if not yet quite completely vanquished. The tables have turned sufficiently, I believe, to warrant looking at the whole issue afresh. It is not simply *chutzpah* to ask an empiricist-oriented question: What, after all, is the *evidence* warranting belief in mathematical *abstracta*, numbers, sets, functions, etc.? Is there really any compelling or even good evidence? In what follows, I want to concentrate on this question, and on an argument to the conclusion that we really should not put much credence in a face value, platonist interpretation of mathematics, at least given our present evidence.

Before proceeding, let me emphasize the epistemic character of the question and the argument to be presented, and the contrast with Goodman's nominalism. As said, for Goodman, the concept of *set* raised problems of intelligibility over "generation" of infinitely many objects out of the finite. But once we abandon the "generation" metaphor and take set theory (or theories) at face value, intelligibility is no longer the problem. Rather, the problem is straightforwardly epistemic: what reason is there to believe that there really are such things as required by the theory (theories) on a face value, platonistic interpretation?

5.3   The Argument

The argument is qualitative and imprecise, but it may be powerful, nevertheless. It assumes the following.

---

[3] For a good survey, see Burgess and Rosen [1997].
[4] See, e.g., Harrington, et al. [1985] and Simpson [1999].

(1) Synthetic claims of existence should not be given great credence without strong evidential support.

(2) Existence claims for mathematical *abstracta,* especially the Axiom of Infinity, that there exists an infinite set or even plurality, are synthetic.

(3) There is no good direct evidence for such existence claims. For instance, contra early efforts of Maddy, we do not perceive sets, and, in any case, not infinite sets, which classical analysis requires.

(4) The best indirect evidence comes via scientific indispensability arguments. However, alternative programs such as predicativism, Chihara's semantic modal-nominalism, and my modal-structuralism separately and cumulatively undercut such arguments. (For certain portions of science, Field's nominalistic constructions also undercut them.)

(5) The fact that current science takes mathematics for granted is not good evidence that premise (1) is suspended in the case of mathematics when viewed from a naturalistic-scientific standpoint.

The conclusion is, therefore, as follows:

(6) The evidence currently available for mathematical *abstracta* is quite weak, and so we should not put much credence in a face value, platonist reading.

Before discussing the particular steps, note that the conclusion concerns only a particular *reading* of mathematical theories. It certainly does not say that the evidence for *mathematics* or any particular portion of it is weak. Some nominalists, such as Field, do not make this distinction, letting mathematics stand or fall with a face value, platonist reading of it. But here I stand firmly – or should I say, laugh heartily? – with David Lewis when he writes,

I'm moved to laughter at the thought of how *presumptuous* it would be to reject mathematics for philosophical reasons. How would *you* like the job of telling the mathematicians that they must change their ways ... now that *philosophy* has discovered that there are no classes? Can you tell them, with a straight face, to follow philosophical argument wherever it leads? If they challenge your credentials, will you boast of philosophy's other great discoveries: that motion is impossible, that a Being than which no greater can be conceived cannot be conceived not to exist, that it is unthinkable that anything exists outside the mind, that time is unreal, that no theory has ever been made at all probable by evidence (but on the other hand that an empirically ideal theory cannot possibly be false), that it is a wide-open scientific question whether anyone has ever believed anything, and so on, and on, *ad nauseam?*

Not me!" (Lewis [1991], p. 59)

5.4   The Existence of Numbers and Infinite Sets Is Not "Analytic"

Perhaps the first premise, that credibility depends on evidence, will be granted, at least for the sake of argument. Resistance can be expected, however, to premise 2, that the existence claims of mathematics are not analytic. By this I understand essentially what Carnap meant, "guaranteed true solely in virtue of linguistic meanings." (I hasten to add, however, that I do not follow Carnap in his application of this concept to mathematics. The mature Carnap [1956], as I understand him, insisted that a question such as "Are there infinite sets?" had to be designated as either internal to a framework or external to any framework. As internal, it is decided, usually trivially, by appeal to an assumption constitutive of the framework. I am not asking whether the Axiom of Infinity is true according to the framework of, say, Zermelo set theory! As external, it is cognitively meaningless (so not "analytic"), and can only be understood pragmatically: "Ought we to speak this way?," or something of this nature. I am not asking that question either.)

The resistance to premise 2 comes from the neo-Fregean efforts (by Wright, Hale, and other authors) to give a principle such as "Hume's Principle" (better called "Cantor's Principle") an analytic reading:

"The number of F = The number of G iff the F are in 1–1 correspondence with the G."

Sometimes appeal is made to a "syntactic priority thesis," guaranteeing that apparently referring terms (singular terms, on syntactic grounds) occurring in (the right kind of) true sentences (true, by ordinary lights) are genuinely referring, guaranteeing the existence sought.[5] My biggest problem with this is the need for the weasely qualification, *"the right kind* of true sentences." Without it, there are hordes of counterexamples: "Thor was the Nordic god of thunder," "Odysseus charted a safe course between Scylla and Charybdis," etc. But how does one distinguish the relevant from irrelevant truths? Surely not on any straightforward syntactic grounds. Consider this instance of the neo-Fregean's favorite construction:

"The god of person P = The god of person Q iff P and Q are coreligionists."

No matter that this may be a novel construction. So was Hume's (rather Cantor's) at one time, and, anyway, the neo-Fregeans insist on the stipulative character of Cantor's principle. We can perfectly well introduce this new one and use it. We can go around saying, "Those two people have the same god, but those two have different ones," etc. We will be understood by those who know our lingo to have spoken truly, by theists and atheists alike. (Perhaps we will need to spell out our criteria of 'coreligionist', but that is not the problem.) I see no non-circular way of restricting the "syntactic priority thesis" for it to have a prayer.

Here is perhaps a decisive reason why the existence of the natural numbers as objects cannot be analytic. Suppose the contrary. Then, presumably, if number theory itself is to be fully justified without appeal to evidence (beyond our use of language – and this is the motivating idea of invoking analyticity to block our

---

[5] See Wright [1983].

argument from the start), it will also be analytic that 0 is not a successor, that every number has a successor, that distinct numbers have distinct successors, and that anything true of 0 and of the successor of anything it is true of is true of all the numbers. (More formally, Wright et al. show how the Dedekind–Peano axioms can be derived from Cantor's Principle. Certain instances of second-order logical comprehension are needed, and the neo-Fregean would have to argue that these too are analytic.) But then, it appears, the consistency of Peano arithmetic (PA) (first order, say) is guaranteed equally well: surely the rules of inference of first-order logic preserve analyticity, and equally surely it is not analytic that $0 = 1$! *So the statement that PA is formally consistent is guaranteed entirely by the meanings of words: we understand those meanings, so there is no epistemic problem whatever about this formal consistency.* And will not a parallel argument show the same for full classical analysis? Will you tell this to Hilbert who formulated the problem of formal consistency and worked hard to solve it, to Gödel who ingeniously proved that a Hilbertian solution is impossible, to Gentzen who pioneered solutions utilizing transfinite methods, and to all their successors? Can you do it with a straight face? If they challenge your credentials, will you boast of philosophy's other great discoveries: that it is unthinkable that anything exists outside the mind, that all questions not decidable by empirical methods or logical analysis are meaningless, that it is an illusion that we can even *understand* untestable statements about the past (e.g., that 12 years ago to the day there were or were not exactly 27 paper clips on my desk) as determinately either true or false, that the very notion of a mind-independent real world is incoherent because there are many conceptual and symbolic systems, and so on, and on, *ad nauseam?*

Not me!

5.5   Sets in Space-Time? Decomposing Apples

Everyday discourse is so rife with talk of sets of material objects that one is naturally led to ask whether abstract mathematics might be grounded in such talk, and whether, moreover, we might be said to interact with and even perceive such physically constituted sets. Maddy [1990] investigated this sympathetically. Responding to criticism of Chihara and others, that we simply are not aware of a set with, say, just a given apple in it over and above the apple, Maddy suggested that material objects be identified with their singletons. But this could only be maintained if an apple "has an unambiguous number property," viz., being one. This is not true, however, she says of "the physical mass which makes up the apple"; but in nearly the same breath she writes that that physical mass "is one apple, many cells, more molecules, even more atoms, and so on." Whether this 'is' is that of identity or of predication, the mass is after all being identified with the apple; but perhaps the thought could be reworded. Regardless, there is a serious problem here: if the apple is indeed different from the whole of physical stuff making it up and located exactly where it is, how can it be a physical object at all? If it is only an apple-under-a-description or an apple-in-relation-to-a-mind, or if it is an individual concept, etc., then it would seem Maddy has departed from physicalism quite radically, contrary to the whole spirit of her approach. Alternatively, Maddy could grant what any good nominalist takes for granted, that indeed an apple = the whole made up of any exhaustive, "intact decomposition" of it into parts, i.e. where the parts must be specified as suitably temporally bounded to accord with the actual (contingent) history of the particular apple (which of course can persist while losing its stem, or its skin, etc.). But then it can no longer be maintained that the apple is "unambiguously one"; just as Frege [1884] insisted so forcefully in the *Grundlagen*, such an object only bears a number relative to a sortal predicate or concept or class. For each intact decomposition, there is the number of its parts, all equally applicable as truly describing the apple, each in its own special sense. So how can the apple be identified with its singleton, as the latter surely is unambiguously one, i.e. one-membered? Moreover, for each intact decomposition there is the set of its parts, and this set is, according to Maddy, "located exactly where the apple is" (since, by hypothesis, the apple *is* the whole of those parts and so must be located exactly where they are). It seems that we have literally thousands of sets (even perceivable ones in the sense that the parts are perceivable) located in the same space-time region. What help, one wonders, is location to the epistemology of these sets?

In any case, the leap from finite sets of material things to infinite sets is huge, and it was never supposed that we somehow perceive infinite sets. But without them, we do not have classical analysis, much less modern functional analysis, topology, differential geometry, etc. At best, the evidence for infinite sets can be indirect.

5.6   Appeals to Indispensability

This is a large topic, and I must summarize. While I do think that indispensability arguments along Quinean–Putnamean lines are cogent (despite objections due to Maddy [1992] and Sober [1993], which Colyvan [1999] and I [1999] independently have sought to answer), it seems that their scope is severely limited. Nominalism apart, the set-theoretic strength of systems adequate to formalizing the vast bulk of scientifically applicable mathematics is surprisingly meager. It is remarkable how much can be carried out in a subsystem of analysis (second-order arithmetic) defined by the so-called arithmetical comprehension axiom scheme (ACA) (recognizing only sets of numbers definable by formulas with number but no set quantifiers). It is difficult to come up with clear examples of mathematical results, genuinely needed in physics, including theoretical physics, that elude predicativist analysis. (It should be emphasized that the background logic of the relevant systems is classical. The case that constructive analysis based on intuitionistic logic is *in*sufficient is much more readily made, in my view, although even here, it is non-trivial to come up with convincing examples.) What this means, essentially, is that one gets by with countable infinities: mathematically speaking, natural numbers and predicatively definable sets of these, along with lots of clever coding tricks, are all that one needs. Appealing to Feferman's systems of flexible types, one can even formulate general results involving reference to sets of sets of numbers, sets of sets of these, etc., without transcending the predicativist framework.

Some time ago, Charles Chihara [1973] recognized that sets can be eliminated from predicativist mathematics in favor of semantic machinery of *satisfaction* of formulas by already accepted objects. Instead of quantifying over sets of natural numbers, say, satisfying some conditions, one can quantify over the formulas that define those sets and assert corresponding conditions of them. This is a crucial step towards a version of nominalistic mathematics, but of course one must somehow provide for a countable infinity of formulas (equivalently for the natural numbers, as either can code for the other). More recently, Chihara [1990] has articulated in detail how this can be done within a logic of "constructibility quantifiers," which builds in "possible existence," understood without introducing possible worlds. At the same time, however, Chihara went way beyond predicativism, allowing for a full semantical recovery of the simple theory of types. This complicates matters *vis à vis* indispensability arguments, for, as just indicated, only a tiny fragment of this theory is actually required for most scientifically applicable mathematics. Thus, the reliance on modality actually required for this much mathematics is not nearly so heavy as might be thought. So long as one is prepared to grant the possibility of, say, a countable infinity of strokes, one has all that one needs, ontologically speaking.

Thus far we can already say that a system of modal-semantical predicative type theory provides a nominalistic recovery of (most, if not quite all) scientifically applicable mathematics. In this sense, even sets of natural numbers need not be recognized to carry out even theoretical physics, in principle. (Of course, no one is suggesting that such a system be instituted in practice.) Chihara's nominalism is neither "hermeneutical" (providing "the deeper, genuine meanings" of mathematical discourse) nor "revolutionary" (in proposing a radical revision of established practice). As I suggested in Hellman [1998], reproduced here as Chapter 6, these are not the only alternatives. Rational reconstructions for epistemological-metascientific purposes should also be considered.

Let us pause briefly to note a comparison with Field's work. His *Science without Numbers* [1980], especially when taken together with Burgess [1984], showed how classical field theories in physics can be carried out along the lines of synthetic geometry, replacing analytical mathematics with relations on space-time obeying suitable axioms. Some efforts have been made to extend such methods to quantum mechanics, but I do not believe successfully, for the simple reason that there are not enough actual physical propensities in nature to support the Hilbert space structure of states (probability measures on propositions). In contrast, modal-predicativism recovers the mathematics of quantum mechanics almost directly from standard treatments. Moreover, the ordinary distinction between truth and falsity is respected for all the mathematics represented, in contrast to Field's stance that all existential claims are false because there are no numbers, etc.! (Compare Lewis' *Credo.*) Finally, note that, whereas Field's original program did not employ modality, even this advantage was not maintained when it came to recovering the metalogic required to sustain the program. A notion of logical possibility was taken as primitive to avoid platonist appeals to models.

Despite the success of predicative modal-nominalism in undercutting claims of indispensability of mathematical *abstracta* for science, one would like to come closer to the content of mathematics as it is normally understood and practiced. Mathematicians simply do not think and reason in terms of satisfaction of open sentences by syntactic objects. Rather they think in terms of structures, such as normed spaces, metric spaces, topological spaces, differentiable manifolds, groups, rings, fields, and so forth. Axioms serve as defining conditions, expressing structures of interest, as well as starting points in logical reasoning. Applications to the physical world are readily understood in terms of partial, approximate manifestation of various types of mathematical structure. All this is lost, or at best deeply hidden, in a modal-nominalist linguistic type theory.

It can be returned to center stage, however, in a *modal-structuralist interpretation* of mathematical theories. Here one seeks to express, in nominalistically acceptable terms if possible, that certain structures, implicitly defined by suitable axioms, are logically possible, and then to understand the theorems of the branch of mathematics in question as asserting what necessarily would hold in any such structure that there might be. The machinery of second-order logic is a natural framework for describing structures central to mathematics, such as many of those listed above and, crucially, the natural number structure, the classical continuum, the complex plane, and segments of the cumulative hierarchy of set theory (say Zermelo–Fraenkel with Choice).

But how can second-order logic be utilized in the interests of nominalism? If we restrict ourselves to *monadic* second-order logic, we can answer (with

Boolos[6]): by reading it *plurally*. The idea is to stand on their head standard examples intended to show that ordinary discourse contains implicit, hidden commitments to sets or classes. The oft-cited example of Geach, "Some critics admire only one another," has no direct first-order representation, but is easily rendered in second-order notation, introducing a class of critics. Instead, Boolos suggested, we may appeal to a prior, independent understanding of such sentences to give a plural reading of monadic second-order quantfiers. As Lewis pointed out in a persuasive discussion,[7] we readily affirm without contradiction that there are all and only the non-self-membered classes. No hidden commitment to a class of such is involved. Existential plural quantifiers seem to arise quite naturally in English. But universal plural quantification, though perhaps less common, is also naturally expressible: "Any" followed by suitable plural constructions seems to work, as in

"Any people whatever, if placed in cramped quarters with one another, will squabble."

To express the principle of mathematical induction, we may say,

"Any objects which include 0 and which include the successor number they include, include all the numbers."

And predicative instances of second-order logical comprehension become trivially true, for example,

$\exists C \, \forall x (x \in C \equiv x$ is a planet),

which is understood as

"There are some things such that each of them is a planet and each planet is one of them,"

or, more succinctly,

"There are the planets."

I think I understand all such examples without ascending a level in type above the objects plurally quantified over. (Boolos' motivation was to allow for second-order formulation of ZF set theory without commitment to proper classes. He wished to avoid "ultimate types" while having the advantages of second-order formulations. But his method can be applied at any level, including the ground level.)

This brings monadic second-order logic within the purview of nominalism; but mathematical structures require functions and relations essentially. How can *polyadic* second-order quantification be introduced? One way – the one

---

[6] See Boolos [1985], reprinted as essay 5, pp. 73–87 in Boolos [1998].
[7] Lewis [1991], pp. 62–71.

I initially pursued in *Mathematics without Numbers* [Hellman, 1989] – is to resort to brute force: since only conceptual possibilities matter, one can postulate enough extraneous atoms to code ordered pairs of given individuals; relations are then coded as wholes of pairs. But more recently, more general and elegant methods were developed by Burgess, Hazen, and Lewis [1991], ingeniously combining mereology with plural quantification. It needs to be assumed that there are infinitely many atoms (or individuals behaving enough like atoms), but, for modal-structural mathematics it suffices to entertain this merely hypothetically. Then (following Burgess' construction) unordered pairs of atoms (diatomic wholes) can code a 1–1 correspondence between all the atoms and some of them, and some other diatoms can code a second 1–1 correspondence between all the atoms and others entirely discrete from (not overlapping) the first targets. (To insure this much, one must assume a kind of partition of all the atoms into three disjoint pieces, (a part of) each equinumerous with the remaining two, something which can be expressed with plural quantification and mereology.) Then one can make sense of the "first image" of any given individual, i.e. the sum (fusion) of the images of its atoms under the first 1–1 correspondence, and the "second image" of a given individual, the sum of the images of its atoms under the second 1–1 correspondence. These images are always discrete from one another. The ordered pair of two, possibly overlapping, individuals is then defined as the sum of the first image of the one to be first and the second image of the one to be second. Relations are then available by quantifying plurally over ordered pairs. All this is relative to a given trisection of the given infinitude of atoms, and it is cumbersome to write it all out. But it works. The upshot is a nominalistic reduction of arbitrary *n*-adic relations in the language of mereology and (monadic) plural quantification. What this means for mathematics is that any structure characterized by *n*-adic relations or functions on a domain can be described within this nominalistic framework, without ever officially invoking sets or classes. In effect, full polyadic second-order logic becomes available.

The extent of the mathematics so representable depends essentially on the cardinality of the initial atoms (or surrogates) entertained as possible. It turns out that, if just a countable infinity is entertained, the resources of plural quantification and mereology take one all the way to full, polyadic *third-order* number theory, equivalently the full, classical theory of sets, relations, and functions of real numbers. There is no restriction to definability of reals or sets or functions of reals. All the impredicative constructions of classical theory carry over intact. The amount of ordinary mathematics thus recovered is truly vast, surely adequate to the presently known needs of the sciences. Yet one can go even farther, arguably without transcending nominalistic conceptions: one can postulate an initial continuum of atoms, at least as a conceptual possibility. (So if space or time or space-time as conceived in classical physics is regarded as possible, this would suffice.) One then attains the level of classical *fourth-order* number theory, the theory of sets, functions and relations of sets, functions and relations of reals. Again, there is full respect for the truth-determinateness of the classical theory, and there are no predicativist limitations. Any mathematical structure definable in second-order logic over a domain of the cardinality of the continuum is thus within nominalistic reach; indeed, the cardinality may be even that of the power set of the continuum. (Utilizing a real pairing function, sets and relations of reals can be coded as certain fusions of reals. Then plural quantification takes us one type-level further, to relational structures on sets etc. of reals.)

It is, of course, possible that in the future we will discover problems, even in physics, which require even richer mathematics for their solution. Harvey Friedman, who has pioneered work on "necessary uses of abstract set theory" in the spirit of Gödel's program, has indeed found arguably genuine mathematical problems pertaining to low level mathematical objects (e.g., functions on n-tuples of integers) whose solution "lies beyond mathematics," i.e. cannot be decided in standard set theory, but requires for example the 1-consistency of Mahlo cardinals or even larger ones. So far, these remain purely mathematical, but who can say that physics might not one day stumble upon such a problem? From a structural point of view, if the problem concerns real numbers or functions of reals or objects codable by reals, then the answer is already determinate within nominalistically describable structures in the sense just outlined. However, stronger axioms pertaining to much larger structures may be required to *derive* the correct answer. *Then* we would have the prospect of an indispensability argument for something genuinely set-theoretic; *then* we might have an argument that nominalism is really inadequate for the natural sciences.[8] But I do not think that we have any such argument today.

5.7   Conclusion

In their book on nominalistic reconstruction programs, Burgess and Rosen concede essentially this much. But they take nominalism to task anyway for not respecting good scientific methods, for in effect placing "philosophy first" rather than "last, if at all" (in Stewart Shapiro's apt phrases), indeed for advocating a revolution with potentially crippling consequences for future science! Confronted with the argument of the present paper, what would they say? Perhaps that the first premise is suspended when it comes to framework principles as centrally entrenched as mathematics is in relation to natural

---

[8] I say "the prospect" of an argument because one must distinguish between statements of consistency of strong axioms of infinity – statements codable in number theory – from the truth of such axioms. It will require further argument to pass from (indispensability of) the former, which Friedman establishes in certain cases (for pure mathematics), to the latter, as Feferman has stressed.

science. Surely it could not be that the *practical* indispensability of "platonistic mathematics" suffices to confer strong evidential support, in spite of its in-principle, theoretical dispensability!

In any case, I would insist that nominalistic *reconstructions* need be neither hermeneutical nor revolutionary but can be – and in most of the cases in question *are* – *preservationist* while attempting to solve or avoid certain epistemological, metamathematical, or metascientific problems not (or not yet) treated within science itself.[9] (It is true that Field's program is exceptional in not preserving mathematical truth. But even this program seeks to preserve the reasonableness of a lot of mathematical-scientific practice.) Like the formalisms of modern proof theory in this respect, they sit side-by-side, not in competition with, mathematics as standardly practiced. *Of course*, natural science takes mathematics for granted and uses it opportunistically without questioning its foundations or its interpretation. But this suggests to me that it does not worry about how to interpret mathematics *at all*, not that it accepts in any considered way a face value, literal, platonist *reading* of mathematics. Even working classical mathematicians do not universally agree on such a reading. Most scientists, I would wager, have never even thought about the issue. We should not, therefore, even say that "platonistic mathematics" is practically indispensable; at best we may be able to say that the compact, standard languages of mathematics are practically indispensable. This does not tell us that any particular reading or interpretation of such languages is required. And it certainly does not support the claim that a methodological principle such as premise (1) of the argument above is somehow "suspended."

Nominalism has no wish to interfere with scientific or mathematical practice. Its interests lie elsewhere. Suppose a good, historical/scientific case were made that genuine scientific or mathematical progress would be adversely affected were scientists or mathematicians even *to read* the likes of Chihara, Feferman, Field, Goodman, Hellman, et al., let alone be persuaded by what any of them has written. Then better that they should not read them! What has that got to do with whether or not there is good epistemic reason to believe in a literal interpretation of set theory?

References

Boolos, G. [1985] "Nominalist Platonism," *Philosophical Review* **94**: 327–344.
Boolos, G. [1998] *Logic, Logic, and Logic*, Jeffrey, R., ed. (Cambridge, MA: Harvard University Press).
Burgess, J. [1984] "Synthetic mechanics," *Journal of Philosophical Logic* **13**: 379–395.

---

[9] See Hellman [1998], Chapter 6 in this volume.

Burgess, J. and Rosen, G. [1997] *A Subject with No Object: Strategies for Nominalistic Interpretation of Mathematics* (Oxford: Oxford University Press).
Burgess, J., Hazen, A., and Lewis, D. [1991] "Appendix," in Lewis, D., *Parts of Classes* (Oxford: Blackwell), pp. 121–149.
Carnap, R. [1956] "Empiricism, semantics, and ontology," in *Meaning and Necessity* (Chicago, IL: University of Chicago Press), pp. 205–221.
Chihara, C. [1973] *Ontology and the Vicious Circle Principle* (Ithaca, NY: Cornell University Press).
Chihara, C. [1990] *Constructibility and Mathematical Existence* (Oxford: Oxford University Press).
Colyvan, M. [1999] "Contrastive empiricism and indispensability," *Erkenntnis* **51**: 323–332.
Feferman, S. [1998] *In the Light of Logic* (Oxford: Oxford University Press).
Field, H. [1980] *Science without Numbers* (Princeton, NJ: Princeton University Press).
Frege, G. [1884] *The Foundations of Arithmetic* (Oxford: Blackwell, 1950), translation by J. L. Austin of *Die Grundlagen der Arithmetik* (Breslau: Wilhelm Koebner, 1884).
Goodman, N. [1972] "A world of individuals," in *Problems and Projects* (Indianapolis, IN: Bobbs-Merrill), pp. 155–172.
Goodman, N. [1977] *The Structure of Appearance*, 3rd edn. (Dordrecht: Reidel).
Goodman, N. and Quine, W. V. [1947] "Steps toward a constructive nominalism," *Journal of Symbolic Logic* **12**: 105–122.
Harrington, L. A., Morley, M., Scedrov, A., and Simpson, S. G. (eds.) [1985] *Harvey Friedman's Research on the Foundations of Mathematics* (Amsterdam: North-Holland).
Hellman, G. [1989] *Mathematics without Numbers: Towards a Modal-Structural Interpretation* (Oxford: Oxford University Press).
Hellman, G. [1998] "Maoist mathematics? Critical study of John Burgess and Gideon Rosen, *A Subject with No Object: Strategies for Nominalist Interpretation of Mathematics* (Oxford, 1997)," *Philosophia Mathematica* **6**(3): 357–368.
Hellman, G. [1999] "Some ins and outs of indispensability," in Cantini, A., et al., eds., *Logic and Foundations of Mathematics* (Dordrecht: Kluwer), pp. 25–39.
Lewis, D. [1991] *Parts of Classes* (Oxford: Blackwell).
Maddy, P. [1990] *Realism in Mathematics* (Oxford: Oxford University Press).
Maddy, P. [1992] "Indispensability and practice," *Journal of Philosophy* **89**: 275–289.
Richman, F. [1996] "Interview with a constructive mathematician," *Modern Logic* **6**: 247–271.
Simpson, S. [1999] *Subsystems of Second Order Arithmetic* (Berlin: Springer).
Sober, E. [1993] "Mathematics and indispensability," *Philosophical Review* **102**: 35–57.
Wright, C. [1983] *Frege's Conception of Numbers as Objects* (Aberdeen: Aberdeen University Press).

6   Maoist Mathematics?

Critical Study of John Burgess and Gideon Rosen,
*A Subject with No Object: Strategies for Nominalistic
Interpretation of Mathematics* (Oxford, 1997)

This book has many virtues. It is concentrated on fundamental questions in the philosophy of mathematics, which it explores with an open mind – or even two open minds; it is richly informed and informative in its clear exposition of the details of nominalistic reconstruction programs, indeed the whole extant gamut of them, some themselves usefully reconstructed; it concludes with a novel insight into the unsuspected value of these programs (to be explained below); and, of special immediate relevance, it is remarkably balanced in its argumentation and self-contained, even to the point of containing its own review! Not *verbatim*, of course, but implicitly, as a scattered whole, merely awaiting a judicious selection and assembly, with occasional textually inspired critical commentary. Here follows an attempt at such.

Recent decades have witnessed a considerable interest in alternatives to the standard mathematical framework of axiomatic set theory. One type of restrictive program goes by the name "constructivism" which typically sets a stricter standard of proof than classical mathematics, especially proofs of existence which must provide instances. While epistemic concerns clearly motivate constructivism, there is not generally an objection to abstract ontology *per se*. This is the preoccupation of a second, very different type of restrictive approach, "nominalism," which, as Burgess and Rosen (hereinafter BR) point out, arose not within the mathematical community proper but among philosophers, and is motivated largely by the difficulty of fitting orthodox mathematics into a general philosophical account of the nature of knowledge. The difficulty largely arises from the fact that the special, 'abstract' objects apparently assumed to exist by orthodox mathematics – numbers, functions, sets, and their ilk – are so very different from ordinary, 'concrete' objects. Nominalism denies the existence of any such abstract objects. That is its negative side. Its positive side is a programme for reinterpreting accepted mathematics so as to purge it of even the appearance of reference to numbers, functions, sets, or the like – so as to preserve the subject while banishing its objects. (p. vii)

There have been many diverse proposals for how to carry this out, each bringing with it its own bag of novel or non-standard technical tools and tricks, its own specific philosophical concerns and aims, and its own degree and kind of success. This poses a considerable challenge to the interested student or professional who would seek to understand and assess the contributions of these reconstruction programs, for not only are the materials scattered in numerous books and articles but they cry out for a unified and clarifying treatment that "begins at the beginning," and goes on in an orderly fashion (even if it does not quite get to the end when it stops!). Fortunately, that is precisely what the authors of this study have provided.

The book falls naturally into three parts. The first is a substantial introduction concerned with the philosophical motivations of nominalism and the reconstruction programs. Second is the central core which undertakes a comparative survey of those programs, seeking to clear up technical issues to the extent currently possible, concentrating on three approaches, a "geometric strategy," inspired by work of Hartry Field, a "purely modal strategy," inspired by work of Charles Chihara, and a "mixed modal strategy," inspired by work of Geoffrey Hellman. This is followed by a "potpourri of other strategies" and a useful guide to the literature. Finally there is a provisional critical assessment. As the authors state early on, they are by no means nominalists or nominalistically inclined; indeed they devote quite a bit of effort to questioning widely shared assumptions motivating nominalism, and, toward the end, they mount a rather severe critique of the programs on scientific grounds. Nevertheless they find considerable value in these programs, especially viewed collectively, from a naturalistic perspective, value that can be appreciated by anyone with serious epistemological interests, nominalist or not.

Let us turn now to expositing the main ideas, leading up to the final assessment. As the passage I quoted from the Preface above indicates, the authors appreciate the positive, "preservationist" thrust of nominalist reconstruction programs, although to be sure, what that comes to specifically varies with the program. This turns out, I believe, to be the most crucial point on which the authors' critical assessment of nominalism turns, and on which my own critical assessment of their critical assessment will turn below. That passage also locates the principal motivation for nominalism in epistemic concerns, especially the concern to overcome the apparent discontinuity between mathematical objects and the natural world as they figure in a naturalistic account of our knowledge. Appreciation of this turns out to raise the second most crucial point of contention in assessing the authors' assessment of the nominalist reconstruction programs.

At the heart of the authors' analysis is an initial twofold classification of nominalisms and nominalist programs, the broadest of all the many distinctions needed to describe them. On the one hand there is "what may be called the

revolutionary conception [whose] goal is reconstruction or *revision*: the production of novel mathematical and scientific theories *to replace* current theories." (p. 6, my italics) Frequently this proceeds by "assigning novel meanings to the words of current theories." In contrast, there is

what may be called the **hermeneutic** conception, [according to which] 'All anyone really means – all the words really mean – is . . . ' (here again giving the reconstrual or reinterpretation). Reconstrual . . . is taken to be an *analysis* of what really 'deep down' the words of current theories have meant all along, despite appearances . . . It is taken to be a means to the end of substantiating the claim that nominalist disbelief in numbers and their ilk is in the fullest sense *compatible with* belief in current mathematics and science. (pp. 6–7, my italics)

BR then tell us they will "ignore the hermeneutic position," except in certain sections toward the end of the book devoted to it. This strongly suggests that BR view the reconstruction programs they describe in detail in Part II as belonging on the "revolutionary" side of the dichotomy they adopt. As will emerge, the core of their critical assessment of these programs flows quite naturally from this way of categorizing them. Moreover, as BR take pains to point out, reading those programs according to the hermeneutic conception – the only alternative the authors leave room for – is quite implausible on linguistic grounds. Surely, ordinary mathematical discourse about, say, real numbers and functions of them is not literally, *accurately rendered* as describing intrinsic relations among space-time points (*à la* Field's program), or as describing possibilities of constructing certain open-sentence inscriptions (*à la* Chihara's later program), or by means of lengthy modal conditionals concerning what would hold in arbitrary structures of the appropriate type (*à la* Hellman's program). While such appeals to ordinary linguistic intuitions cannot settle all questions about "deep structural meaning," as BR explicitly recognize in their Conclusion, it certainly seems as if these programs are Maoist, at least in supporting a cultural revolution!

Having presented this basic dichotomy, the authors continue their Introduction with an orderly discussion, first, of "What Is Nominalism," second, of "Why Nominalism?," and third, of "Why Reconstrual?" These are aimed, respectively, at highlighting the importance of three fundamental questions which the authors feel have not been adequately addressed in the literature on nominalism, namely, (i) What is an abstract entity? (ii) Why should one disbelieve in abstract entities? and (iii) Why should one who disbelieves in abstract entities seek to reconstrue theories that involve them? Why not simply treat such theories instrumentally, for instance? There is much to be learned from the three sections that follow, each of which raises significant challenges to commonly held views. While BR do not pretend to arrive at final answers, their examination of the issues with its balanced dialectic will be found instructive and useful by anyone interested in the central questions of metaphysics and epistemology, regardless of their interest in learning about the detailed formulations of nominalist reconstruction programs and their logics.

Especially valuable, in this reviewer's opinion, is the discussion of the "abstract/concrete" distinction, or better, distinctions which, as BR expose, citing work of David Lewis, cut across one another in complex ways. As Nelson Goodman [1977] emphasized long ago, "abstract" has multiple meanings, and a nominalist who renounces classes, for instance, need not eschew experiential qualities or other entities that may also be labelled "abstract." Despite these complexities, nominalists agree on a range of examples of "abstract entities," paradigmatically from mathematics, and they agree that "their" causal inefficacy is at the root of the problem "they cause" for naturalized epistemology.

The discussion of this problem, or cluster of problems, in connection with "causal theories of knowledge," however, seems at times in danger of veering off course. The central nominalist worry is over how to provide a naturalistic explanation of how anyone knows of mathematical objects. This stands regardless of the difficulties in providing an adequate *conceptual analysis* of "know," which the nominalist need not pretend to have. Furthermore, the discussion of Field's "refined version" of the epistemological argument against mathematical platonism does not seem to do justice to the nominalist's challenge. It may be true as well as amusing that, after certain reductive manoeuvres, the alleged problem arising within platonism of explaining "the pervasive correlation" between what mathematicians believe about mathematical objects and what is true about them reduces to the dubious demand for an explanation of a mere *conjunction* of two propositions, that "the full cumulative hierarchy of sets exists" *and* that this is believed. Still, one feels that the point is being missed. If one considers mathematics as practiced, prior to regimentation in a single system such as set theory, a genuine problem of accounting for reliability *would* seem to arise from consideration of *multiple systems* or *mathematical structures*, which seem to be "free for the thinking up," structures which the platonist regards as *all* existing (provided no contradiction is encountered in the axioms or defining conditions laid down). On the other hand, even if one concentrates on mathematics as codified, say in set theory, surely a central nominalist worry – one which platonists may share – is that we do *not* in fact seem to have knowledge of the first conjunct, that "the full cumulative hierarchy of sets exists," precisely because this seems so far removed from experience and from anything we encounter in natural science generally, or in ordinary mathematics for that matter. Belief in such a thing is an integral part of *extraordinary* mathematics, not of ordinary mathematics. This suggests that something has gone wrong in the reduction of the "pervasive correlation." The focus should really be on weak set theories which suffice for ordinary mathematics and scientific applications. But since there are many weak set theories around for codifying ordinary mathematics, especially if room is to be left for unforeseeable future applications, there seems to be more to Field's reliability argument than the authors recognize.

This brings us to the central core of the book, which is a detailed presentation of the three principal nominalist reconstruction programs mentioned above along with several variants. No attempt will be made here even to summarize the rich and extensive content of this part, except to say that a great service has been performed for both students and professionals interested in this subject. The formal essence of the programs is clearly laid out in each case, with just enough detail to give the reader a real sense of how the program in question works but not so much as to obscure the broader picture. Especially valuable is the authors' detailed listing of the requisite modal operators and quantifiers used in the second and third strategies, culled from scattered sources and organized into suitable systems not found elsewhere in the extensive literature on modal logic. In addition, there are several pedagogically useful sections, including historical background to the geometric strategy, description of some non-nominalistic, though not yet physical, geometric implications of higher set theory (turning on determinacy postulates), and a section developing in detail the instructive analogy between temporal and modal idioms.

The upshot of this development of nominalist reconstruction programs can perhaps be summarized as follows. The geometric strategy, dispensing with mathematicalia in favor of space-time points and regions, is successful in preserving physical applications of mathematics in classical field theories, although it does not preserve the truth of pure mathematics at all, unless it is coded in spatial terms. (This step was not part of Field's original program, but one might add it. Note, however, that the modal, i.e., non-contingent, status of pure mathematics would not thereby be preserved, although its truth would be.) Whether it can be extended to preserve applications in general relativity or quantum mechanics are open questions. In the first case, the absence of a synthetic (semi-)Riemannian geometry is cited. And in the second, the abstract character of the states as elements of Hilbert spaces is cited. In a rare omission, David Malament's [1982] review of Hartry Field's [1980] *Science without Numbers,* which first made this point, is not cited but should have been.

The purely modal strategy, so-called because its logic employs only modal extensions of standard quantifier logic, dispenses with mathematicalia in favor of the possibility of concreta, such as numerals or open sentences, standing in appropriate relations to one another and to other physical entities. It succeeds in preserving scientific applications of mathematics, at least all those that can be carried out in classical analysis; indeed it corresponds closely in mathematical strength to the simple theory of types, and so preserves roughly this much pure mathematics as well. Just how far it extends depends on how rich the possibilities of constructing open sentences are permitted to be. This is left rather open-ended in Chihara [1990].[1] Finally, although BR do not go into this in any depth, the purely modal as well as the mixed modal strategy, unlike the geometric one, does also preserve the non-contingent status of pure mathematics: if a suitable array of concreta is possible, it is necessarily so, according to the modal logics favored in these approaches.

The mixed modal strategy is so called because its logic employs in addition to modal extensions certain other extensions, especially what BR call "plethynticology," the logic of plural quantifiers introduced by George Boolos, which together with mereology serves to recover second-order logic. This provides a rich framework for carrying out modal-structuralist interpretations of mathematical theories. Given an axiom of infinity, to the effect that it is possible that there exist countably infinitely many atomic individuals, indeed pure and applied mathematics through third-order number theory can be recovered. Stronger hypotheses of infinity lead to higher orders. Certain special modal operators, for actuality and consequentiality, are useful in describing applications of mathematics to actual systems, and, while the results are more long-winded than standard platonist formulations, there is little doubt that platonist ontological commitments are dispensable for scientific purposes. As BR indicate, the combined use of plural quantification and mereology to extend the reach of the mixed modal strategy was not available as of Hellman [1989], but it is incorporated in more recent developments of his approach,[2] not available to BR in writing the book under review. Key technical insights appeared first in Burgess, Hazen, and Lewis [1991], in which it was shown how to achieve the effect of ordered pairing with plural quantification and mereology but no further primitives. The possibility of an infinity of individuals is needed, but it is needed anyway for applications of analysis.

In sum, the assumption of a special mathematical ontology is not really needed for scientific applications of mathematics or for most ordinary pure mathematics, and there are two, three, perhaps a dozen systematic alternatives to the standard platonist reading of mathematics to show this. Does this mean that the cultural revolution is at hand, that adherents of the old ways will have to be exposed and re-educated, and, if necessary, publicly denounced for their "reactionary tendencies"? This brings us to the concluding section of the book in which the authors provide their provisional assessment of the nominalist reconstruction programs. This is, philosophically, probably the most interesting and controversial part of the book. Proponents of the reconstruction programs will no doubt have their own reactions and may formulate their own responses. In the remainder of this review, I will summarize what I take to be the authors'

---

[1] See also my review, Hellman [1993].
[2] See Hellman [1996], reproduced as Chapter 1 in this volume.

key points and indicate briefly the lines along which it seems to me a nominalistically inclined response may effectively proceed.

Toward the end of their Introduction, the authors raise the question whether, should a nominalistic reconstruction of a scientific theory (or of a purely mathematical one) succeed, in the sense that it *"could be adopted"* in place of the original, *"should it be?"* (p. 63). Further, they assert that the famous Quine–Putnam anti-nominalist "indispensability argument" makes a major concession to nominalism in granting that "if nominalistic alternatives to standard scientific theories could be developed, then they *should be adopted"* (p. 64, my emphasis). This leads to their exploration in the Conclusion of the "scientific merits" of the various nominalistic reconstructions set out in the central portions of the book. Immediately they stress that this question will be considered only from a naturalistic, as opposed to a "first-philosophical" perspective, a perfectly reasonable choice given the naturalistic views and motivations behind the nominalist programs themselves. Next they urge that the question of scientific merits is really not for philosophers ultimately to answer but is rather one for "the scientific community." Indeed, they write (somewhat chidingly, as I read it),

the true test would be to send in the nominalistic reconstruction to a mathematics or physics journal and see whether it is published, and if so how it is received . . . a test to which reconstructive nominalists have been unwilling to submit (and prudently so, in view of [the omission of substantial parts of pure mathematics by the reconstructions and their admitted impracticality for much scientific work]). (p. 206)

And, in the same passage, they ask, "in what sense can philosophers *proposing a revision of science* claim to be judging by scientific standards?" (my emphasis).

Having thus modestly disqualified themselves from passing final judgment on such matters, the authors develop a provisional assessment based on a bit of "descriptive methodology, a branch of naturalized epistemology," beginning with "a somewhat heterogeneous list of features that tend to be implicitly used in science as standards for judging when a theory is a *better choice* than its rivals" (p. 209, my emphasis). The first three concern empirical accuracy and internal rigor and consistency. The final four play a critical role: (iv) economy of assumptions, (v) coherence with familiar established theories, with minimality of amendment, (vi) perspicuity of the basic notions and assumptions, and (vii) fruitfulness, or capacity for being extended to answer new questions. Then follows a two-pronged exploration leading to a critique of the nominalist reconstructions. First, it is argued that "to give a high score to any of the nominalistic reconstructions would require one to discount factors (v)–(vii) almost entirely" in favor of (iv), concentrated on just one kind of economy, ontological. The authors seem to allow room for a distinction between theory and practice; but, they suggest, it would hardly be "sane to suggest that purely theoretical explanations of [astronomical and financial phenomena of concern respectively to applied physicists and economists] *should be couched in synthetic or modal language*" (p. 211, my emphasis). It is recognized, moreover, that "science itself observes a division of labor," that, for example, physicists are not held to the same standard of rigor as mathematicians, who in turn leave some issues to logicians, etc. "There may well be much to be said for a line of response something like this," it is conceded, but nominalists "have not themselves much said it": they have not presented studies of division of labor in the scientific community and then cited these "as warrant *for discarding* pure mathematics and ignoring familiarity, perspicuity, and fruitfulness" (p. 213, my emphasis).

Second, there follows an interesting discussion, valuable in its own right, of "Occam's razor" as it has actually figured in science. While acknowledging that they are not able to offer decisive, expert opinion on the matter, BR are unable to find a clear case in scientific or mathematical history in which one theory was preferred to another purely or largely for reasons of ontological economy. The case of Newtonian absolute space versus Leibnizian relationism bears further scrutiny than the authors provide, I believe. But they make a good case that at least such a basis for "theory choice" must be quite exceptional, although the force of this may seem somewhat muted in light of the multifaceted nature of "theory choice," which suggests that "clear cases" are going to be the exception anyway. Ultimately, the authors suggest, revolutionary nominalism is on shaky ground at best with regard to the scientific merits of its programs.

The real value of these programs is to be found elsewhere, BR propose, namely in the role they can play in an important part of naturalized epistemology, that is, in helping sort out the contributions to our science, respectively, of the world and of our own intelligence and cognitive apparatus. "Using [an alternative theory] might be quite inconvenient or even unfeasible for us; but provided it would in principle be possible for intelligences unlike us … comparison of the theory with our actual scientific theories would help give a sense of what and how much our character has contributed to shaping the latter." Even theories "inferior by our standards" to our actual theories can thereby "advance the philosophical understanding of the character of science" (p. 243). Finally, nominalism in its recent manifestations should serve to instruct constructivists: "It is all very well to say that producing a constructivist physics would be a philosophically more profound achievement than producing a nominalist physics; but the production of a constructivist physics is a hope for the future, while the production of a nominalist physics is an actual accomplishment, at least to the extent surveyed in this book" (p. 244). To this I say, Amen.

The critique of the nominalist reconstructions on their "scientific merits," however, I find deeply problematic. From the places in the above quoted passages where I supplied italics, the most serious problem can be discerned. Should these reconstructions be viewed as "up for adoption" by the scientific community, as candidates for being "chosen" as preferable on scientific grounds to current theories? Should they be viewed as really proposing "a revision" of current science? In what sense? Should their authors be regarded as advising physicists or economists what language they should really employ? Is anyone giving reasons "for discarding pure mathematics"? Is the only conceivable alternative to "yes" answers to these questions the pigeonhole of "hermeneutical nominalism" which claims to be providing deep-structural translations of ordinary mathematical-scientific discourse?

It seems to me that none of these questions should be answered in the affirmative and that at the root of the problem is the very dichotomy between "revolutionary" and "hermeneutical" nominalism with which the authors began. For this dichotomy leaves no room for what I should have thought was the proper category for most of these programs, namely a kind of "rational reconstruction." Consider proof theory in the logical foundations of mathematics. No one seriously maintains that mathematicians ought to write out their proofs in any of the formal systems developed by logicians, or even that they should spend time checking whether this can be done in any given case. On the other hand, no one proposes proofs in formal systems as accurate translations of actual mathematical proofs, although certain *key aspects* of the latter are surely being codified and respected. Rather the logicians are providing standards that coexist side by side with mathematical practice and which are designed to help answer *metamathematical* questions that naturally arise when one views the practice in rational terms. What do proofs rest on, ultimately, and what do they accomplish? Why should theorems be believed? And so on. I suggest that something similar can be said for most, if not all, of the nominalist reconstruction programs. Their purpose is to help answer certain *metamathematical* or *metascientific* questions, not normally entertained in pure and applied mathematical work proper. How can the essential mathematical content and results of mathematics be understood so that a naturalized epistemology of science and mathematics can proceed smoothly? Cannot this content be understood independently of platonist ontology? How, if possible, can the seemingly embarrassing questions associated with the platonist picture be blocked, *while respecting and preserving the reasonableness of ordinary practice, including the use of ordinary theories?* Of course, one may attempt to disarm these questions while remaining platonistic, as for instance Tait has done; and one may ask whether the nominalist reconstructions really can serve naturalized epistemology as intended, or whether they raise intractable problems of their own regarding knowledge of space-time or of possibilities, etc. *But that is to lock horns properly over the key philosophical and metascientific questions; it is not to raise the false spectre of Maoist mathematics!*

In my review [Hellman, 1993] of Charles Chihara [1990], I raised the question, "Could it be that the 'problems of platonism' are so serious that, short of a successful nominalization strategy, transfinite set theory would have to be renounced?" I am no longer sure what I meant by "renounced" – perhaps "as a rational pursuit." But Chihara's own view, which I went on to cite, is instructive: regardless of its dispensability for science, there is *no objection* to transfinite set theory at all, but only to what some mathematicians and philosophers have said about such a theory. So if his own constructibility theory can be extended so as to preserve axioms and theorems of transfinite set theory, it would be put up "for adoption" only in the sense that mathematicians could refer anyone to it who tried to distract them with ontological and epistemological questions raised by platonism. Similarly with regard to my own modal-structural approach to set theory. And even Field's program, while it is not *as* preservationist as the others in that it does not respect mathematical truth even at the level of the natural numbers, still can be understood as respecting the rationality of a portion of mathematical practice, including the use of standard theories. For they provide manageable (and perhaps indispensable) proofs of nominalistically statable conclusions about the physical world and are fruitful in a variety of ways. BR quote David Lewis as saying, "Even if we reject mathematics gently – explaining how it can be a most useful fiction, 'good without being true' – we still reject it, and that's still absurd" (see p. 34). As wonderful as Lewis' *Credo* is, it highlights the question without answering it: What do you mean by these sweeping words, "adopt," "accept," "reject," "revise," etc., especially in connection with something so broad and multi-faceted as "mathematics"? BR have framed their whole debate in these terms and, if I am right, associated as they are too closely with social and political phenomena, they simply cannot do the philosophical work demanded of them.

At the outset, I quipped that *A Subject with No Object* in essence contains its own review. Much of what I have just written can be read as drawing out implications of the authors' own initial characterization of the nominalists' central purpose, "to preserve the subject of mathematics while banishing its objects." Now if the subject is *pure* mathematics, this does not apply to Field's original program, although it may if the subject is mathematized physics. But either way, it does apply to the programs of Chihara and Hellman and all the variants and related approaches reviewed by BR. Perhaps all the talk of "revolution," "rejection," etc., resulted from taking Field's stance toward pure mathematics as representative. In any case, that it clearly is not.

Finally, we should pick up the themes of naturalistic epistemology and the division of intellectual labor. For, once the nominalists' concerns are located in the metamathematical and metascientific, as just indicated, the whole evaluation of the reconstructions in terms of descriptive methodological criteria must be suitably adjusted. If a reconstruction is not intended as a *replacement* or even *revision* of ongoing theory, if it is intended rather as a coexisting proposal for understanding such theory or its accomplishments, preserving its substance while facilitating accommodation within a naturalistic epistemology, then it would indeed be fairly preposterous for the author of such a reconstruction to submit it to a physics or mathematics journal. Physics and mathematics proper are not the right scientific subjects; rather the subject is that of *epistemology*, fledgling though it certainly is as a science. Then, when *desiderata* such as "tractability," "fruitfulness," "consonance with other successful theories" are considered, it will be in comparison to other proposals on the same level or to other epistemologies, not in comparison to ongoing physics and mathematics.

Of course, it may be objected that the "science" of epistemology is too young for any really telling comparisons to be made. Indeed! We are in speculative territory, not knowing just where the gold may lie. As the authors themselves state early on, however,

> naturalism ... is not to be confused with crass "scientism" ... For in the first place, the various forms of enquiry we call science do not speak to every question ... And in the second place, science is not a closed guild with rigid criteria of membership. (p. 65)

And surely "science" cannot be confined to "today's science." Thus, it seems that, once the proper distinction between levels of questions is drawn and the relevant branch of science correctly identified, much of the authors' critique simply dissolves. Why should an historical survey of "theory choice" in the contexts of physics and mathematics, proper, or even within today's linguistics and psychology, be expected to tell us very much with respect to the weighing of theoretical virtues in the context of future epistemology? To be sure, an emphasis on restricting ontology in physics or mathematics proper would probably have proved detrimental if not disastrous in many cases. Is not that what we should expect? But why should that carry over to new cognitive, epistemic proto-sciences whose problems and aims are utterly different?

It is ironic that the authors have not seen matters in this way, since in the final part of their assessment they turn precisely in the direction I am suggesting. The irony is even greater in light of their actual proposal. For if, indeed, the nominalist reconstruction programs help us sort out our own contribution from "reality's," does that not strongly suggest that platonist commitments, if they are forced on us, are realy forced *by* us and are not best taken as an accurate description of reality? What more nominalistic conclusion could any nominalist desire?

Even should my critical remarks hold up, it should be clear that this book is of great value and interest and that, on the whole, it exemplifies philosophy practiced at its best.

References

Burgess, J., Hazen, A., and Lewis, D. [1991] "Appendix," in Lewis, D., *Parts of Classes* (Oxford: Blackwell), pp. 121–149.

Chihara, C. S. [1990] *Constructibility and Mathematical Existence* (Oxford: Oxford University Press).

Field, H. [1980] *Science without Numbers* (Princeton, NJ: Princeton University Press).

Goodman, N. [1977] *The Structure of Appearance*, 3rd edn. (Dordrecht: Reidel).

Hellman, G. [1989] *Mathematics without Numbers: Towards a Modal-Structural Interpretation* (Oxford: Oxford University Press).

Hellman, G. [1993] Book review: Charles S. Chihara, *Constructibility and Mathematical Existence* (Oxford University Press, 1990)," *Philosophia Mathematica* **1**(1): 75–88.

Hellman, G. [1996] "Structuralism without structures," *Philosophia Mathematica* **4**(2): 100–123.

Malament, D. [1982] Review of Field's *Science without Numbers*, *Journal of Philosophy* **79**: 523–534.

Part II

Predicative Mathematics and Beyond

7   Predicative Foundations of Arithmetic

Solomon Feferman and Geoffrey Hellman

Introduction

Predicative mathematics in the sense originating with Poincaré and Weyl begins by taking the natural number system for granted, proceeding immediately to real analysis and related fields. On the other hand, from a logicist or set-theoretic standpoint, this appears problematic, for, as the story is usually told, impredicative principles seem to play an essential role in the foundations of arithmetic itself.[1] It is the main purpose of this paper to show that this appearance is illusory: as will emerge, a predicatively acceptable axiomatization of the natural number system can be formulated, and both the existence of structures of the relevant type and the categoricity of the relevant axioms can be proved in a predicatively acceptable way.

Indeed, there are three aspects of the set-theoretical foundations of arithmetic which might appear to involve impredicative principles in an essential way, either explicitly or implicitly. These may be listed as follows.

(a) **Axiomatization**: The axiom of *mathematical induction* ((N-III) below) is *prima facie* in full second-order form.

(b) **Categoricity**: As usually formulated, the proof of categoricity employs an impredicative construction, namely in the specification of an isomorphism between two models (of arithmetic with the full second-order induction axiom) as the intersection of all one-one maps between their domains preserving the zeros and succession.

(c) **Existence**: To construct a set satisfying the axioms of (b), assuming one has a set satisfying the usual first-order axioms on successor, one takes the intersection of all sets closed under successor; this is an impredicative definition.

Concerning these points, we make the following remarks:

---

[1] For a careful discussion in support of this conclusion, see Parsons [1983]. We shall comment on this below, in the final Discussion section.

Ad (a): As is well known, no first-order axiomatization is categorical. The axiom (N-III) is on its face second-order, so the usual conclusion is that we need "full" second-order logic to underlie the use of (N-III).

Ad (b): This kind of proof is standard and is given, for example, in Shapiro [1991], pp. 82–83.

Ad (c): The question of existence is raised in a logicist or modified logicist (set-theoretical) approach to the foundations of arithmetic, where one is also required to "construct" the natural numbers. This played an important role for Dedekind [1888], whose "existence proof" calls for special comment and reformulation (cf., e.g., Hellman [1989], Ch. 1). Predicativists such as Poincaré and Weyl, however, did not require this step (cf. Feferman [1988]).

Thus, we may ask, to what extent can (a)–(c) be accounted for on predicatively acceptable grounds?

Re (a): Unrestricted second-order variables may be considered to range over predicatively defined classes satisfying weak, predicatively acceptable closure conditions.

Re (b): We assume that the notion of *finite set* is predicatively understood, governed by some elementary closure conditions.[2] Instead of defining the isomorphism between two "*N*-structures" (as Dedekind's "simply infinite systems" or the set theorist's "ω-sequences" will be called below) as the *intersection of* (*certain*) *classes,* we obtain it as the *union of* (*certain*) *finite sets.*[3]

---

[2] It is well known that a categorical theory for the natural number system can be expressed in (monadic) weak second-order logic in its semantical sense: following Monk [1976], pp. 488–489, letting the variable A range over finite sets of individuals, we can extend Robinson's system Q of arithmetic with the axiom

$$\forall x \, \exists A [x \in A \ \& \ \forall y (y' \in A \rightarrow y \in A)].$$

("Every individual belongs to a finite set closed under predecessor.") However, since finite set variables are the only higher-order variables available in this system, neither (the full statement of) mathematical induction nor categoricity can even be expressed, so it is clearly inadequate as a predicativist framework for arithmetic. (Monk's point in considering this system is to demonstrate that even such a minimal enrichment of first-order logic is non-compact and not recursively axiomatizable.)

The systems presented here extend an axiomatizable fragment of weak second-order logic so that mathematical induction, categoricity, and notions of "infinite" and "Dedekind infinite" can be expressed. It then turns out that important metatheorems are provable from within these predicatively justifiable systems.

[3] Interestingly enough, this was also Dedekind's procedure, cf. his "126 Theorem of the definition by induction" (Dedekind [1888], pp. 85–86) used to prove categoricity, his major "132 Theorem," although indeed Dedekind did not formalize the comprehension principles he employed. Of course, Dedekind worked entirely with the notion of "Dedekind-finite," which involves quantification over general functions, and so a formal version of his proof of Theorem 126 would ultimately appeal to impredicative comprehension.

Re (c): Even if we pursue the logicist challenge to "construct" the natural numbers, this can in fact be carried out within a predicatively justified system.

In sum, the moral is that appearances are deceiving!

Our work will be carried out within a three-sorted system, EFSC, standing for "Elementary Theory of Finite Sets and Classes," and a certain extension EFSC*. The three sorts are (general) *individuals*, *finite sets*, and *classes*. The notion of *N-structure* can be formulated in the language of EFSC, and it is proved there that any two *N*-structures are isomorphic (Section 7.1). EFSC* is obtained from EFSC by adding an axiom to the effect that any (truly) finite set is Dedekind finite. EFSC* proves the existence of an *N*-structure (Section 7.3). It will be shown that EFSC* contains the first-order system of Peano arithmetic, PA (under suitable definition of its basic notions) and is a conservative extension of PA (Section 7.4). The system PA is a part of what is predicatively acceptable, granted the conception of the totality of natural numbers (cf. Feferman [1964, 1968]).[4]

7.1   The System EFSC

The *language* $\mathcal{L}$ *(EFSC)* contains the following.

> *Individual variables:* a, b, c, u, v, w, x, y, z, with or without subscripts.
>
> *Finite set variables:* A, B, C, with or without subscripts.
>
> *Class variables:* $\mathbf{X}$, $\mathbf{Y}$, $\mathbf{Z}$, with or without subscripts.
>
> *Operation symbol:* (,).
>
> *Individual terms*, s, t, ... are generated from the individual variables by the operation (,).
>
> *Atomic formulas:* $s = t$, $s \in A$, $s \in \mathbf{X}$.
>
> *Formulas* ϕ, ψ, ... are generated from atomic formulas by &, and $\vee$, $\neg$, $\rightarrow$ and $\forall$, $\exists$ applied to any one of the three sorts of variables.
>
> *WS-formulas: Predicative* or *weak second-order formulas* are those in which there are no *bound* class variables.

The underlying logic is classical three-sorted (first-order) predicate calculus with equality (in the first sort).

Definition: $\mathbf{X} \subseteq \mathbf{Y} \equiv^{df} \forall x[x \in \mathbf{X} \rightarrow x \in \mathbf{Y}]$, $\mathbf{X} = \mathbf{Y} \equiv^{df} \mathbf{X} \subseteq \mathbf{Y} \ \& \ \mathbf{Y} \subseteq \mathbf{X}$. Similarly defined are $\mathbf{X} \subseteq A$, $A \subseteq \mathbf{X}$, $\mathbf{X} = A$, $A \subseteq B$, $A = B$.

---

[4] There is a radical form of predicativism which does not accept the natural numbers as a "completed totality," i.e., over which unbounded quantification has a definite truth-functional value. This is the sense of Nelson [1986]; Nelson's system is *much* weaker than PRA, Primitive Recursive Arithmetic, whereas RA itself is already acceptable to finitists.

The *Axioms* of EFSC are (WS-CA), (Sep), (FS-I), (FS-II), and (P-I), (P-II), as follows:

$$(\text{WS-CA}) \qquad \exists \mathbf{X} \forall x [x \in \mathbf{X} \leftrightarrow \phi],$$

where ϕ is a WS-formula lacking free $\mathbf{X}$ (WS-CA is for weak second-order comprehension axiom).

Notation: We write $\{x|\phi\}$ for an $\mathbf{X}$ such that $x \in \mathbf{X} \leftrightarrow \phi$. Since X is determined up to definitional =, this is an inessential extension of the symbolism. As special cases, we define: $\mathbf{X} \cap \mathbf{Y} = \{x|x \in \mathbf{X} \ \& \ x \in \mathbf{Y}\}$, $\mathbf{X} \cup \mathbf{Y} = \{x|x \in \mathbf{X} \vee x \in \mathbf{Y}\}$, $-\mathbf{X} = \{x|x \notin \mathbf{X}\}$, $\mathbf{X} \times \mathbf{Y} = \{z| \exists x, y(x \in \mathbf{X} \ \& \ y \in \mathbf{Y} \ \& \ z = (x, y))\}$, $\mathbf{V} = \{x|x = x\}$, $\Lambda = \{x|x \neq x\}$, $A \cap \mathbf{X} = \{x|x \in A \ \& \ x \in \mathbf{X}\}$, $\{a\} = \{x |x = a\}$, etc.

$$(\text{Sep}) \quad \forall A \exists B \forall x [x \in B \leftrightarrow x \in A \ \& \ \phi],$$

ϕ aWS-formula, 'B' not free in ϕ.

(FS-I)(*Empty*)	$\exists B \forall x[x \notin A]$.
(FS-II)(*Adjunction*)	$\forall a \forall A \exists B \forall x[x \in B \leftrightarrow x \in A \vee x = a]$.
(P-I)	$(x_1, x_2) = (y_1, y_2) \leftrightarrow x_1 = y_1 \ \& \ x_2 = y_2$.
(P-II)	$\exists u \forall x, y[(x, y) \neq u]$.

Remarks on the axioms
1. Define $Fin\mathbf{X} \equiv \exists A[\mathbf{X} = A]$, "$\mathbf{X}$ is (truly) finite." Then under (WS-CA), the Separation axiom scheme (Sep) is equivalent to $\forall A, \mathbf{X}[\mathbf{X} \subseteq A \rightarrow Fin(\mathbf{X})]$. However, we can consider variant formulations of EFSC without class variables. Then we need (Sep) as above. (See Remark 7 and Metatheorem 1, below.)
2. Under (WS-CA), the first Finite Set Axiom (FS-I) is equivalent to $Fin(\Lambda)$.
3. Under (WS-CA), (FS-II) is equivalent to
   $\forall A, a[Fin(A \cup \{a\})]$ or $\forall \mathbf{X}, a[Fin(\mathbf{X}) \rightarrow Fin(\mathbf{X} \cup \{a\})]$.
4. We use pairs so as to define relations, functions, etc., in terms of classes. Alternatively, we can dispense with pairs by introducing more sorts of variables for binary, ternary, ... relations, and for (truly) finite relations of each "arity."
5. The axiom (P-II) says that there are "urelements" under pairing.
6. The system EFSC has a certain analogy to the NBG (von Neumann–Bernays–Gödel) theory of sets and classes. In fact, there is a simple interpretation of EFSC in NBG: both individual and finite set variables are taken to range over all sets in NBG, and (x, y) is defined as usual, $(x, y) = \{\{x\}, \{x, y\}\}$. We can take u in (P-II) to be the empty set $\mathcal{A}$. Class variables are taken to range over classes in NBG. This

interpretation does not require the axiom of infinity. Note that NBG – $\{Inf\}$ has a standard model in which the set variables range over the hereditarily finite sets V_ω i.e., $\bigcup_{n<\omega} V_n$, where $V_o = \Lambda$ and $V_{n+1} = V_n \cup \mathcal{P}(V_n)$, and the class variables range over the definable subsets of V_ω.

7. Let EFS be EFSC without (WS-CA) in the language without class variables.

Metatheorem 1: *EFSC is a conservative extension of EFS.*

Proof (model-theoretic). By Gödel's completeness theorem, it suffices to show how any model $\mathcal{M}$ of EFS can be expanded to a model $\mathcal{M}'$ of EFSC. We can simply take the range of the class variables to be all the WS-definable subsets of $\mathcal{M}$ (from parameters for individuals and "finite" sets in $\mathcal{M}$).

A proof-theoretic argument can also be given. This is analogous to the conservation result of NBG over ZF, which was established by proof-theoretic means in Shoenfield [1954].

8. As will be seen, the system EFSC allows for a natural development of arithmetic and a general proof of categoricity, since it allows for quantification over general structures. (In particular, individuals may be any objects whatever, and are not restricted to objects in the range of numerical quantifiers.) The alternative approach of adopting number-theoretic axioms directly along with those of EFSC just presented – a system that we have dubbed "Predicative Dedekind Arithmetic" (PDA) – has the awkwardness of being able to treat only structures with "numbers" as individuals. An advantage of PDA, however, that we forego in the present approach, is that the implication from (true) finiteness to Dedekind finiteness (TF $\rightarrow$ DF) – adopted as a further axiom (Card) below – can be derived in PDA from the more intuitive axiom that "truly finite sets are bounded" (in the natural ordering of natural numbers). That implication (TF $\rightarrow$ DF), however, is elementary enough for present purposes, and so we shall pursue the present systems, in which the number-theoretic structures of interest may be both characterized and proved to exist.

Definition: $Func(\mathbf{X}) \equiv \forall x, y_1, y_2[(x, y_1) \in \mathbf{X} \; \& \; (x, y_2) \in \mathbf{X} \rightarrow y_1 = y_2]$. We use $f, g\ h$ with or without subscripts to range over functions.

Define $Dom(f) = \{x | \exists y[(x, y) \in f]\}$; $Ran(f) = \{y | \exists x[(x, y) \in f]\}$; we write $f(x) = y$ for $(x, y) \in f$; $f: \mathbf{X} \rightarrow \mathbf{Y}$ means $Dom(f) = \mathbf{X} \; \& \; Ran(f) \subseteq \mathbf{Y}$. We shall use functional notation freely in the following.

Definition: $(x_1, \ldots, x_n, x_{n+1}) = ((x_1, \ldots, x_n), x_{n+1})$;

$$\mathbf{X}^{n+1} = \mathbf{X}^n \times \mathbf{X}.$$

Thus, functions of n arguments may be reduced to unary functions on Cartesian powers.

Definition: A *pre-N-structure* is a triple $\mathcal{M} = \langle \mathbf{M}, a, g \rangle$, where

$a \in \mathbf{M}$, $g \colon \mathbf{M} \to \mathbf{M}$, and

(N-I) $\forall x \in \mathbf{M}[g(x) \neq a]$

(N-II) $\forall x, y \in \mathbf{M}[g(x) = g(y) \to x = y]$.

Definition: An *N-structure* is a pre-*N*-structure $\langle \mathbf{M}, a, g \rangle$ such that

$$(\text{N-III})\,(Induction) \quad \begin{aligned} &\forall \mathbf{X} \subseteq \mathbf{M}[a \in \mathbf{X} \ \& \\ &\forall x(x \in \mathbf{X} \to g(x) \in \mathbf{X}) \to \mathbf{M} \subseteq \mathbf{X}]. \end{aligned}$$

By (WS-CA), the scheme of induction for WS-formulas, relativized to any *N*-structure, can be derived from (N-III). In particular, for any *N*-structure $\langle \mathbf{N}, 0, ' \rangle$, one can prove

$$(\text{Pr})\ (Predecessors)\colon x \neq 0 \to \exists y(x = y').$$

Definition: Let $\mathcal{M} = \langle \mathbf{M}, a, g \rangle$ be a pre-*N*-structure; we define

$$y \leqslant_{\mathcal{M}} x \equiv \forall A[x \in A \ \& \ \forall z(g(z) \in A \to z \in A) \to y \in A].$$

Where no confusion will arise, the subscript $\mathcal{M}$ may be dropped. It will also contribute to intuitive clarity to work with the notation $\langle \mathbf{N}, 0, ' \rangle$ for an arbitrary but fixed pre-*N*-structure.

Theorem 2: *EFSC proves the following for any pre-N-structure* $\langle \mathbf{N}, 0, ' \rangle$, *where* $x, y \in \mathbf{N}$:

(i) $x \leqslant x$;

(ii) $w \leqslant y \leqslant x \to w \leqslant x$;

(iii) $y' \leqslant x \to y \leqslant x$;

(iv) $y \leqslant 0 \leftrightarrow y = 0$;

(v) $y \leqslant x' \leftrightarrow y \leqslant x \vee y = x'$.

Proof. (i), (ii), (iii), and $\leftarrow$ of (iv) and (v) are immediate.

(iv) $\to$: Suppose $y \leqslant 0$. Let $A = \{0\}$, which exists by (FS-I), (FS-II), i.e., $\forall z(z \in A \leftrightarrow z = 0)$. Then $0 \in A$ and $\forall z(z' \in A \to z \in A)$, since $z' \neq 0$ (by (N-I)). So, by hypothesis, $y \in A$, whence $y = 0$.

(v) $\to$: Suppose $y \leqslant x'$ but not $y \leqslant x$, to show $y = x'$. By hypothesis, $\exists A[x \in A \ \& \ \forall z(z' \in A \to z \in A) \ \& \ y \notin A]$. Let $B = A \cup \{x'\}$, i.e., $z \in B \leftrightarrow z \in A \vee z = x'$. Then $x' \in B \ \& \ \forall z (z' \in B \to z \in B)$. Since $y \leqslant x'$, it follows that $y \in B$. But $y \notin A$, so $y = x'$.

Let $\langle \mathbf{N}, 0, ' \rangle$ be an *N*-structure and let *M*, *a*, and *g* be arbitrary satisfying $a \in \mathbf{M}$ and $g \colon \mathbf{M} \to \mathbf{M}$. We seek to introduce f satisfying the recursion equations

$$(1) \qquad \begin{cases} f(0) = a \\ f(x') = g(f(x)). \end{cases}$$

Let $Rec(A, g, a, x)$ be the following formula, which expresses that the finite binary relation A is the graph of f restricted to $\{z | z \leqslant x\}$:

$$
\begin{aligned}
Rec(A, g, a, x) \equiv\ & \forall z, w[A(z, w) \to z \leqslant x]\ \& \\
& \forall w[A(0, w) \leftrightarrow w = a]\ \& \\
& \forall z < x \forall w[A(z', w) \leftrightarrow \exists u(A(z, u)\ \& \\
& g(u) = w)].
\end{aligned}
$$

(Here and in the following two theorems, it simplifies notation to use x, y, z as individual variables ranging over **N** and u, v, w as individual variables ranging over **M**.)

Theorem 3: *(EFSQ)*
   (i) $Rec(A, g, a, x) \to \forall z[z \leqslant x \to \exists! w A(z, w)]$;
  (ii) $Rec(A, g, a, x)\ \&\ Rec(B, g, a, x) \to A = B$;
 (iii) $\forall x \exists A Rec(A, g, a, x)$;
 (iv) $Rec(B, g, a, x')\ \&\ \forall z, w[A(z, w) \leftrightarrow B(z, w)\ \&\ z \leqslant x] \to Rec(A, g, a, x)$;
  (v) $Rec(A, g, a, x)\ \&\ Rec(B, g, a, y)\ \&\ x \leqslant y \to A \subseteq B$.

Proof. (i) is by induction (in **N**) on z. For (ii), by induction on z, $\forall w[A(z, w)] \leftrightarrow B(z, w)]$. (iii) is by induction on x (for any given g, a), invoking (Sep), (FS-I), (FS-II). If $Rec(A, g, a, x)$ and $A(x, u)$, then $Rec(B, g, a, x')$, where $B = A \cup \{(x', g(u))\}$. (iv) is immediate. (v) is by induction on x.

Now we can define f satisfying (1) by a WS-formula:

$$(2)\quad (f(x) = u) \equiv \exists A[Rec(A, g, a, x)\ \&\ A(x, u)].$$

The preceding theorem then yields the following.

Theorem 4: *(EFSC). Let $\langle \mathbf{N}, 0,' \rangle$ be an N-structure and let* **M**, a, g *satisfy $a \in \mathbf{M}$ and g:* $\mathbf{M} \to \mathbf{M}$; *then*

  (i) $\forall x \exists! u \exists A[Rec(A, g, a, x)\ \&\ A(x, y)]$;
 (ii) *for f defined by (2), the equations (1) hold, i.e.,*
    $\exists f\{f: \mathbf{N} \to \mathbf{M}\ \&\ f(0) = a\ \&\ \forall x \in \mathbf{N}[f(x') = g(f(x))]\}$.

Theorem 5: *(Categoricity) (EFSC). If $\langle \mathbf{N}, 0,' \rangle$ and $\langle \mathbf{M}, a, g \rangle$ are any two N-structures, then $\langle \mathbf{N}, 0,' \rangle \cong \langle \mathbf{M}, a, g \rangle$, that is, there exists a bijection f between* **N** *and* **M** *such that*

  (i) $f(0) = a$;
 (ii) $f(x') = f(f(x))$.

Proof. By Theorem 4(ii), there exists f from **N** into **M** satisfying (i) and (ii), i.e., preserving the "zeros" and "succession." It remains to show that f is a bijection between **N** and **M**. By induction on x (in **N**), it follows from the suppositions on **M** and g and (i) and (ii) that $f(x) \in \mathbf{M}$. Further, we prove

$\forall y[f(x) = f(y) \rightarrow x = y]$, also by induction on x. Setting $x = 0$, $\forall y[f(0) = f(y) \rightarrow 0 = y]$ is equivalent to $\forall y(f(y) = a \rightarrow y = 0)$. Now if $y \neq 0$, find z such that $y = z'$ (by (Pr) above); then $f(y) = f(z') = g(f(z))$. But $g(f(z)) \neq a$, so we have a contradiction. Now suppose $\forall y(f(x) = f(y) \rightarrow x = y)$; we must show $\forall y[f(x') = f(y) \rightarrow x' = y]$, that is $\forall y(g(f(x)) = f(y) \rightarrow x' = y]$. Here if $y = 0$, we again get a contradiction. So $y = z'$ and $f(y) = f(z') = f(f(z))$. But then $f(x) = f(z)$, whence $x = z$, and then $x' = z' = y$. Finally, to show $\forall u \in \mathbf{M} \exists x(f(x) = u)$, we proceed by induction in $\mathbf{M}$, using $\mathbf{X} = \{w | w \in \mathbf{M} \ \& \ \exists x(f(x) = w)\}$. Clearly $a \in \mathbf{X}$. If $u \in \mathbf{X}$, then $\exists x(f(x) = u)$, so $g(x) = f(x') = g(u)$. Hence $\mathbf{M} \subseteq \mathbf{X}$. q.e.d.

For purposes of comparing EFSC with well-known systems, the following theorem establishing closure under primitive recursion on any N-structure will also be useful.

Theorem 6: *(EFSC). If $\langle \mathbf{N}, 0, ' \rangle$ is any $\mathbf{N}$-structure and $h: \mathbf{N}^n \rightarrow \mathbf{N}$, $g: \mathbf{N}^{n+2} \rightarrow \mathbf{N}$, then there exists $f: \mathbf{N}^{n+1} \rightarrow \mathbf{N}$ such that:*

$$f(x_1, \ldots, x_n, 0) = h(x_1, \ldots x_n)$$
$$f(x_1, \ldots, x_n, x') = g(x_1, \ldots, x_n, x, f(x_1, \ldots, x_n, x)).$$

Proof. Exactly analogous to that of Theorem 3 above.

7.2   Existence of Special Pre-*N*-Structures in EFSC

Definition: Let 0 be some fixed urelement under pairing, i.e., suppose

$$(\text{P-II})_0 \qquad \forall x, y[(x, y) \neq 0].$$

Then define $x' =^{df} (x, 0)$.

Lemma: (i) $\forall x[x' \neq 0]$;
(ii) $\forall x, y[x' = y' \rightarrow x = y]$.

Proof. Immediate, by (P-I), (P-II)$_0$.

Corollary: $\langle \mathbf{V}, 0, ' \rangle$ *is a pre-N-structure.*

Proof. Immediate, by the definition of pre-*N*-structure and the lemma.

Now the question is whether we can construct a pre-*N*-structure satisfying the induction axiom (N-III). The obvious way to do this would be to take

$$\mathbf{M} = \{x | \ \forall \mathbf{X}[0 \in \mathbf{X} \ \& \ \forall y(y \in \mathbf{X} \rightarrow y' \in \mathbf{X}) \rightarrow x \in \mathbf{X}]\}.$$

However, in EFSC we cannot infer the existence of such $\mathbf{M}$ using (WS-CA) – indeed, this is a *prima facie* impredicative definition. We can, however, construct a pre-N-structure satisfying some special conditions which hold in any N-structure; this will serve as a preliminary to the construction of an N-structure in the stronger theory EFSC* of Section 7.3.

Definition: $\leqslant$ is the $\leqslant_\mathcal{V}$-*relation* for the structure $\mathcal{V} = \langle \mathbf{V}, 0,' \rangle$.

For any x, let $Pd_\leqslant(x) =^{df} \{y | y \leqslant x\}$.

By Theorem 2, $\leqslant$ is reflexive and transitive, $Pd_\leqslant(0) = \{0\}$, $Pd_\leqslant(x') = Pd_\leqslant(x) \cup \{x'\}$.

Definition: $\mathbf{M}$ is *special* just in case:

(i) $0 \in \mathbf{M}$;
(ii) $x \in \mathbf{M} \rightarrow x' \in \mathbf{M}$;
(iii) $x \in \mathbf{M}$ & $y \leqslant x \rightarrow y \in \mathbf{M}$;
(iv) $x \in \mathbf{M} \rightarrow Fin(Pd_\leqslant(x))$;
(v) $x \in \mathbf{M}$ & $x \neq 0 \rightarrow \exists y \in \mathbf{M}[x = y']$.

If $\mathbf{M}$ is special, then the substructure $\langle \mathbf{M}, 0,' \rangle$ of $\langle \mathbf{V}, 0,' \rangle$ is called a *special pre-N-structure*.

Theorem 7: *(EFSC). There exist special pre-N-structures.*

Proof. Let $\mathbf{M} = \{x | Fin(Pd_\leqslant(x))$ & $\forall u[u \leqslant x \rightarrow u = 0 \lor \exists w(u = w')]\}$.

Then (i) $0 \in \mathbf{M}$, because $Pd_\leqslant(0) = \{0\}$ and $\forall u[u \leqslant 0 \rightarrow u = 0]$.

(ii) Suppose $x \in \mathbf{M}$. Then $x' \in \mathbf{M}$ because $Pd_\leqslant(x') = Pd \leqslant (x) \cup \{x'\}$, so $Fin(Pd_\leqslant(x)) \rightarrow Fin(Pd \leqslant (x'))$, and $\forall u[u \leqslant x' \rightarrow u = 0 \lor \exists w(u = w')]$ since $u \leqslant x' \rightarrow u \leqslant x \lor u = x'$.

(iii) $x \in \mathbf{M}$ & $y \leqslant x \rightarrow y \in \mathbf{M}$, for first $Pd_\leqslant(y) \subseteq Pd \leqslant (x)$, so $Fin(Pd_\leqslant(x)) \rightarrow Fin(Pd_\leqslant(y))$ (cf. Remark 1 on the axioms, above). Now if $u \leqslant y$, then $u \leqslant x$, so $u = 0 \lor \exists w(u = w')$.

(iv) $x \in \mathbf{M} \rightarrow Fin(Pd_\leqslant(x))$, by construction of $\mathbf{M}$.

(v) $x \in \mathbf{M}$ & $x \neq 0 \rightarrow \exists y \in \mathbf{M}[x = y']$, for $x \leqslant x$, so $x = 0 \lor \exists y[x = y']$, by construction of $\mathbf{M}$. Suppose $x \neq 0$, and let y be such that $x = y'$. Since $y \leqslant y' = x$, we have $y \in \mathbf{M}$, by (iii).

7.3   Existence of N-Structures

The theory EFSC* will be obtained from EFSC by adding the axiom that every (truly) finite set is Dedekind finite.

Definition: $DedFin(\mathbf{X}) \equiv \forall f[f: \mathbf{X} \rightarrow \mathbf{X} \ \& \ f$ is one-one $\rightarrow Ran(f) = \mathbf{X}]$, where "one-one" is defined in the usual way.

(Card) *(Cardinality Axiom):* $\forall A[DedFin(A)]$.

It follows from (Card) that $\forall \mathbf{X}[Fin(\mathbf{X}) \rightarrow DedFin(\mathbf{X})]$ (this is what we called TF $\rightarrow$ DF above), and conversely, so this is equivalent to (Card).

We define EFSC* to be EFSC + (Card).

Lemma: *(EFSC*). Let $\langle \mathbf{N}, 0,' \rangle$ be a special pre-N-structure, then*

(i) $\forall x \in \mathbf{N} \ \exists! \ \mathbf{Y} \subset \mathbf{N} \ \forall z \in \mathbf{N}[z \in \mathbf{Y} \leftrightarrow z \leq x]$;

(ii) *(DFIS)* $\forall x \in \mathbf{N} \ \forall \mathbf{Y} \subseteq \mathbf{N}[\forall z \in \mathbf{N}(z \in \mathbf{Y} \leftrightarrow z \leqslant x) \rightarrow DedFin(\mathbf{Y})]$ *(DFIS is for "Dedekind finite initial segments").*

Proof. Immediate, by (WS-CA), the definition of "special" and (Card).

Theorem 8: *(EFSC*). Suppose $\langle \mathbf{N}, 0,' \rangle$ is a special pre-N-structure; then it is an N-structure.*

Proof. What must be shown is that $\langle \mathbf{N}, 0,' \rangle$ satisfies mathematical induction, (N-III). Suppose Induction fails, i.e., for some $\mathbf{X} \subseteq \mathbf{N}$ we have $0 \in \mathbf{X}, \forall y(y \in \mathbf{X} \rightarrow y' \in \mathbf{X})$ but for some z, $z \notin \mathbf{X}$. Introduce the predecessor function $p_{\leqslant z}(y)$ defined for $y \leqslant z$ and $y \neq 0$, i.e., $(p_{\leqslant z}(y) = u) \equiv (y \leqslant z \ \& \ y \neq 0 \ \& \ u' = y)$. Next introduce the set

$$\mathbf{Y} = \{u | u \leqslant z \ \& \ u \notin \mathbf{X}\}.$$

Clearly, $\mathbf{Y}$ is Dedekind-infinite: by (Pr) (Predecessors), $p \leqslant z$ is a one-one function on $\mathbf{Y}(0 \notin \mathbf{Y}$ by hypothesis that $0 \in \mathbf{X})$, with values in $\mathbf{Y}$ (by Theorem 2(iii), and contraposing the second hypothesis of Induction), but $z \notin Range(p_{\leqslant z})$ by stipulation of $p_{\leqslant z}$. It follows that $Pd_{\leqslant}(z) \equiv \{u | u \leqslant z\}$ is also Dedekind-infinite (by a general argument: define the witnessing function on $Pd_{\leqslant}(z)$ to be $p_{\leqslant}z \cup$ the identity on $Pd_{\leqslant}(z) - \mathbf{Y}$). This contradicts the Lemma (ii) (DFIS). q.e.d.[5]

---

[5] This theorem together with Theorem 5 (Categoricity) realizes in effect a suggestion attributed to Michael Dummett for characterizing the natural numbers in a predicatively acceptable way, as those individuals x belonging to every class containing 0 and closed under successor applied to individuals distinct from x (i.e., $\forall \mathbf{X}[0 \in \mathbf{X} \ \& \ \forall y(y \in \mathbf{X} \ \& \ y \neq x \rightarrow y' \in \mathbf{X}) \rightarrow x \in \mathbf{X}]$, which specifies the initial segment inclusively up to x), a construction which works as well when the class variable is replaced throughout with a finite-set variable. For a discussion, see Isaacson [1987], pp. 155–156).

Clearly we disagree with Isaacson's diagnosis of the situation, that "the weak second-order definition does not fare significantly better on the score of avoiding impredicativity than the one based on full second-order logic," (p. 156). The reason given is that "an exact representation of the natural number sequence must occur as elements of the domain [of the second-order quantifiers]." But, as maintained here, this assumption – of the existence of finite initial segments

The following metatheorem is now a direct consequence of Theorem 8.

Corollary: *The system PA of Peano arithmetic is interpretable in EFSC*.*

Proof. Fix an N-structure, $\langle \mathbf{N}, 0,' \rangle$, e.g. as defined in Section 7.2 above. By Theorem 6 (Section 7.1), all primitive recursive functions are definable on $\mathbf{N}$, satisfying their defining equations. In particular, the recursion equations for addition and multiplication hold in $\langle \mathbf{N}, 0,' \rangle$. Further, any formula $\psi(x, y_1, \ldots, y_n)$ in $\mathcal{L}(PA)$ with parameters $y_1, \ldots, y_n \in \mathbf{N}$ defines a class, $\mathbf{X} = \{x \mid x \in \mathbf{N} \ \& \ \psi^{(\mathbf{N})}(x, y_1, \ldots, y_n)\}$, where in $\psi^{(\mathbf{N})}$ all (first-order) quantifiers are relativized to $\mathbf{N}$. Hence, by (N-III),

$$\langle N, 0,', +, \cdot \rangle \vDash \forall y_1, \ldots \ldots, y_n [\psi(0, y_1, \ldots \ldots y_n) \ \&$$
$$\forall z(\psi(z, y_1, \ldots, y_n) \to \psi(z', y_1, \ldots, y_n)) \to$$
$$\forall x \psi(x, y_1, \ldots, y_n)].$$

This completes the proof.

7.4   Proof-Theoretic Strength

Metatheorem 9: *EFSC* is of the same proof-theoretic strength as PA, and it is a conservative extension of PA under the interpretation in the preceding Corollary (to Theorem 8).*

Proof. By the Corollary, EFSC* is at least as strong as PA. In the other direction, EFSC* is interpretable in the second-order extension ACA_0 of PA (ACA stands for "arithmetic comprehension axiom" and the subscript 0 indicates that induction is taken as the second-order axiom $\forall \mathbf{X}[0 \in \mathbf{X} \ \& \ \forall x(x \in \mathbf{X} \to x' \in \mathbf{X}) \to \forall x(x \in \mathbf{X})]).$[6] Individuals are interpreted as natural numbers, pairing is a primitive recursive function with $(x, y) \neq 0$, finite sets are interpreted as the codes by numbers of finite sets (with the primitive recursive $\in$-relation – the empty set is coded by 0, the finite non-empty set $\{x_1, \ldots, x_k\}$ with $x_1 < \ldots < x_k$ is represented by $2^{x_k} + \ldots + 2^{x_1}$), and classes are interpreted as sets of natural numbers in the second-order language $\mathcal{L}(\mathrm{ACA}_0)$. Under this translation, the axioms of EFSC* are derivable as theorems in ACA_0. For example, for (Card), one proves in ACA_0 that every finite set is equinumerous with an initial segment.

(of isomorphisms between ω-sequences (cf. Theorem 3(iii)), as well as of the individual ω-sequences) – is clearly predicatively justified. Moreover, there is a vast difference between presupposing infinitely many finite sets and presupposing an infinite one, especially the very one you are trying to "introduce"! The predicativist can afford to talk like a platonist about (hereditarily) finite sets, but not about the infinite.

[6] For information on ACA_0 and related subsystems of analysis (of a hierarchy explored by Harvey Friedman and others), see, e.g., Simpson [1987].

One then proves by induction on k that every initial segment $(0, k)$ is Dedekind finite. Now it is well known that ACA_0 is a conservative extension of PA (either by a direct model-theoretic argument or by a proof-theoretic result similar to that of Shoenfield [1954]). Thus, following through the above interpretation of EFSC* in ACA_0, we infer that EFSC* is conservative over PA.

Now, by Feferman [1964] or [1968], PA is a small part of what is directly predicatively acceptable. Thus, EFSC* is what Feferman has called a predicatively reducible system, that is, one which can be reduced by finitary proof-theoretic methods to a directly predicatively justified system.

Remark: Reflecting on the last two theorems, we see that, although EFSC* proves the existence of a (standard) model of PA, this cannot – in virtue of the second Gödel incompleteness theorem – be converted into a consistency proof within EFSC*. Although the relativization to an N-structure of each PA axiom (and indeed PA theorem) can be proved, one by one, in EFSC*, there is no way to introduce *satisfaction* and hence no way to prove a general *soundness* theorem (to the effect that any PA theorem *holds* in any N-structure). Indeed, the introduction of *satisfaction* would require impredicative construction, beyond the power of EFSC*. (Again, the analogy with NBG set theory and ZF is a good one: NBG "proves the existence of a (standard) ZF model, $\langle \mathbf{V}, \in \rangle$" – again in the sense of proving the relativization to V of each ZF theorem – but, without impredicative class construction, it cannot introduce satisfaction for unbounded set theoretic formulas, and, of course, cannot prove the consistency of ZF.)

We close this section with two questions and a brief discussion.

Question 1: What is the exact proof-theoretic strength of EFSC? N.B.: The proof of the categoricity theorem (Section 7.1) uses only a small part of EFSC. This part should be interpretable in the system designated by Friedman RCA_0, i.e., the weak fragment of ACA_0 which uses only (relative) Δ_1^0-comprehension. By Friedman's work (cf. Simpson [1987] or Sieg [1985]) RCA_0 is a conservative extension of PRA (primitive recursive arithmetic).

Question 2: What is a nice axiomatization of a subsystem of EFSC (i) in which categoricity is provable, and (ii) which is equivalent in strength to PRA?

Discussion: Even without claiming a strongest possible result, we have seen how the machinery of (a fragment of) weak second-order logic can be exploited to derive both the categoricity of number-theoretic axioms and the existence of arithmetical structures. In both cases, the central idea is that initial segments be finite (as in Theorem 3(iii) and Condition (iv) in the notion of a special pre-N-structure), allowing for a predicative construction of isomorphisms (Theorems 4 and 5) and of special pre-N-structures (see the construction of **M** in the proof of Theorem 7), which then are proved to satisfy full (predicative) induction (Theorem 8). Clearly this procedure bears on the position of Poincaré, according to which induction plays a primitive role in our conception of the natural numbers. As the above demonstrates, however, there is a viable alternative which begins with "Dedekind-finite initial segments" and which actually allows for a "predicative logicist" construction of the natural number system.

The argument that mathematical induction necessarily involves impredicativity (given, e.g., by Parsons [1983], p. 137) turns on the evident circularity of attempting to guarantee induction by introducing the natural numbers as "those objects obtainable from 0 by iterating the successor operation *an arbitrary finite number of times.*" There is, however, no circularity in beginning with the notion of "finite set" as governed by the axioms of EFSC* (including the link with Dedekind-finitude), which are articulable as above, prior to the construction of the natural number system. (It was indeed one of Dedekind's [1888] principal insights that the concept of finitude *can* be introduced prior to the natural numbers.) Moreover, from the general predicativist standpoint there is independent justification for proceeding in this manner: predicativism restricts itself to domains in which every object is explicitly describable by symbolic expressions of a predicativist language; so the finite subsets of any such domain **D** are also explicitly describable, by means of disjunctions of the form $x = c_1 \lor x = c_2 \lor \ldots x = c_n$ (where the c_i, represent designators (in the predicativist language) of objects in **D**). Again, this is conceptually independent of the natural number sequence.

It should be stressed in this connection that intelligibility of the notion "finite set" does not depend on a prior grasp of the structure of finite sets (of a given domain, ordered by inclusion), which is indeed a fairly complicated infinitistic object. It is, however, the infinitistic *structure* we call "the natural number sequence" that we seek to ground. Moreover, on a structuralist view, one can say (with Dedekind) that the individual numbers have no identity apart from position in this structure; but this contrasts with the self-standing character of "finite set" we have just noted.

It is true, however, that the weak second-order language is employed in these constructions in an essential way: if one tried to bypass it by substituting the explicitly defined notion "**X** is Dedekind-finite" for "**X** is finite" (i.e.,

$\exists A \; \forall \; x(x \in \mathbf{X} \leftrightarrow x \in A)$) throughout, one would find oneself appealing to impredicative comprehension, due to the quantifier over general functions in the definition of "Dedekind-finite."

Thus, we can see a far-reaching trade-off between predicativism – or *predicative logicism* – and *classical logicism*. Classical logicism provides a complete analysis of the concepts "finite," "infinite," and "cardinal number," but at the price of *impredicative comprehension* with all of its attendant "metaphysical" commitments. Predicativism avoids the latter but must presuppose the concept of "finite" in some form or other. However, as the above demonstrates, it can do this in a natural way *without thereby taking the natural number system as given*. On the contrary, it can exploit its assumptions to recover the essential core of Dedekind's analysis.

References

Dedekind, R. [1888] "The nature and meaning of numbers," reprinted in Beman, W. W., ed., *Essays on the Theory of Numbers* (New York: Dover, 1963), pp. 31–115, translated from the German original, *Was sind und was sollen die Zahlen?* (Brunswick: Vieweg, 1888).

Feferman, S. [1964] "Systems of predicative analysis," *Journal of Symbolic Logic* **29**: 1–30.

Feferman, S. [1968] "Autonomous transfinite progressions and the extent of predicative mathematics," in *Logic, Methodology, and Philosophy of Science III* (Amsterdam: North Holland), pp. 121–135.

Feferman, S. [1988] "Weyl vindicated: *Das Kontinuum* 70 years later," in *Temi e Prospettive della Logica e della Filosofia della Scienza Contemporanee* (Bologna: CLUEB), pp. 59–93.

Hellman, G. [1989] *Mathematics without Numbers: Towards a Modal-Structural Interpretation* (Oxford: Oxford University Press).

Isaacson, D. [1987] "Arithmetical truth and hidden higher-order concepts," in Paris Logic Group, eds., *Logic Colloquium '85* (Amsterdam: North-Holland,), pp. 147–169.

Monk, J. D. [1976] *Mathematical Logic* (New York: Springer,).

Nelson, E. [1986] *Predicative Arithmetic* (Princeton, NJ: Princeton University Press,).

Parsons, C. [1983] "The impredicativity of induction," in *How Many Questions? Essays in Honor of S. Morgenbesser* (Indianapolis, IN: Hackett), pp. 132–154.

Shapiro, S. [1991] *Foundations without Foundationalism: The Case for Second-Order Logic* (Oxford: Oxford University Press).

Shoenfield, J. [1954] "A relative consistency proof," *Journal of Symbolic Logic* **19**: 21–28.

Sieg, W. [1985] "Fragments of arithmetic," *Annals of Pure and Applied Logic* **28**: 33–72.

Simpson, S. G. [1987] "Subsystems of Z_2 and reverse mathematics," in Takeuti, G., ed., *Proof Theory*, 2nd edn. (Amsterdam: North-Holland), pp. 432–446.

8 Challenges to Predicative Foundations of Arithmetic

Solomon Feferman and Geoffrey Hellman

> The White Rabbit put on his spectacles. "Where shall I begin, please your Majesty?" he asked. "Begin at the beginning," the King said gravely, "and go on till you come to the end: then stop." Lewis Carroll, *Alice in Wonderland*

This is a sequel to our article "Predicative foundations of arithmetic" (Feferman and Hellman [1995], reproduced as Chapter 7 in this volume), referred to in the following as PFA; here we review and clarify what was accomplished in PFA, present some improvements and extensions, and respond to several challenges. The classic challenge to a program of the sort exemplified by PFA was issued by Charles Parsons in a 1983 paper, subsequently revised and expanded as Parsons [1992]. Another critique is due to Daniel Isaacson [1987]. Most recently, Alexander George and Daniel Velleman [1998] have examined PFA closely in the context of a general discussion of different philosophical approaches to the foundations of arithmetic.

The plan of the present paper is as follows. Section 8.1 reviews the notions and results of PFA, in a bit less formal terms than there and without the supporting proofs, and presents an improvement communicated to us by Peter Aczel. Then, Section 8.2 elaborates on the structuralist perspective that guided PFA. It is in Section 8.3 that we take up the challenge of Parsons. Finally, Section 8.4 deals with the challenges of George and Velleman, and thereby, that of Isaacson as well. The paper concludes with an Appendix by Geoffrey Hellman, which verifies the predicativity, in the sense of PFA, of a suggestion credited to Michael Dummett for another definition of the natural number concept.

8.1 Review

In essence, what PFA accomplished was to provide a formal context based on the notions of finite set and predicative class and on *prima facie* evident principles for such, in which could be established the existence and categoricity

117

of a natural number structure. The following reviews, in looser formal terms
than PFA, the notions and results therein prior to any discussion of their
philosophical significance. Three formal systems were introduced in PFA,
denoted EFS, EFSC, and EFSC*. All are formulated within classical logic.
The language L(EFS), has two kinds of variables:

> *Individual variables* : $a,\ b,\ c,\ u,\ v,\ w,\ x,\ y,\ z,\ldots,$ and
> *Finite set variables* : $A,B,C,F,G,H,\ldots$

The intended interpretation is that the latter range over *finite sets of individuals*.
There is one binary operation symbol (,) for a *pairing function* on individuals,
and *individual terms s, t,* ... are generated from the individual variables by
means of this operation. We have two relation symbols, '=' and '∈', by means
of which *atomic formulas* of the form $s = t$ and $s \in A$ are obtained. *Formulas* φ,
ψ, ... are generated from these by the propositional operations '¬', '&', '∨',
'→', and by the quantifiers '∀' and '∃' applied to either kind of variable. The
language L(EFSC), which is the same as that of EFSC*, adds a third kind of
variable:

Class variables: **X, Y, Z,** ...[1]

In this extended language, we also have a membership relation between
individuals and classes, giving further atomic formulas of the form $s \in$ **X**.
Then formulas in L(EFSC) are generated as before, allowing, in addition,
quantification over classes. A formula of this extended language is said to be
weak second-order if it contains no bound class variables. The intended range
of the class variables is the collection of weak second-order definable classes of
individuals. We could consider finite sets to be among the classes, but did not
make that identification in PFA. Instead, we write $A =$ **X** if A and **X** have the
same extension. Similarly, we explain when a class is a subclass of a set, and so
on. A class **X** is said to be *finite* and we write Fin(**X**) if $\exists A(A = $ **X**$)$.

The *Axioms of EFS* are denoted (Sep), (FS-I), (FS-II), (P-I), and (P-II), and
are explained as follows. The separation scheme (Sep) asserts that any definable subset of a finite set is finite; that is, for each formula φ of EFS,
$\{x \in A \mid \varphi(x)\}$ is a finite set B when A is a given finite set. The axiom (FS-I)
asserts the existence of an empty (finite) set, and (FS-II) tells us that if A is
a finite set and a is any individual then $A \cup \{a\}$ is a finite set. The pairing
axioms (P-I) and (P-II), respectively, say that pairing is one-one and that there
is an urelement under pairing; it is convenient to introduce the symbol 0 for an
individual that is not a pair.

The *Axioms of EFSC* augment those of EFS by the scheme (WS-CA) for
weak second-order comprehension axiom, which tells us that $\{x \mid \varphi(x)\}$ is

---

[1] The class variables are given in boldface, to distinguish them from the finite set variables.

a class $\mathbf{X}$ for any weak second-order φ. In this language, we allow the formula φ in (Sep) to contain free class variables; then it can be replaced by the assertion that any subclass of a finite set is finite. The following theorem (numbered 1 in PFA) is easily proved by a model-theoretic argument, but can also be given a finitary proof-theoretic argument.

Metatheorem: *EFSC is a conservative extension of EFS.*

In the language of EFSC, (binary) relations are identified with classes of ordered pairs, and functions, for which we use the letters $\mathbf{f}$, $\mathbf{g}$, ..., [2] are identified with many-one relations; n-ary functions reduce to unary functions of n-tuples. Then we can formulate the notion of *Dedekind finite class* as being an $\mathbf{X}$ such that there is no one-one map from $\mathbf{X}$ to a proper subclass of $\mathbf{X}$. By the axiom (Card) is meant the statement that every (truly) finite class is Dedekind finite. The *Axioms of EFSC** are then the same as those of EFSC, with the additional axiom (Card).

Now, working in EFSC, we defined a triple $\langle \mathbf{M}, a, \mathbf{g} \rangle$ to be a *pre-N-structure* if it satisfies the following two conditions:

(N-I) $\forall x \in \mathbf{M} \, [\mathbf{g}(x) \neq a]$, and

(N-II) $\forall x, y \in \mathbf{M} \, [\mathbf{g}(x) = \mathbf{g}(y) \rightarrow x = y]$.

These are the usual first two Peano axioms when a is 0 and $\mathbf{g}$ is the successor operation. By an *N-structure* is meant a pre-N-structure that satisfies the *axiom of induction* in the form

(N-III) $\forall \mathbf{X} \subseteq \mathbf{M} \, [a \in \mathbf{X} \, \& \, \forall x (x \in \mathbf{X} \rightarrow g(x) \in \mathbf{X}) \rightarrow \mathbf{X} = \mathbf{M}]$.

It is proved in EFSC that we can define functions by primitive recursion on any N-structure; the idea is simply to obtain such as the union of finite approximations. This union is thus definable in a weak second-order way. From that, we readily obtain the following theorem (numbered 5 in PFA).

Theorem: *(Categoricity, in EFSC). Any two N-structures are isomorphic.*

Now, to obtain the existence of N-structures, in PFA we began with a specific pre-N-structure $\langle \mathbf{V}, 0, \mathbf{s} \rangle$, where $\mathbf{V} = \{x | x = x\}$ and $\mathbf{s}(x) = x' = (x, 0)$; that this satisfies (N-I) and (N-II) is readily seen from the axioms (P-II) and (P-I), respectively. Next, define

---

[2] As a point of difference with PFA, function variables here are given in boldface to indicate that they are treated as special kinds of classes.

$$\text{Clos}^-(A) \leftrightarrow \forall x[x' \in A \to x \in A], \tag{1}$$

and

$$y \leq x \leftrightarrow \forall A[x \in A \,\&\, \text{Clos}^-(A) \to y \in A]. \tag{2}$$

In words, $\text{Clos}^-(A)$ is read as saying that A is closed under the predecessor operation (when applicable), and so, $y \leq x$ holds if y belongs to every finite set that contains x and is closed under the predecessor operation. Let

$$\text{Pd}(x) = \{y | y \leq x\}. \tag{3}$$

The next step in PFA was to cut down the structure $\langle \mathbf{V},\ 0,\ \mathbf{s}\rangle$ to a special pre-N-structure:

$$\mathbf{M} = \{x | \text{Fin}(\text{Pd}(x)) \,\&\, \forall y\,[\,y \leq x \to y = 0 \lor \exists z(y = z')]\}. \tag{4}$$

This led to the following theorem (numbered 8 in PFA):

Theorem: *(Existence, in EFSC*).* $\langle \mathbf{M},\ 0,\ \mathbf{s}\rangle$ *is an N-structure.*

To summarize: in PFA, categoricity of N-structures was established in EFSC and existence in EFSC*. Following publication of this work, we learned from Peter Aczel of a simple improvement of the latter result obtained by taking in place of $\mathbf{M}$ the following class:

$$\mathbf{N} = \{x | \text{Fin}(\text{Pd}(x)) \,\&\, 0 \leq x\}. \tag{5}$$

Theorem: *(Aczel). EFSC proves that* $\langle N,\ 0,\ s\rangle$ *is an N-structure.*

We provide the proof of this here, using facts established in Theorem 2 of PFA.
  (i) $0 \in \mathbf{N}$, because $\text{Pd}(0) = \{0\}$ and $0 \leq 0$.
  (ii) $x \in \mathbf{N} \to x' \in \mathbf{N}$, because $\text{Pd}(x') = \text{Pd}(x) \cup \{x'\}$, and $0 \leq x \to 0 \leq x'$.
  (iii) If $\mathbf{X}$ is any subclass of $\mathbf{N}$ and $0 \in \mathbf{X} \land \forall y[\,y \in \mathbf{X} \to y' \in \mathbf{X}\,]$, then $\mathbf{X} = \mathbf{N}$. For, suppose that there is some $x \in \mathbf{N}$ with $x \notin \mathbf{X}$. Let $A = \{y\,|\,y \leq x \,\&\, y \notin \mathbf{X}\}$; A is finite since it is a subclass of the finite set $\text{Pd}(x)$. Moreover, A is closed under the predecessor operation, and so, A contains every $y \leq x$; in particular, $0 \in A$, which contradicts $0 \in \mathbf{X}$.
The theorem follows from (i)–(iii), since the axioms (N-I) and (N-II) hold on $\mathbf{V}$ and hence on $\mathbf{N}$.

It was proved in PFA that EFSC* is of the same (proof-theoretic) strength as the system PA of Peano axioms and is a conservative extension of the latter under a suitable interpretation. The argument was that EFSC* is interpretable in the system ACA_0, which is a well-known second-order conservative extension of PA based on the arithmetical comprehension axiom scheme together with induction axiom in the form (N-III). Conversely, we can develop PA in

EFSC* using closure under primitive recursion on any N-structure. Since any first-order formula of arithmetic so interpreted then defines a class, we obtain the full induction scheme for PA in EFSC*. Now, using the preceding result, the whole argument applies *mutatis mutandis* to obtain the following.

Metatheorem: *(Aczel). EFSC is of the same (proof-theoretic) strength as PA and is a conservative extension of PA under the interpretation of the latter in EFSC.*

This result also served to answer Question 1 on p. 13 of PFA.

Incidentally, it may be seen that the definition of $\mathbf{N}$ in (5) above is equivalent to the following:

$$x \in \mathbf{N} \leftrightarrow \forall A[x \in A \,\&\, \mathrm{Clos}^-(A) \to 0 \in A] \,\&\, \exists A[x \in A \,\&\, \mathrm{Clos}^-(A)]. \quad (6)$$

For, the first conjunct here is equivalent to the statement that $0 \leq x$, and the second to $\mathrm{Fin}(\mathrm{Pd}(x))$. In this form, Aczel's definition is simply the same as the one proposed by George [1987], p. 515.[3] Part of the progress that is achieved by this work in our framework is to bring out clearly the assumptions about finite sets that are needed for it and that are *prima facie* evident for that notion.

There is one further improvement in our work to mention. It emerged from correspondence with Alexander George and Daniel Velleman that the remark in footnote 5 on p. 16 of PFA asserting a relationship of our work with a definition of the natural numbers credited to Dummett was obscure. The exact situation has now been clarified by Geoffrey Hellman in the Appendix to this paper, where it is shown that Dummet's definition also yields an N-structure, provably in EFSC.

8.2   The Structuralist Standpoint and "Constructing the Natural Numbers"

In developing predicative foundations of arithmetic, we have been proceeding from a structuralist standpoint, one that each of us has pursued independently in other contexts. In general terms, structuralism has been described by one of us as the view that "mathematics is the free exploration of structural possibilities, pursued by (more or less) rigorous deductive means" (Hellman [1989], p. 6), along with the claim that,

---

[3] That, in turn, was a modification of a definition of the natural numbers proposed by Quine [1961] using only the first conjunct in (6), which is adequate when read in strong second-order form, but not when read in weak second-order form; cf. George [1987], p. 515, and George and Velleman [1998], n.10.

In mathematics, it is not particular objects which matter but rather certain 'structural' properties and relations, both within and among relevant totalities. (Hellman [1996], p. 101, reproduced here as Chapter 1)

Such general formulations raise questions of scope, for it seems that there must be exceptional mathematical concepts requiring a non-structural or pre-structural understanding so that prior sense can be made of "items *in* a structure," *substructure,* and other concepts required for structuralism to get started.[4] For present purposes, however, this question need not be taken up in a general way, as we may work within a more specialized form of structuralism, one explicitly concerned with number systems. As the other of us has put it:

The first task of any general foundational scheme for mathematics is to establish the number systems. In both the extensional and intensional approach this is done from the modem *structuralist* point of view. The structuralist viewpoint as regards the basic number systems is that it is not the specific nature of the individual objects which is of the essence, but rather the isomorphism type of the structure of which they form a part. Each structure $\mathcal{A}$ is to be characterized up to isomorphism by a structural property P which, logically, may be of first order or of higher order. (Feferman [1985], p. 48)

So long as this is understood, we may work with a system such as EFSC, leaving open whether this itself is to be embedded in a more general structuralist framework or whether it is thought of as standing on its own.

The central point here is that what we are seeking to define in a predicatively acceptable way is not, strictly speaking, the predicate '*natural number*' *simpliciter*, but rather the predicate '*natural-number-type structure*'. That is, we seek to characterize what it is to be a *structure* of this particular type – what Dedekind [1888] called "*simply infinite systems*" and what set-theorists call "*ω-sequences*" – and also to prove that, mathematically, such structures exist. Once this has been accomplished, we may then, as *a façon de parler*, identify the elements of a particular such structure as "the natural numbers," employing standard numerals and designations of functions and relations, but this is essentially for mathematical convenience. Officially, we *eliminate* the predicate 'is a natural number' in its absolute sense and speak instead of what holds in any natural-number-type structure. And thanks to our (limited) second-order logical machinery, we can render arithmetical statements directly, relativized to structures, as illustrated by the conditions (N-I)–(N-III) (Section 8.1, above); there is no need to introduce a relation of *satisfaction* between structures and sentences.

This standpoint has some implications worth noting. First, since no absolute meaning is being assigned to 'natural number', the same goes for 'non-number'. While of course a good definition of 'natural-number-type structure'

---

[4] For a good discussion of this and related issues, see Parsons [1990].

must rule out anything that does not qualify as such a structure, there is simply no problem of "excluding non-numbers" such as Julius Caesar (on standard platonist conceptions). This notorious Fregean problem simply does not arise in the structuralist setting. Rather than having to answer the question, "Is Julius Caesar a number?" (and presumably get the right answer), we sidestep it entirely. We even regard it as misleading to ask, "Might Julius Caesar be or have been a number?" for this still employs 'number' in an absolute sense. Of course, Julius Caesar might have been – and presumably is, in a mathematical sense – a member of many natural-number-type structures. On the other hand, we can make sense of standard, mathematically sensible statements such as "3/5 is not a natural number" by writing out "In any structure for the rationals with a substructure for the natural numbers (identified in the usual way), the object denoted '3/5' does not belong to the domain of the latter." And, of course, many elliptical references to "the natural numbers" are harmless.

More significantly, the whole question of circularity in "constructions of the natural numbers" must be looked at afresh. In contrast to 'natural number', 'natural-number-type structure' is an infinitistic concept in the straightforward sense that any instance of such a structure has an infinite domain with (at least) a successor-type operation defined on it. While it might well appear circular to define 'natural number' in terms of a predicate applying to just finite objects – for example, finite sets or sequences from some chosen domain – since it might seem obvious that such objects can do the duty of natural numbers, nevertheless if one succeeds in building up an *infinite structure* of just the right sort from finite objects, using acceptable methods of construction, and then proves by acceptable means that one has succeeded, *prima facie* one has done as much as could reasonably be demanded.

In predicative foundations, it is quite natural to take the notion 'finite set' as given, governed by elementary closure conditions as in EFSC. The cogency of this can be seen as follows. Within the definitionist framework, a predicatively acceptable domain is one in which each item is specified by a designator, say in a mathematical language. Hence any finite subset of the domain is specifiable outright by a disjunction of the form $x = d_1 \lor x = d_2 \lor \ldots \lor x = d_k$, where each d_i is a designator of an object in the domain. Thus, the finite sets correspond to finite lists of designators, and it is reasonable for the definitionist to take *this* notion – "finite list of quasi-concrete objects" – as understood. The claim is, along Hilbertian lines, that this does not depend on a grasp of the *infinite structure of natural numbers*, nor does it depend on an explicit understanding of the even more complex infinite structure of finite subsets ordered, say, by inclusion. Once given such a starting point, the closure conditions of EFSC are then evident.

There is a further related point of comparison between the concepts 'finite set' and 'natural number' that is relevant to our project. Given an infinite domain X of objects, we think of a finite set A of Xs as fully determined by its members. Although certain relations to other finite sets of Xs are also evident for us – for example, adjoining any new element to A yields a finite set – the identity of A as a *finite set* is not conceived as depending on its position in an infinite structure of finite sets of Xs. Yet this "self-standing" character of finite sets is not shared by natural numbers, even on platonist views. To identify a natural number is to identify its position in an infinite structure. Even on a set-theoretic construction, while the sets taken as numbers are of course determined *as sets* by their members, they are not determined *as numbers* until their position in a sequence is determined. Such considerations lead naturally to the structuralist project of PFA.

The significance of these points has perhaps not been sufficiently appreciated because, historically, structuralism has not been articulated independently of platonism. If one succeeds in defining 'natural number' platonistically, say as Frege or Russell did, or as Zermelo or von Neumann did, so that the natural numbers are identified uniquely with particular abstract objects, then, since the whole sequence of natural numbers thus defined together with arithmetic functions and relations are unproblematic as objects in such frameworks, it is a trivial matter to pass to an explicit definition of 'natural-number-type structure': one simply specifies as such a structure any that is isomorphic to the original, privileged one. Then, clearly all the work has gone into the original definition of 'natural number', and questions of circularity are directed there. However, the approach of PFA is different, sharing more with Hilbert's conception of mathematical axioms and reference than with Frege's.[5] For we bypass construction of 'the natural numbers' as particular objects and proceed directly to the infinitistic concept, 'natural-number-type structure' (much as Dedekind [1888] proceeded directly to define 'simply infinite system'). Then, in proving the existence of such structures, we introduce a certain sequence of finite objects available within our framework. Collecting these is predicatively unproblematic, for they are specified as having finitely many earlier elements (including an initial one), not as fulfilling mathematical induction. That they satisfy induction is then proved as a theorem.[6]

---

[5] For a valuable discussion of Hilbert's structuralist views of axioms and reference in mathematics and the contrast with Frege's views, see Hallett [1990].

[6] Our construction thus improves on Dedekind's, for he relied, for a Dedekind-infinite system, on a totality – of "all things which can be objects of my thought" (Dedekind [1888], Theorem 66) – which, even apart from its unmathematical character, is unacceptable to a predicativist on logical grounds, for, presumably, such a totality would contain itself! Furthermore, for a *simply infinite system*, he then relied on a subtotality impredicatively specified as the intersection of all subtotalities containing an initial element and closed under the given function (Theorems 72 and 44). But it is noteworthy that the particular example that Dedekind sought to invoke to insure non-vacuity of his definitions was not identified as "the numbers." As it happened, Dedekind did go on to speak of such abstract particulars, but that is another story, and, in any case, it is a further move that we have not been tempted to make.

Despite this result and the related ones established in PFA – especially the categoricity of our characterization and the proof-theoretic conservativeness of our system over PA – questions have been raised, implicitly and explicitly, concerning circularity and possible hidden impredicativity in the constructions. In the remaining two sections, we will address these specifically.

8.3   Parsons' Challenge

In his stimulating paper "The Impredicativity of Induction" (I of I in the following), Charles Parsons [1992] takes up a number of issues in his typically thoughtful and thorough manner. Our main purpose here is to address the points most directly related as a challenge to what PFA was intended to accomplish, namely, a predicative foundation of the structure of natural numbers, given the notion of finite set of individuals.[7] But it is necessary, first, to make some distinctions in regard to the idea of predicativity. To begin, a putative definition of an object c is said to be *impredicative* if it makes use of bound variables whose range includes c as one of its possible values.[8] Such bound variables may appear attached to quantifiers, or as the variable of abstraction in definitions of sets or functions, or as the variable in a unique description operator, and so on. We do not agree with the position ascribed to Poincaré and Weyl,[9] that impredicative definitions are *prima facie* viciously circular and to be avoided. For example, we regard the number associated with the Waring problem for cubes – defined as the least positive integer n such that every sufficiently large integer is a sum of, at most, n positive cubes – as a perfectly meaningful and non-circular description of a specific integer; it is known that $n \leq 7$, but beyond that, the exact value of n is not known. While this definition would generally be considered non-constructive, and is impredicative according to the general idea given above, from a classical predicative point of view it is not viciously circular, since we are convinced by predicative arguments that such a number exists and must have an alternative predicative definition, be it 7 or a smaller integer. So, for us, the issue is to determine when there is a predicative warrant for accepting a *prima facie* impredicative definition. That cannot be answered without saying what constitutes a predicative proof of existence of objects of one kind or another. Moreover, the above explanation of what it is about the form of a putative definition that makes it impredicative does not tell us what

---

[7] Parsons' paper appeared well before PFA, and so, the challenge was not issued to *it* but rather to the kind of program that it exemplifies. That challenge was addressed briefly in the final discussion section of PFA, pp. 14–15, but is expanded on substantially here.

[8] The informal explanation of what constitutes an impredicative definition varies from author to author. A representative collection of quotations is given by George [1987]; the explanation given in the text here is closest to that taken by George from an article of Hintikka [1956].

[9] Cf. I of I, pp. 152–153 and p. 159, n.24.

constitutes a *predicative definition*, because it only tells us what should *not* appear in it, and nothing about what (notions, names, etc.) *may* appear in it. Since the latter have to be, in some sense, prior to the object being defined, and since it is not asserted in explaining what is to be avoided just what that is, an answer to this necessarily makes of predicativity a *relative* rather than an *absolute* notion.

Considerations such as this led Kreisel to propose a formal notion of predicative provability *given the natural numbers*, and that was characterized in precise proof-theoretical terms independently (and in agreement with each other) by Feferman [1964] and Schütte [1965]. Speaking informally, that characterization takes for granted the notions and laws of classical logic as applied to definitions and statements involving, to begin with, only the natural numbers as the range of bound variables in definitions of sets of natural numbers, and then admits, successively, definitions employing variables for sets ranging over collections of sets that have been comprehended predicatively.[10] The details need not concern us; suffice it to say that Parsons, among others, has found this analysis of predicativity given the natural numbers to be persuasive (I of I, p. 150). However, as he suggests in the latter part of I of I, he also finds it reasonable to ascribe the term 'predicative' to the use of certain generalized inductive definitions that breach the bounds of the Feferman–Schütte characterization. There is no contradiction here from our point of view; the latter simply shifts what the notion of predicativity is taken relative to. One might go further and consider a notion of predicativity relative to the structure of real numbers, if one regarded that structure as well determined, and so on to higher levels of set theory. Though the idea is clear enough, none of these has been studied and characterized in precise proof-theoretical terms.[11]

Now, finally, we return to the program of PFA. There, the aim is to consider what can be done predicatively in the foundations of arithmetic relative to the notion of a finite set of individuals, where the individuals themselves may have some structure as built up by ordered pairs.[12] Philosophically, the significance of this is that we have a prior conception of finite set that does not require the understanding of the natural-number system, and for this notion we

---

[10] To be more precise, this is spelled out by means of an autonomous transfinite progression of ramified systems, where autonomy is a bootstrap condition that restricts one to those transfinite levels that have a prior predicative justification; cf. Feferman [1964].

[11] The relative notion of predicativity is recast by Feferman [1996] in terms of a formal notion of the *unfolding* of a schematic theory, which is supposed to tell us what more should be accepted once we have accepted basic notions and principles.

[12] Parsons has an interesting discussion in I of I (pp. 143–145), of what is reasonable to assume about the range of first-order variables in proposed definitions of the natural numbers. We believe that the assumptions (P-I) and (P-II) are innocuous, in the sense that the notion of ordered pair is a prerequisite to an understanding of any abstract mathematics.

have some evident closure principles, which are simply expressed by the axioms (Sep), (FS-I), and (FS-II) of PFA. We do not regard the success of the program PFA to be necessary for the acceptance of the natural-number system, but believe that its success, if granted, is of philosophical interest.

The challenge raised by Parsons in I of I begins with the evident impredicativity of Frege's definition of the natural numbers, in the form

$$(\text{Frege-N}) \quad Na \leftrightarrow \forall P\{P0 \,\&\, \forall x(Px \rightarrow P(Sx)) \rightarrow Pa\},$$

where the variable P is supposed to range over "arbitrary" second-order entities in some sense or other (Fregean concepts, predicates, propositional functions, sets, classes, attributes, etc.), including, among others, the entity N supposedly being defined. But Parsons enlarges on what constitutes the impredicativity of Frege's definitions in that he says that, to use it to derive induction in the form (say) of a rule,

$$(\text{Ind-Rule}) \quad \frac{\varphi(0), Na \rightarrow [\varphi(a) \rightarrow \varphi(a')], Nt}{\varphi(t)}$$

we must allow instantiation of the variable P in (Frege-N) by formulas $\varphi(x)$ which may contain the predicate N. In this sense, the focus of Parsons' discussion is on the *impredicativity of induction*, rather than the *prima facie* impredicativity of the putative definition (Frege-N). He expands the implications of this still further as follows:

The thesis of the present note is that the impredicativity that arises from Frege's attempt to reduce induction to a definition is not a mere artifact of Frege's strategy of reduction. As Michael Dummett observed some years ago, the impredicativity – though not necessarily impredicative second-order logic – remains if we regard induction in a looser way as part of the explanation of the term 'natural number'. If one explains the notion of natural number in such a way that induction falls out of the explanation, then one will be left with a similar impredicativity. (I of I, p. 141; the reference is to Dummett [1978], p. 199)

Perhaps what we were up to in PFA is orthogonal to the issue as posed in this way by Parsons, but let us see what we can do to relate the two. First, as explained in Section 8.2, what we are *not* after is a definition of the notion of natural number in the traditional sense in which this is conceived, but rather it is to establish the existence (and uniqueness, up to isomorphism) of a *natural-number-structure, or* N-structure (as it was abbreviated, PFA, i.e., Feferman and Hellman [1995]). Second, induction in the form of the principle (N-III) of Section 8.1, above, is taken to be *part* of what constitutes an N-structure. We agree with Parsons (I of I, p. 145) that "[s]tated as a general principle, induction is about 'all predicates'," but we do not agree with the conclusion that he draws (ibid.) that "[i]nduction is thus inherently impredicative, because . . . we cannot apply it without taking predicates

involving quantification over [the domain of natural numbers] as instances."
Rather, our position is that our – or, perhaps better, Aczel's – proof of the existence
(and categoricity) of an N-structure is predicative, given the notion of arbitrary
finite set of individuals, and thence in any such structure we may apply induction
to any formula that is recognized to define a class in our framework, including
formulas that refer to the particular definition of our N-structure. Specifically,
within EFSC, these are the weak second-order formulas, in which only quantification over individuals and finite sets is permitted. Of course, if we want to apply
induction to more general classes of formulas in our system, or to formulas in more
extensive systems, the question of predicativity has to be re-examined on a case-
by-case basis. For example, if we expand the system EFSC by a principle that says
that in any N-structure we may apply induction to *arbitrary* formulas of L(EFSC),
the resulting system EFSC + FI is no longer evidently predicative, given the notion
of finite set, but it is so nonetheless. The reason is that EFSC + FI can be
interpreted in the system ACA with full second-order induction – which is
predicative given the natural numbers according to the Feferman–Schütte characterization. And since, on our analysis, the natural numbers are predicative, given
the finite sets, this also justifies EFSC + FI on that same basis. Naturally, one may
expect that if the language is expanded by introducing terms for impredicatively
defined sets (specified by suitable instances of the comprehension axiom), or if one
adds impredicative higher-type or set-theoretical concepts, then the expanded
instances of induction that become available will take us beyond the predicative,
whether considered relative to the natural numbers or to finite sets.[13] But this
cannot be counted as an objection to what is accomplished in PFA. It is not the
general principle of induction that is impredicative, but only various of its
instances; and those instances that Parsons argues to be impredicative, in the
above quotation, are not examples of such, granted the notion of finite set.

Now, finally, and relatedly, we take up the objection that Parsons raises in I of
I, pp. 146–168, to the predicativity of Alexander George's [1987] revision of
Quine's definition of the natural numbers using quantification over finite sets,
which is equivalent to Aczel's definition of an N-structure as we pointed out in
Section 8.1, above. Of this he says: "To the claim that the Quinean definition of
the natural numbers is predicative, one can also reply that it is so only because
the notion of finite set is assumed." Indeed, as the above discussion affirms, we
could not agree more. But the reason for his objection then is that "[o]nce one
allows oneself the notion of finite set, it seems one should be allowed to use
some basic forms of reasoning concerning finite sets," and in particular
(according to Parsons) of induction and recursion on finite sets, which would
then allow one to define the natural numbers as the cardinal numbers of finite

---

[13] Addition of higher types or even set-theoretical language does not *per se* force us into
impredicative territory; cf. Feferman [1977].

sets. But it is just this that we do *not* assume in EFSC (or EFSC*); no assumptions are made on finite sets besides the closure principles (Sep), (FS-I), and FS-II) (and (Card) in the case of EFSC*). Of course, within our system, once we have an N-structure, we can formally define what it means to be a finite set by saying that it is in one-one correspondence with an initial segment of that structure, and then derive principles of induction and recursion for *that* notion. But we cannot prove that these exhaust the range of the finite-set variables.

8.4   The Challenge of George and Velleman

In their paper, "Two conceptions of natural number" (TC in the following), Alexander George and Daniel Velleman [1998] take up the PFA constructions in connection with two main conceptions of natural number, which they describe as "pare down" (PD) and "build up" (BU) corresponding to two ways of characterizing the minimal closure of a set A under an operation f. On the PD approach, this is defined explicitly as the intersection of all sets including A and closed under f. In the case of the natural numbers, this corresponds to the definitions given by Dedekind, Frege, and Russell, essentially as the intersection of all classes containing zero and closed under successor. In contrast, the BU approach provides an inductive definition, illustrated in the case of the natural numbers by clauses such as

(1) 0 is a natural number, and

(2) if n is a natural number, then so is $S(n)$,

together with an *extremal clause,* which says that natural numbers are only those objects generated by these rules. As their discussion brings out, the PD approach comports with a platonist view, according to which impredicative definitions are legitimate means of picking out independently existing sets, whereas the BU approach comports with a constructivist view that rejects the platonist stance and impredicative definitions in favor of rules for generating the intended set of objects. Not surprisingly, neither camp is satisfied with the other's approach, the constructivist rejecting the PD approach as just indicated, but the platonist also rejecting the BU approach as failing properly to define the intended class by failing explicitly to capture the required notion of "finite iteration" of the rules of construction. Furthermore, neither camp is impressed with the other's critique. And so the impasse persists.

The question arises for George and Velleman: To which type of definition should that of PFA be assimilated? As they recognize, it seeks to avoid impredicativity and so surely should not be thought of as a PD definition. On the other hand, in PFA, "the completed infinite" is recognized; moreover (although George and Velleman do not highlight this), an *explicit* definition of "*natural-number-type structure*" is provided, not merely an inductive or recursive description of "natural numbers," and so, assimilation of PFA to the

BU approach is misleading. Here we would suggest that a new, third category of definition be recognized, one that combines the explicitness demanded by PD with the predicative methods demanded by BU; it might be called "predicative structuralist" (PS), if one wants a two-letter label. But before recognizing a qualitatively new product, we want to be sure that at least the labeling is honest and accurate.

In notes, George and Velleman raise questions on this score. The essential worry seems to be that the construction in PFA (or its simplification by Aczel) succeeds only if the range of the finite-set quantifiers is restricted to truly finite sets; otherwise, "non-standard numbers" will not be excluded. But, for some reason, any effort to impose this restriction must appear circular or involve some hidden impredicativity. They put it this way:

> As Daniel Isaacson [1987] suggests, the predicativist definition will be successful only if (i) the second-order quantifier in the definition ranges over a domain that includes all finite initial segments of N and (ii) the domain contains no infinite sets. He concludes that the definition therefore "does not fare significantly better on the score of avoiding impredicativity than the one based on full second-order logic" (p. 156). Feferman and Hellman argue in response (1995, note 5, p. 16) that the existence of the required finite initial segments can be justified predicatively, but it seems to us that they have failed to answer part (ii) of Isaacson's objection, namely that infinite sets must be excluded from the domain of quantification. As we saw earlier, it is this exclusion of infinite sets from the second-order domain that guarantees that Feferman and Hellman's definition will capture *only* natural numbers. In fact, the difficulty here is in effect the same as the difficulty that the platonist finds with the BU definition; it is not the inclusion of desired elements in the domain that causes problems, but rather the exclusion of unwanted elements. (TC, n.9)

Now an adequate response to this requires distinguishing what may be called "external" and "internal" viewpoints concerning formalization of mathematics. From an external standpoint, one views a formalization from the outside and asks whether and how non-standard models of axioms or defining conditions can be ruled out. Here the metamathematical facts are clear. So long as one works with a consistent formal system based on a (possibly many-sorted) first-order logic, or indeed any logic that is compact, non-standard models of arithmetic are inevitable. *But this is true even if an impredicative definition of "N-structure" is given.* Even a PD definition in ZFC is subject to this limitation and will have realizations in which "numbers" with infinitely many predecessors appear. No extent of analysis of 'finite' or 'standard number', and so on, can overcome this limitation. What this shows is that the problem of "excluding non-standard models" in this sense is "orthogonal," so to speak, to the problem of predicativity. All formal definitions are in the same boat, and the only recourse, from the external vantage point, is somehow to transcend the framework of first-order logic. Let us return to this momentarily.

Alternatively, one can look at matters from an *internal* point of view. One accepts the inevitability of non-standard models of theories built on formal logic, but then one attempts to lay down axioms that are intuitively evident of the informal notions one is trying to capture, and then one seeks to prove the strongest theorems that one can, which, on their ordinary informal interpretation, express interesting and desirable results. Thus, one can lay down closure conditions, as in PFA, that are evident of finite sets, and, although they can hold of other collections as well, the theorems that one proves, such as mathematical induction in specified pre-N-structures, establish desired results even if they can be non-standardly interpreted. (Bear in mind that every mathematical result about the continuum, say, recovered in ZFC has non-standard interpretations.) Indeed, on this score, a good case can be made that the predicativist can prove results on the existence and uniqueness of natural-number-type structures that are just as decisive as those the classicist can prove. Let us return to this after elaborating a bit further about what can be said on behalf of PFA and the improvements described in Section 8.1 from the *external* viewpoint.

To effect the desired "exclusion of infinite sets" that can lead to "non-standard numbers," that is, elements of N-structures with infinitely many predecessors, one takes the bull by the horns, so to speak: the exclusion is imposed by fiat in the metalanguage by stipulating that we are only concerned with interpretations in which the range of the finite-set quantifiers contains only finite sets. 'Finite' is taken as absolute. This is the framework of "weak second-order logic" in its semantic sense. As is well known, it is non-compact and not recursively axiomatizable, but this is offset by gains in expressive power, exploited in PFA. For now one can collect items of a pre-N-structure that correspond to genuinely finite initial segments of a linear ordering, and this suffices to characterize N-structures.

There is a limited analogy with the classicist's approach via PD definitions, for example, those of Dedekind, Frege, and Russell, formalized say in second-order notation; for these characterize N-structures only if non-standard, less-than-full ranges of the second-order quantifiers are excluded (so that second-order monadic quantifiers must range over *all* subsets of the domain, precluding Henkin models). The problem of non-standard models is overcome by moving to non-compact, non-axiomatizable "full second-order logic." But the analogy is only partial. For, whereas the classical logicist excludes non-full interpretations on the basis of a claim to understand "*all subsets* of an infinite set," the predicative logicist merely excludes infinite sets from the range of finite-set quantifiers on the basis of a claim to understand "*all finite subsets.*" If the objection is that this is illegitimate because 'finite' "is as much in need of analysis as the concept 'natural number' " (TC, n.9), then it is appropriate to refer back to Section 8.2, above, and the whole case for grounding the infinitistic notion of "natural-number-type structure" on elementary assumptions on finite objects, together with the point made earlier (Section 8.3) that nowhere do we have to invoke finite-set induction in order to prove any of our theorems, including the theorem that says that mathematical induction holds in any special pre-N-structure. (*Mutatis mutandis* for the Aczel theorem.) Indeed, since induction is essential to the natural-number concept and to reasoning "about the natural numbers," the very fact that finite-set induction is *not* needed to recover this much counts in favor of the view that 'finite set' is actually *less* in need of analysis than 'natural number'.

Moreover, on the question of existence, there is a fundamental disanalogy between the PD and the PS approaches. For, as George and Velleman bring out, the impredicative definitions of the logicists still must presuppose existence of the minimal closure, and this is an additional assumption, not guaranteed merely by the restriction to full interpretations. There still must *be* some full interpretation of the right sort, that is, containing the real minimal closure. In contrast, the predicative constructions of PFA, Aczel, and the Appendix below yield the desired classes by a restricted comprehension principle, WS–CA. Given finite sets as objects, such a principle is justified much as arithmetical comprehension is; one can even eliminate talk of classes of individuals in favor of satisfaction of formulas, since these contain only bound individual and finite-set variables but no bound class variables.

Thus, the predicative logicist accompanies the platonist classicist only a relatively small step beyond first-order logic; then construction takes over on the new higher ground, while the platonist continues ascending, eventually into the clouds.

Consistently with this external view, one can, however, also pursue the internalist course of proving desirable theorems. Here, perhaps surprisingly, the predicativist is able to recover predicativist analogues of well-known classical results. The proofs of categoricity or unicity of N-structures and of mathematical induction in the pre-N-structures of PFA, Aczel, and the Appendix already illustrate this. But one can go further and prove theorems that, informally understood, say explicitly that N-structures cannot contain any non-standard elements. The idea is to formalize the following, familiar reasoning. Let $\langle \mathbf{M}, 0, ' \rangle$ be an N-structure. Induction implies than any non-empty class (subclass of $\mathbf{M}$) closed downward under predecessor, $p(x)$, contains 0. Consider the class of non-standard numbers (of $\mathbf{M}$); call it $\mathbf{K}$. If $z \in \mathbf{K}$, then also $p(z) \in \mathbf{K}$ (contraposing the Adjunction axiom); therefore, if $\mathbf{K}$ is non-empty, it contains 0, a blatant contradiction. (Put positively, 0 is standard and if z is standard, so is z', and so, all members of $\mathbf{M}$ are standard.) Elements with infinitely many predecessors are ruled out directly by Induction.

But in what system is the above reasoning carried out? If we attempt to formalize it in EFSC, expressing "x is non-standard" by "$\forall A (A \neq \{y : y \leq x\})$," we immediately contradict the definition of **M**! On the other hand, we cannot simply plug in "$\{y : y \leq x\}$ is Dedekind-infinite" or any other second-order analysis of "infinite" involving general class or function quantifiers, for then we would not be able predicatively to form the class **K**. However, there is an alternative method that gets around this. For here we may appeal to the metatheorem mentioned in Section 8.3. If we add to EFSC the axiom schema known as "full induction" (FI), that is, induction for *arbitrary* second-order formulas, the resulting system, EFSC + FI, is interpretable in the subsystem of PA^2 known as ACA. This also contains FI and, moreover, is a predicatively acceptable system relative to the natural numbers (on the Feferman–Schütte characterization) as noted in Section 8.3. But, as was also observed there, since the natural numbers or N-structures are predicative given the finite sets, EFSC + FI is also predicatively acceptable relative to the finite sets. Although it cannot prove the existence of subclasses of an N-structure defined by formulas with class quantifiers, it can prove that induction holds directly for any formula that platonistically defines a subclass, as it were. In particular, now one can formalize the above induction ruling out non-standard numbers, using, in place of $x \in \mathbf{K}$, a second-order formula $\varphi(x)$ to express "x has infinitely many predecessors," for example, "the predecessors of x form a class, a subclass of which is in one-one correspondence with an unbounded subclass of **M**"; or it could just as well be "the predecessors of x form a Dedekind-infinite class." The predicativist, as well as the classicist, regards these as good formalizations of the intended notion. Thus, the reasoning is formalizable in a predicatively acceptable extension of EFSC without appealing to the special finite-set variables and without any circular or impredicative reference to the class **K**.[14]

---

[14] There is some irony in the fact that George and Velleman, after claiming (TC, n.10) that the Aczel construction cannot rule out non-standard numbers without a circular appeal to "the complement of N," present an argument of their own for the predicative acceptability of an extension of EFSC* in which the full induction schema is derivable. (See their note 14.) They argue for a direct extension to include the separation schema for finite sets with arbitrary second-order formulas. This is closely related to the fact that, in a weak subsystem of analysis, FI is equivalent to the so-called "bounded comprehension scheme,"

$$\forall n \exists X \forall m \; (m \in X \leftrightarrow m < n \;\&\; \varphi(m)),$$

where $\varphi(m)$ is any formula of second-order arithmetic (lacking free 'X') (see Simpson [1985], p. 150). This corresponds to the separation scheme for finite sets with arbitrary second-order formulas. We prefer the direct route to full induction via ACA and proof theory, since it is predicatively problematic to say that an arbitrary formula "specifies unambiguously which elements of the [given] finite set are to be included in a subset" (TC, n.14). It then turns out that their proposed stronger separation scheme is derivable from full induction, and so inherits a predicative justification after all. In any case, once full induction is available, the reasoning that N-structures are truly standard is predicatively formalizable without appeal to finite-set variables, even while employing a standard logicist analysis of 'finite' or 'infinite' as just indicated.

Looking at the contrapositive, one sees that one has thus *derived* the consequence of the axiom (Card) directly relevant to ruling out non-standard members of N-structures, viz., the statement that the predecessors of any such element form a Dedekind-finite set,

$$\forall x[\, x \in \mathbf{M} \to \mathrm{DedFin}(\mathrm{Pd}(x))].$$

This follows straightforwardly by induction on the formula, $\varphi(z)$, expressing $\mathrm{DedFin}(\mathrm{Pd}(z))$. Again, we need not be able to collect all elements satisfying this formula in order to reason with it by mathematical induction.

Thus, "non-standard numbers" are ruled out as decisively as they can be. From the external standpoint, they are excluded by the semantics of weak second-order logic, which, as has been argued, is a good framework for elementary predicative mathematics. From an internal perspective, without falling back on special finite-set variables, we can employ standard, logicist *analyses* of 'finite', 'infinite', and so forth, and derive theorems in predicatively acceptable systems that directly express the desired exclusion. This may seem like "having one's cake and eating it at the same time." But really it is more like having two desserts.

Acknowledgement: This paper was written while the first author was a Fellow at the Center for Advanced Study in the Behavioral Sciences (Stanford, CA) whose facilities and support, under grants from the Andrew W. Mellon Foundation and the National Science Foundation, have been greatly appreciated.

References

Dedekind, R. [1888] "The nature and meaning of numbers," reprinted in Beman, W. W., ed., *Essays on the Theory of Numbers* (New York: Dover, 1963), pp. 31–115, translated from the German original, *Was sind und was sollen die Zahlen?* (Brunswick: Vieweg, 1888).

Dummett, M. [1978] "The philosophical significance of Gödel's theorem," in *Truth and Other Enigmas* (London: Duckworth), pp. 186–201 (first published in 1963).

Feferman, S. [1964] "Systems of predicative analysis," *Journal of Symbolic Logic* **29**: 1–30.

Feferman, S. [1977] "Theories of finite type related to mathematical practice," in Barwise, J., ed., *Handbook of Mathematical Logic* (Amsterdam: North-Holland), pp. 913–971.

Feferman, S. [1985] "Intensionality in mathematics," *Journal of Philosophical Logic* **14**: 41–55.

Feferman, S. [1996] "Gödel's program for new axioms: why, where, how and what?" in Hájek, P., ed., *Gödel '96*, Lecture Notes in Logic 6 (Berlin: Springer), pp. 3–22.

Challenges to Predicative Foundations of Arithmetic

Feferman, S., and Hellman, G. [1995] "Predicative foundations of arithmetic," *Journal of Philosophical Logic* **24**: 1–17.

George, A. [1987] "The imprecision of impredicativity," *Mind* **96**: 514–518.

George, A., and Velleman, D. [1998] "Two conceptions of natural number," in Dales, G. and Olivieri, G., eds., *Truth in Mathematics* (Oxford: Oxford University Press), pp. 311–327.

Hallett, M. [1990] "Physicalism, reductionism, and Hilbert," in Irvine, A. D., ed., *Physicalism in Mathematics* (Dordrecht: Kluwer), pp. 183–257.

Hellman, G. [1989] *Mathematics without Numbers: Towards a Modal-Structural Interpretation* (Oxford: Oxford University Press).

Hellman, G. [1996] "Structuralism without structures," *Philosophia Mathematica* **4**: 100–123.

Hintikka, J. [1956] "Identity, variables, and impredicative definitions," *Journal of Symbolic Logic* **21**: 225–245.

Isaacson, D. [1987] "Arithmetical truth and hidden higher-order concepts," in Paris Logic Group, eds., *Logic Colloquium '85* (Amsterdam: North-Holland), pp. 147–169.

Parsons, C. [1990] "The structuralist view of mathematical objects," *Synthese* **84**: 303–346.

Parsons, C. [1992] "The impredicativity of induction," in Detlefsen, M., ed., *Proof, Logic and Formalization* (London: Routledge), pp. 139–161 (revised and expanded version of a 1983 paper).

Quine, W. V. O. [1961] "A basis for number theory in finite classes," *Bulletin of the American Mathematical Society* **67**: 391–392.

Schütte, K. [1965] "Eine Grenze für die Beweisbarkeit der transfiniten Induktion in der verzweigten Typenlogik," *Archiv für Mathematische Logik und Grundlagenforschung* **7**: 45–60.

Simpson, S. [1985] "Friedman's research on subsystems of second-order arithmetic," in Harrington, L. A., Morley, M., Scedrov, A., and Simpson, S. G., eds., *Harvey Friedman's Research on the Foundations of Mathematics* (Amsterdam: North-Holland), pp. 137–159.

Wang, H. [1963] "Eighty years of foundational studies," reprinted in *A Survey of Mathematical Logic* (Amsterdam: North-Holland), pp. 34–56.

Appendix Realizing Dummett's Approach in EFSC

Geoffrey Hellman

Here the notation of Section 8.1 will be followed, including the use of F, G, H, as finite-set variables. Let 0 and $'$, respectively, be the initial element and successor-like function of a given pre-N-structure. Our principal aim is to prove the following.

Theorem: *(EFSC)*

Let

$$\mathbf{M} =^{df} \{x: \exists F(0 \in F \ \& \ \forall y(y \in F \ \& \ y \neq x \rightarrow y' \in F) \ \& \ x \in F \ \& \ \forall G[[0 \in G \ \& \\ \forall y(y \in G \ \& \ y \neq x \rightarrow y' \in G)\} \rightarrow F \subseteq G])\},$$

that is, x belongs to a (the) minimal finite set containing 0 and "closed upward except at x." Then $\langle \mathbf{M}, 0, ' \rangle$ is an N-structure.

Remark: Note that this definition of **M** incorporates both an existential condition and a universal one, corresponding to the conditions Wang attributes to Dummett (Wang [1963]; cf. TC, n.10).

The proof is simplified by adopting the following abbreviations, which also brings out the relationship between this theorem and that of Aczel:

$$\mathrm{Clos}_d(F, [z, x]) \equiv z \in F \ \& \ \forall y \ (y \in F \ \& \ y \neq x \rightarrow y' \in F),$$

read as "*F* is closed upward from *z* to *x*" (the subscript *d* is for Dummett). Next define

$$z \leq_d x \text{ by } F \equiv \mathrm{Clos}_d(F, [z, x]) \ \& \ x \in F \ \& \ \forall \ G(\mathrm{Clos}_d(G, [z, x]) \rightarrow F \subseteq G).$$

This can be read as "*F* witnesses $z \leq_d x$." Trivially, if both F_1 and F_2 witness $z \leq_d x$, then $F_1 = F_2$ (extensionally). Now, define

$$z \leq_d x \equiv \exists F(z \leq_d x \text{ by } F).$$

Now, **M** in the Theorem can be defined by $\mathbf{M} = \{x : 0 \leq_d x\}$. For purposes of comparison, recall the Aczel construction, $\mathbf{M} = \{x : \mathrm{Fin}(\mathrm{Pd}(x)) \ \& \ 0 \leq x\}$, where '$\leq$' is the ordering introduced in PFA, as in Section 8.1, above. We now proceed to the proof of the theorem.

Proof: Let F_x denote the unique *F* that witnesses $0 \leq_d x$. Then, to say that $x \in \mathbf{M}$ is to say that F_x exists. We have

(i) $0 \in \mathbf{M}$, to wit $\{0\}$ as F_0.

(ii) $z \in \mathbf{M} \rightarrow z' \in \mathbf{M}$.

   Given F_z, set $F_{z'} = F_z \cup \{z'\}$. We must show that this works. $F_z \cup \{z'\}$ is finite, by adjunction (FS-II). $0 \in F_z \cup \{z'\}$ and $z' \in F_z \cup \{z'\}$. Now if $y = z$, then trivially $y' \in F_z \cup \{z'\}$; and if $y \neq z$, then if $y \in F_z \cup \{z'\} \ \& \ y \neq z'$, then $y \in F_z$, and by hypothesis then so is y', whence $y' \in F_z \cup \{z'\}$. Thus, $\mathrm{Clos}_d (F_z \cup \{z'\}, [0, z'])$. It remains to prove minimality.

   Suppose $\exists u \in F_z \cup \{z'\}$ such that $u \notin G$, some *G* such that $\mathrm{Clos}_d (G, [0, z'])$. Consider $G - \{z'\}$. If $y \neq z \ \& \ y \in G - \{z'\}$, $y \neq z'$ either, so $y' \in G - \{z'\}$ by hypothesis on *G*. So, $\mathrm{Clos}_d(G - \{z'\}, [0, z])$, and so, $F_z \subseteq G - \{z'\}$, by hypothesis on F_z. Thus, $z \in G - \{z'\}$, and so, $u \neq z'$, by the closure condition on G that forces $z' \in G$. Therefore, $u \in F_z$ but, by hypothesis that $u \notin G$, $u \notin G - \{z'\}$ either, contradicting the minimality of F_z. This completes the proof of minimality of $F_z \cup \{z'\}$ and of step (ii).

(iii) Induction, (N-III): Let $\mathbf{X}$ be a class such that $0 \in \mathbf{X}$ and $y \in \mathbf{X} \rightarrow y' \in \mathbf{X}$, all $y \in \mathbf{M}$.

To prove: $z \in \mathbf{M} \rightarrow z \in \mathbf{X}$.

We will prove $F_z \subseteq \mathbf{X}$, which suffices since $z \in F_z$ and indeed $z \in \bigcap [G : \mathrm{Clos}_d(G, [0, z])]$. Let $H = G \cap \mathbf{X}$, for some such G (existence by F_z itself). H is finite by (WS-Sep). We have $\mathrm{Clos}_d(H, [0, x])$ by the closure conditions on G and $\mathbf{X}$. Therefore, $F_z \subseteq H = G \cap \mathbf{X}$, whence $F_z \subseteq \mathbf{X}$.   ∎

By virtue of the unicity of N-structures ("categoricity," Theorem 5 of PFA), an N-structure can be represented as of the form $\langle \mathbf{M}, 0,' \rangle$ of Theorem 1, that is, the domain $\mathbf{N}$ of any N-structure $= \{x \in \mathbf{N} : 0 \leq_d x\}$, where 0 here is the initial element of $\mathbf{N}$ and $\leq_d$ is defined over $\mathbf{N}$ via the successor-type relation on $\mathbf{N}$.

As expected, the ordering $\leq_d$ is closely related to '$\leq$' of PFA. This is spelled out in the following.

Theorem: *(EFSC)*

(1) *In any pre-N-structure,*

   $z \leq_d x \rightarrow z \leq x.$

(2) *In any N-structure $\langle \mathbf{M}, 0,' \rangle$ defined as in Theorem 1, and hence in any N-structure,*

   $z \leq_d x \leftrightarrow z \leq x.$

Proof: (1) and $\rightarrow$ of (2): Suppose the implication fails, that is, that $z \leq_d x$ but $\exists A(x \in A \,\&\, \forall y(y' \in A \rightarrow y \in A) \,\&\, z \notin A)$. (So, $z \neq x$, and $z' \notin A$, nor is z'', etc.) Let F be the witness to $z \leq_d x$; that is $F = \cap[G : \mathrm{Clos}_d(G, [z, x])]$.

Let $B = \{u : u \notin A \,\&\, u \in F\}$. B is finite, by (WS-Sep), and $z \in B$ and $\forall y(y \in B \,\&\, y \neq x \rightarrow y' \in B)$ that is, $\mathrm{Clos}_d(B[z, x])$. So, by definition of $z \leq_d x$, we have $x \in B$, that is, $x \notin A$, a contradiction. (Remark: Note the similarity to the Aczel proof, except that here we are "stepping forward" instead of "stepping back.")

$\leftarrow$ of (2): Now assume that we are in an N-structure, $\langle \mathbf{M}, 0,' \rangle$, as in Theorem 1. We proceed by induction on z.

(i) For $z = 0$, the implication is trivial.

(ii) Let $z = y$. If $y = x$, the implication is trivial, as then $y' \nleq x$. Let $y \neq x$, and let H_y be the (minimal) witness to $y \leq_d x$, which we can suppose, by inductive hypothesis. We claim that $H_y - \{y\}$ is the minimal witness to $y' \leq_d x$. Since $y \in H_y$ and $y \neq x$, $y' \in H_y$ by $\mathrm{Clos}(H, [y, x])$, and so,

$y' \in H_y - \{y\}$. If $u \in H_y - \{y\}$ and $u \neq x$, then, because $u \neq y$, $u \in H_y$, and so, $u' \in H_y$, whence $u' \in H_y - \{y\}$. (In the last step, we appeal to the minimality of H_y, which implies that $p(y) \notin H_y$, so that $u' \neq y$.) It remains to prove minimality of $H_y - \{y\}$ as witness to $y' \leq_d x$. Let G be such that $Clos_d(G[y', x])$, and suppose $\exists u$ such that $u \in H_y - \{y\}$ but $u \notin G$. Then, $u \in H_y$ but $u \neq y$, so $u \notin G \cup \{y\}$. But, $G \cup \{y\}$ contains y and meets the closure condition for H_y, viz. $Clos_d(G \cup \{y\}, [y, x])$. Therefore, by minimality of H_y, $H_y \subseteq G \cup \{y\}$, contradicting the supposition of u. This proves the minimality of $H_y - \{y\}$ and completes the inductive step. ■

Thus, the Dummett-inspired construction of Theorem 1, as well as the Aczel construction, defines N-structures, provably in EFSC. And the orderings involved, $\leq_d$ and $\leq$, respectively, are extensionally equivalent in these structures.

9 Predicativism as a Philosophical Position

9.1 Predicativity Requirements Versus Predicativist Theses

As is well known, predicative mathematics has long been motivated by skepticism concerning the classical conception of the Cantorian transfinite and, in particular, of the continuum, or the notion of "all subsets of (even) a countably infinite set." Along with constructivism, predicativism regards as suspect talk of functions which cannot even in principle be given a definite mathematical description. Indeed, a basic predicativity requirement is that any recognized mathematical object be presentable by means of a finite string of symbols from a countable language, where this is understood to include formulas defining sets or functions, with quantification in the formulas restricted to "already acceptable" objects. In contrast with intuitionism, however, the natural number system is treated classically, and classical logic is taken as legitimate. The focus, then, is on principles of set existence. In accordance with Russell's "vicious circle principle," sets of natural numbers cannot legitimately be introduced via definitions or formulaic conditions with unrestricted quantifiers over such sets. Such definitions are "impredicative." A prime example, familiar to philosophers, is the Fregean definition of the natural number system itself as the intersection of all classes containing 0 and closed under the successor operation (however '0' and 'successor' are defined). In general, the concept of *minimal closure* of a set under an operation, which abounds in mathematics, must be got at in a more constructive way or done without (to wit, the least upper bound of a bounded set of reals, or the notion of outer measure in measure theory).

Nevertheless, thanks to the pioneering efforts of Weyl and the modern proof theoretic work of Feferman, Schütte, and others, it is now clear that a great deal of ordinary mathematics, including virtually all of nineteenth century analysis and much of twentieth century functional analysis, can be carried out in systems that qualify as "predicative" according to rather precise standards that these proof theorists have forged.[1] Beginning with the natural numbers ($\mathbb{N}$) as given (or, better, with a weak system governing finite sets from a countably generated domain of individuals, as in Feferman and Hellman [1995], reproduced here as Chapter 7; see also Ferreira [1999]), one allows iteration of arithmetical comprehension along countable ordinals themselves capable of predicative construction. (Allowing iteration just along ω gives the sets of natural numbers – or real numbers, under well-known correspondence – familiar from Russell's ramified type theory, taking the natural numbers as individuals and adding the sets of numbers of all finite orders.) It is even fairly settled just how far such iteration may be taken: the countable limit ordinal known as Γ_0 marks the limit of predicative construction in a double sense: One can identify the "predicatively definable" subsets of $\mathbb{N}$ as those occurring in the Gödel constructible hierarchy up to Γ_0, $L\Gamma_0$.[2] Furthermore, formal systems of predicative analysis whose consistency strength can be measured by an ordinal $\alpha < \Gamma_0$ (e.g. proving an encoded statement of transfinite induction up to α) qualify as providing predicatively acceptable proofs of their theorems. Γ_0 itself, however, is, by these same standards, impredicative; embracing it involves, at least indirectly, commitment to genuinely uncountable totalities.[3] Theorems requiring consistency strength measured by Γ_0 (or greater) for their proof are regarded as essentially impredicative.[4]

All this means that we now have a rather precise understanding of the range of "predicative mathematics." When it comes to philosophical theses associated with predicative mathematics, however, the situation is much less clear, as will emerge in the next section.

---

[1] Weyl's program originates in his [1918] publication. For an informative overview and presentation of a modern system realizing Weyl's aims, see Feferman [1988], in Feferman [1998a], Ch. 13. This book also contains an extensive bibliography on predicativity.

[2] For purposes of analysis, one can limit oneself to $\mathbf{R}\Gamma_0$, the ramified analytic hierarchy up to Γ_0, in which, at each stage, only new subsets of $\mathbb{N}$ are added. In $\mathbf{L}\Gamma_0$, higher type objects are added as well, making this a universe for predicative mathematics, as seen from the outside.

[3] For a good discussion of Γ_0 and its significance, see Smorynski [1982].

[4] The subsystem ATR$_0$ of second-order arithmetic highlighted in reverse mathematics has proof theoretic ordinal Γ_0, and so is just at the limit of predicativity. (ATR$_0$ stands for "arithmetic transfinite recursion"; the subscript 0 here and elsewhere indicates that mathematical induction is restricted to those formulas whose extensions can be proved to exist in the system, rather than the full induction scheme.) Although mathematics requiring the full strength of ATR$_0$ is not predicatively provable (e.g. that any pair of countable well-orderings are comparable, or that any uncountable closed or analytic set has a non-empty perfect subset), important classes of consequences are; one speaks of (partial) reducibility to predicativity. See Simpson [1999], also Feferman [1998a]).

9.2   Strong, Limitative Predicativist Theses

It is easiest to begin with well-known views of a negative character, concerning what the predicativist does not accept. (In the following, 'N' represents 'negative'.)

(N1) Reference to "all subsets" of an infinite set is not legitimate, and to be avoided.

(N2) Quantification over arbitrary countable ordinals (like any other purport-edly absolutely uncountable totality) is not legitimate, and to be avoided.

(N3) The classical continuum as a mind-independent, objective reality is rejected and replaced by an open ended series of approximations, each countable from outside but extendable to a richer totality. Talk of "the real numbers" is relative to a background language in which real-number descriptions or definitions are available.

Each such claim is, of course, part and parcel of the predicativist methodo-logical stance. But on what deeper claims or principles do they rest? Is nominalism in play, for example? Not in the traditional sense (rejecting abstract or even purely mathematical entities), surely, since, as we already have noted, reference to "the natural numbers" and classical reasoning about them are regarded as unproblematic; similarly with regard to finite sets of countably many given objects. But perhaps a non-traditional form of nominalism is in the background.

Perhaps, even more fundamentally, there is a worry about the intelligibility of reference to that which is beyond even the possibility of identification by means of mathematico-linguistic expression. Perhaps, like the intuitionist, the predicativist thinks that classicists are deluding themselves in their efforts to communicate about genuinely uncountable totalities, and perhaps, as in the case of some intuitionists, this view can be traced back to a verifiability theory of meaning.

Or perhaps, only somewhat more modestly, an epistemological claim would be made, that rational belief only extends as far as that which is graspable in predicativist terms – in the case of mathematics, objects that are definable/describable in such terms and theorems that are provable in a predicatively justified system.

One may be tempted at this point to enter into a dialogue with a hypothetical predicativist about such claims in the hopes both of clarifying the position(s) and of determining better whether the classicist has reason on her/his side for resisting and transcending the implied predicativist limits to intelligibility, ontology, or rational belief. I will resist this temptation, however, because of a certain Socratic observation that can be made about these *strong, limitative* predicativist theses, which seems to spare us the effort: namely, that *the very effort to articulate such theses, given the precision available to us from all the logical work alluded to in the opening section above, reveals them to be self-defeating: the predicativist implicitly transcends predicativity him/herself in the very formulation of the limitative theses!*[5]

Consider, for example, the strong, limitative semantic thesis:

> (LS) *Commitment to what is not predicatively definable is not genuinely intelligible.*

The problem here is that the term 'predicatively definable' is itself impredicative. There are various ways to see this. Perhaps the crispest demonstration is in terms of ordinals. Suppose the contrary, that 'predicatively definable ordinal' is itself a predicative notion (relative to the natural numbers or finite sets of countably generated individuals). As already mentioned, as a result of proof-theoretic analysis, this is widely agreed to include all ordinals $\alpha < \Gamma_0$. But then the condition 'predicatively definable ordinal' (call it Pred.Df.) should give rise to an acceptable instance of predicative comprehension (say in one of Feferman's systems of flexible or variable types, in which classical constructions can be quite closely imitated). The result would be

$$\alpha \in \Gamma_0 \leftrightarrow \text{Pred.Df.}(\alpha),$$

a direct predicative construction of Γ_0 itself, contrary to its status as the least impredicative ordinal! (Recall that, under the von Neumann definition of 'ordinal', the ordering relation $<$ is just $\in$.)

One could also argue the point in terms of real numbers. By hypothesis, there should only be countably many predicative reals (lest the predicativist find him/herself committed to the uncountable after all). Then an application of Cantor's diagonal argument leads outside the class. But that argument provides, in principle, a definition of a new real that the predicativist should understand.[6] (Such an argument is non-constructive from an intuitionistic standpoint, but, recall, the predicativist accepts classical logic.)

The attempt to assert (LS) is self-defeating from within the predicativist framework.

The situation with a strong, limitative *ontological* thesis is similar. Consider the claim,

> (LO) *Only predicatively specifiable objects or totalities exist (in pure mathematics).*

(The parenthetical qualification is advised by a principle of charity.) It is common practice in predicative mathematics to take a predicatively acceptable specification of an object as sufficient for its (mathematical) existence. So by an

---

[5] We are not attributing the strong, limitative claims to anyone in particular, although scholars may be able to locate them in various places. To our knowledge, however, Feferman's writings would not be among them, and it seems reasonably clear that he would not endorse them.

[6] For a detailed discussion of these points, see Hellman [1994].

argument analogous to the above for (LS), the impredicative ordinal Γ_0 exists, contrary to (LO).

The situation with regard to a strong, limitative *epistemological* thesis is hardly any better. The analogue to (LS) and (LO) would naturally be

> (LE) *Rational belief extends only to predicatively constructible objects (in pure mathematics).*

Again, the converse is normally assumed to hold. Then, again, Γ_0 should be rationally believable, contrary to (LE)

In each case, the conclusion is that the general notion *'predicatively constructible (or specifiable, or definable)'* is itself impredicative, so that the attempt to enunciate the strong, limitative thesis fails from within. On the other hand, from outside, the very expression of (LS) undermines its own content; whereas, in the cases of (LO) and (LE), they are inconsistent with what is normally assumed in formulating them.

9.3   Anti-Platonism?

The above jujitsu-like manoeuvres were made possible, of course, by appealing to the precise explications afforded by modern logic of notions such as "the predicative universe," "limit of predicativity," etc. Suppose, however, that these are avoided in favor of a general expression of skepticism – or *malaise,* one might say – regarding the predominant platonist practice of treating mathematical objects as mind-independent and statements about them as objectively true or false, independently of human capacities to prove them so. Feferman explicitly articulates this, as in the following passage:

I am a convinced antiplatonist in mathematics [and reject the view that] objects such as numbers, sets, functions, and spaces exist independently of human thoughts and constructions, and [that] statements concerning these abstract entities . . . have a truth value independent of our ability to determine them . . . .although this [platonist view] accords with the mental practice of the working mathematician, I find the viewpoint philosophically preposterous. (Feferman [1998a], Preface, p. ix)

We have already suggested that the anti-platonism of predicativism should not be equated with nominalism on account of the predicativist's acceptance of the natural numbers and classical logic. But in this passage, Feferman explicitly lists numbers as non-exempt from anti-platonist scruples. While it is clear that the natural numbers and the integers can be "constructed" (à la Hilbert, via strokes, etc.), how does the use of classical logic square with the expressed rejection of objective truth? Why does the predicativist not go all the way with intuitionism (or, at least, Bishop constructivism, which insists on intuitionistic logic but adds no non-classical axioms to analysis)?

In recent correspondence and presentations, Feferman has revised the formulation of his anti-platonism to allow for objectivity of truth values, at least for arithmetical statements, without an ontology of numbers as mind-independent abstract objects. In fact, just this distinction – between objectivity of truth values ("realism") and abstract ontology ("objects platonism") – lies at the core of an eliminative structuralism I have tried to develop, so, not surprisingly, I regard this as a salutary move.[7] It is sufficient to entertain progressions as coherent, conceptual possibilities in order to support classical arithmetical reasoning. The nature of the constituents of progressions can be left entirely open, depending on background ontology which can be left to vary. And, since we are merely entertaining possibilities of infinities, there is no contingency in our mathematics, as Russell recognized there should not be. Perhaps our own capacity to *imagine* progressions – for example of stars, or space-time regions, or whatever you like – is enough dependence on "human thoughts and constructions" to remain within the spirit of Feferman's anti-platonism, without resorting to intuitionistic logic. So far, the predicativist and the (modal) nominalist are at one. But then the next question would be, *is this a stable position for the predicativist?* What blocks its extension – along a slippery, if steeply ascending, slope – to full classical analysis?

To see how such an extension naturally arises, consider the natural next step for the *nominalist structuralist*: that is to admit *all parts* of any progression already recognized (or hypothetically entertained). This is *far* more liberal than what the predicativist can allow. In case the progression is made up of mutually non-overlapping "atoms" (i.e. lacking proper parts, in any sense one cares to specify), these parts are just wholes of those atoms, and these correspond one-to-one with the non-empty *subsets* of those atoms, which in turn correspond one-to-one with the points of the classical continuum; that is, we have already arrived at the classical real number system. (The structural relations and operations involved can also be introduced by mimicking well-known classical constructions based on integers and rationals, themselves easily coded as items of a given progression.) Presumably, the predicativist objects to most of these wholes of atoms, despite their non-abstractness. They are, after all, no more abstract than the items of the given progression. The claim must be that, regardless of their non-abstractness, such wholes exist *only if describable in predicativist language*. (Never mind that, as already argued, strictly speaking this general position really cannot be fully expressed predicatively.) Is this tenable?

Consider now a progression of non-overlapping little devices, which behave as follows. A quantity of radioactive substance is present and hooked up to

---

[7] See Hellman [1989]. The general view accepting objectivity of truth values for arithmetic *without* numbers as abstract objects has been endorsed by others, e.g. Hofeweber [2000], p. 142.

a switch which is triggered by detection of a decay particle, turning on a little light which remains *on*, say for a fixed time interval (e.g. 10^{-1} seconds).[8] Assume, with our best current physics, that the decay events are entirely random, and that they are, physically and stochastically, entirely independent of one another. (Let these "lights" be arranged along a line in space – assume Minkowski space-time as a background, if you like. Describe the states of the progression in a particular, arbitrarily chosen frame of reference.) At any moment, some lights may be on, but which ones is entirely random, physically speaking, and as time goes on, they keep changing entirely randomly. Call those lights which are simultaneously *on* an "array." (Arrays are just certain of the much-maligned scattered wholes of mereology. Perhaps they will seem less objectionable to some because of the way we like to celebrate Christmas.) It is an entirely objective matter which arrays have occurred within any specified time period. Surely, there can be no expectation that all arrays which have occurred have any particular kind of mathematical description or, indeed, any mathematical description at all. Many of them may be picked out by nothing other than the non-mathematical term "*on* now." Indeed, there can be no basis for excluding any non-empty subset, classically conceived, from among the arrays that may have occurred or may yet occur. What does the predicativist say about these arrays? Surely, their existence cannot depend on any kind of *mathematical* describability; and describability by a non-mathematical index-ical such as "*on* now" imposes no limits whatsoever. It seems to me that such considerations as these should scotch the idea that arbitrary subsets of an infinite set are somehow mind or language dependent in a way in which the items of the given set are not.

9.4   Moderate Theses

Thus far we have argued that strong, limitative claims are predicatively inex-pressible, and we have sought to refute the predicativist view that sets or wholes of integers are somehow mind or language dependent or more so than the integers. What about more modest predicativist claims? Here we confront the opposite sort of problem, namely that such a claim may be obvious, uncon-troversial, and perfectly acceptable to defenders or champions of impredicative mathematics. For example, consider the moderate epistemological claim,

> (ME) *Predicative mathematical results are more secure than impredi-cative ones.*

---

[8] Thus no appeal is being made to the continuity of time (which would raise problems of circularity), and we are not attempting to describe a scenario in which all possible combinations of lights are "run through," even in infinite time. The point at issue is the alleged mind- or language-dependence of the mathematical notion, "subset of integers."

Once it is agreed that consistency strength is one relevant measure of "security," this claim is nearly tautological, as ordinals measuring consistency strength of impredicative systems are greater than or equal to Γ_0. An arch-platonist can cheerfully accept (ME).

However important such claims may be in their own right, this avenue does not seem promising for articulating a distinctive predicativist philosophical view.

9.5   Challenges to Indispensability Claims

I would suggest at this point that the philosophical interest of work on predicativity lies primarily, not in grand philosophical theses that might be thought to lie behind that work, but rather in two other aspects. First, in helping to delineate "what rests on what" – on what mathematical axioms and methods are sufficient and necessary for what results – the study of predicativity, like the work of *reverse mathematics*, just *is* an important chapter in the detailed epistemology of mathematics. An excellent example is the work Feferman [1998b] has done in connection with his system **W**, in honor of Hermann Weyl. This is a system designed for predicative analysis that dispenses with the awkwardness of ramified type theory, allowing for the free introduction of sets and functions of arbitrary finite type over the natural numbers; in it a great deal of classical analysis and modern functional analysis can be reconstructed, with minimal reliance on coding devices; and yet it is of the consistency strength of Peano arithmetic, exploiting only a tiny fraction of the power in principle available to predicative systems. Thus, the power of **W** tells us that all the analytic theorems it recovers are as secure as ordinary arithmetic, even though they may appear to concern highly infinitistic, uncountable objects. Epistemologists of mathematics, I would suggest, ought to be envious of such results.

Second, and more controversially, work on predicativity is interesting in the challenges it raises to programs seeking to make out the indispensability of impredicative mathematics for various purposes, especially for solving combinatorial problems or problems at the level of analysis as ordinarily practiced, or for applications to the sciences. Thus, Feferman has upheld the resources of predicative mathematics as a direct challenge to the Gödel–Friedman program of justifying the Cantorian transfinite, especially large cardinal hypotheses, as necessary for solving problems at the level of the integers or sets thereof, and also as a challenge to arguments in the style of Quine and Putnam seeking indirect support for abstract mathematics through its allegedly indispensable role in formulating and testing scientific theories. Here we shall concentrate on aspects of the Gödel–Friedman program and predicativity.

Now it is well know that certain problems of finite mathematics require axioms well beyond the limits of predicativity for their solution. A striking example is Friedman's finite form of Kruskal's theorem on infimum embeddability of finite trees. To state Kruskal's theorem, define a finite tree T to be *embeddable* into another T' just in case there exists a one-to-one mapping $\varphi : T \rightarrow T'$ such that for any $x, y \in T$, $\varphi(x \wedge y) = \varphi(x) \wedge \varphi(y)$. Kruskal's theorem can be stated as: *There is no infinite set of pairwise non-embeddable finite trees.* Equivalently, as follows:

Kruskal's Theorem: *For any sequence of finite trees, $T_1, T_2, \ldots, T_n, \ldots$ there exist i, j with $i < j$ such that T_i is embeddable into T_j.*

This itself is, of course, not a finitary statement as it quantifies over infinite sequences. However, Friedman discovered a finite form of Kruskal's Theorem, i.e. with quantifiers over only finite objects, but which is not predicatively provable.

Friedman's Finite Form of Kruskal's Theorem: *For any positive integer c there exists a positive integer $n = n_c$ so large that if $T_1, T_2, \ldots, T_n$ is any finite sequence of finite trees with $|T_i| \leq c \cdot i$ for all $i \leq n$, then there exist indices i, j such that $i < j \leq n$ and T_i is embeddable into T_j.*

(The bound on the growth of the finite trees can be improved.) Though finitistic, this theorem is not provable even in the formal system ATR_0.[9] Of course, we already knew from Gödel's work that plenty of finitistc statements are not provable in any given formal system, for example "$\text{Con}(\text{ATR}_0)$" is not provable in ATR_0 unless it is inconsistent. However, the statement "$\text{Con}(\text{ATR}_0)$" is metamathematical and totally different in character from genuine mathematical problems. Like its predecessor, the Paris–Harrington finitization of Ramsey's partition theorem, Friedman's finite form of Kruskal improves on Gödel's results considerably. There is nothing at all metamathematical about it, and it is almost as simple to state as the original Kruskal Theorem.

Nevertheless, Feferman has argued that this example (and others like it) is inconclusive on at least two grounds. First, there is the issue of "naturalness" or "mathematicality." Although not especially complicated as a statement, Friedman's Finite Form (call this FFF) is contrived for logical purposes and, of course, did not arise naturally in mathematical work proper. Second, there is the issue of just what is shown indispensable in such a case: Feferman argues that it is not impredicative set existence *per se*, nor is it platonistic commitment to the uncountable. In fact, Friedman obtained his independence result by proving that

---

[9] See note 4 (pg 140). In relation to ordinary mathematics, this system is very powerful. Indeed, it is surprising how much can already be proved in the far weaker system known as RCA_0 (recursive comprehension axiom).

FFF implies the 1-consistency of the system ATR_0.[10] This is what is logically indispensable, but note that it is at the level of a statement of number theory, not even sets of integers let alone uncountable sets. Moreover, it turns out that FFF can be proved in a system of iterated inductive definition (with intuitionistic logic), which can be given a kind of constructive justification, without anything like appeal to the power set of $\mathbb{N}$. What should we make of these points?

In response to the first, I would caution, "Beware of an intentionalist fallacy!" It is fair enough to scrutinize independent statements for their intrinsic mathematical content; Friedman himself does this and is always seeking improvements. But if the purpose behind the statements is grounds for objection, then, taking the point to its logical conclusion, the search for examples in pursuit of the Gödel–Friedman program would have to be shut down entirely, as anything arrived at would perforce be the work of logicians trying to come up with examples rather than of mathematicians pursuing number theory or analysis!

Concentrating then on intrinsic mathematical content *per se*, there is another response to be made on behalf of Friedman. The original Kruskal Theorem has only one initial universal quantifier over sequences; dropping this yields a "free-variable" formulation, similar in form to "laws of analysis" that constructive mathematics has always wanted to respect (e.g. commutativity of addition of real numbers). But, as Friedman showed, this formulation of Kruskal is already impredicative (providing for well-orderings beyond Γ_0) (see Simpson [1985]). Here we do have a genuine, mathematical theorem (a celebrated one, I am told), without any contrivance – bells, whistles, or frills – of a form the predicativist should seek to recover, but transcending predicative resources. (A similar story can be told about the Graph Minor Theorem of Robertson and Seymour (Friedman, et al. [1987]), suggesting a pattern.)

Once again Feferman can point to the fact that the free-variable formulation of Kruskal's Theorem is provable in a system, though stronger than ATR_0, justified without appealing to the platonist conception of the power set of $\mathbb{N}$. This is true, but the natural response is: "One thing at a time. Wherever you choose to stop along the way to the full, classical continuum, new results will emerge requiring even stronger systems for their proofs; even large cardinal hypotheses will eventually be necessary for a unified treatment of such problems." Indeed, this is the whole thrust of Friedman's more recent Boolean Relation Theory, which describes problems at the level of functions and set of integers not solvable in all of ZFC but which are solvable with large cardinal axioms of Mahlo type. Here is a recent example of this kind of result. Let $A \cup . B$ denote the union of A and B but implying that they are disjoint. A multivariate

---

[10] A system (capable of carrying out elementary arithmetic) is said to be *1-consistent* just in case there is no quantifier free formula $\varphi(x)$ such that the system proves '$\exists x \varphi(x)$' but also proves, for each numeral k, '$\neg\varphi(\kappa)$'.

function f from $\mathbb{N}^k$ into $\mathbb{N}$ has *expansive linear growth* iff there are c, d, $e > 1$ such that for all x in dom(f), if $|x| > e$, then $c|x| \leq f(x) \leq d|x|$, where $|x|$ is the maximum term of the tuple x. For $A \subseteq \mathbb{N}$, we write fA for the set of all values of f at arguments from A. Now consider the following:

Proposition: *For all multivariate f, g from $\mathbb{N}^k$ into $\mathbb{N}$ of expansive linear growth, there exist infinite sets A, B, C contained in $\mathbb{N}$ such that*

$$A \cup . fA \subseteq C \cup . gB$$
$$A \cup . fB \subseteq C \cup . gC.$$

Now let MAH+ denote the theory ZFC + "for all n there exists an n-Mahlo cardinal." Friedman proves the following:

Theorem: *The Proposition is provable in* MAH+ *but not in* ZFC *(assuming* ZFC *is consistent). In fact, the Proposition is not even provable in* ZFC + *the scheme, "there exists an n-Mahlo cardinal," each n, as it proves the 1-consistency of the latter (over the subsystem of second-order arithmetic known as* ACA *(arithmetic comprehension axiom)).*

Remarkably, if the pattern of letters, A, B, C, is altered at all in the statement of the Proposition, it becomes provable in RCA_0, the weakest of the main subsystems of analysis studied in reverse mathematics.

Again, Feferman will claim, correctly, that what is shown indispensable, in a strict, formal sense, by such examples is not large cardinals axioms on a realist conception but, rather, the 1-consistency of some such axioms, which again are just statements at the level of the integers. This was just Feferman's second objection, already mentioned.

This second objection is in line with Feferman's whole strategy in countering "Gödel's doctrine," by which Feferman means Gödel's view that the "true reason" for the incompleteness phenomena is that "the formation of ever higher types can be continued into the transfinite" (Gödel [1931], n. 48a). This, in turn, Feferman interprets as implying that "the unlimited transfinite iteration of the power set operation is necessary to account for finitary mathematics." The strategy is to point out that, although undecidables, for example asserting the consistency of a given formal system, *can* be decided by introducing higher-type objects, for example relevant truth sets, giving the appearance of confirmation of Gödel's doctrine (which, in the case of set theory, seems to justify even the unlimited extension to inaccessible cardinals, which provide for models of ZF, for example), this is illusory. For, instead, one can get by by adding much less, for example a proof-theoretic reflection scheme:

(R) *Prov(A)* → *A, for each sentence A (of the original language in question).*

This expresses confidence in the original system and implies its consistency. But there is no appeal to anything more abstract than whatever is already expressed in *A*.

Now, in response, there is a certain irony here: on the one hand, statements at low levels, for example combinatorial in character, to be derived as *theorems*, are held to a high standard of "mathematicality," as judged by our best mathematical practice. If "metamathematics" creeps in, that is certainly grounds for criticism, and nowadays even for outright dismissal; so much as a hint of "artificiality" is cause to seek improvements. (As already indicated, here Feferman and Friedman are in agreement.) Yet when the question turns to "new *axioms*" for mathematics occasioned by incompleteness, even overtly metamathematical schemata, such as (R), are proposed in place of what one would have thought was far more typical of "mathematics," viz. the recognition of new sorts of objects – for example, ordinal numbers, set-theoretic structures – seen as natural extensions of those already recognized.

A related point can also be raised concerning the unifying, "explanatory role" of new axioms that may be proposed for mathematics, a role explicitly highlighted by Gödel in his famous paper [Gödel 1947], where an analogy was drawn with theoretical physics. Suppose we have a sizeable body of mathematical statements concerning say the integers and sets and functions thereof, and suppose that each of these is independent of ordinary mathematics (say, ZFC), but each can be obtained by adjoining a formal statement of consistency (or 1-consistency) of some stronger set of axioms, say, ZFC + a small large cardinal hypothesis.[11] Indeed, we may assume that, as in reverse mathematics, relative to a weak base theory, each mathematical statement on the list is actually equivalent to some such consistency statement (depending on the given statement). Suppose further, favorably for Feferman's approach, that the statements are even demonstrable in some extension of ordinary mathematics susceptible of a constructive or quasi-constructive justification (say by reducibility, with respect to a relevant class of formulas, to a predicative system or something stronger but still weaker than full, classical set theory.) It appears that, in such a situation, Feferman's strategy of disarming Gödel's doctrine is poised for success; for each of the independent statements, it suffices to assume merely the number-theoretic coded statement of (1-)consistency of the relevant formal extension of ordinary mathematics, without entertaining anything further about the transfinite. Ironically, however, the very conditions of success we have assumed can be used to strengthen the Gödel–Friedman case; for, we may suppose, all of the various

---

[11] "Small" large cardinals are, in a sense, approachable "from below," such as inaccessibles, Mahlo cardinals, etc., and are compatible with Gödel's Axiom of Constructibility, "V = L." "Large" large cardinals, e.g. measurable, certain partition cardinals, strongly compact, etc., are of a different character and incompatible with V = L; see, e.g., Drake [1974].

formal extensions and their (1-)consistency are justified at once by a single, simple large cardinal hypothesis (say, of Mahlo type). By hypothesis, we have some independent reason to believe in the consistency claims (as we have assumed a quasi-constructive justification – although this need not be the only source of confidence in the systems); so it is natural to ask after the *best explanation* for these claims. Why are all these independent, complex combinatorial statements (of reflection, consistency, or 1-consistency, etc.) true? Even if there is a maximal one, corresponding, say, to the largest of the large cardinals mentioned in them, and from which all the others follow, why should this complex, mathematically arcane statement hold? The question seems all the more pressing if one thinks that the large cardinal hypothesis in question is false or incoherent. From the Gödel–Friedman perspective, the large-cardinal hypothesis provides a cogent answer. Consistency emerges as a necessary consequence of the coherence or satisfiability of mathematical concepts (as, of course, Tarskian semantics spells out in detail); it need not be taken as a brute fact. Rather, the large cardinal hypothesis is a natural, *mathematical* one which extends existence claims about set-theoretic structures already familiar from ZF and weaker extensions by ordinary inaccessible cardinals, *and it unifies a whole body of new mathematics, and metamathematical claims as well.*[12]

The Axiom of Replacement of ZF itself is instructive: despite the fact that it takes us far beyond "ordinary mathematical experience," it is well motivated and justified on mathematical grounds, for example, in allowing derivation of a general transfinite recursion scheme, used in a smooth development of ordinal arithmetic. What would we say about a predicativist-inspired alternative to Replacement that replaced it with the claim, in effect, that "every demonstrable, arithmetic (say) consequence of Replacement is true"? This can be stated (as a scheme) in arithmetic, and so avoids any appeal to higher-type objects. Evidently such a trick is universally available to avoid any structures beyond the integers. Suppose some newly introduced structures involving higher types are characterized by a certain condition, call it "plato." We can always fall back on "All arithmetic (say) consequences of plato are true"! But again it raises the question of why such mathematically arcane facts should

---

[12] The reader may recall that Gödel ([1947], p. 477) explicitly considered "verifiable consequences" of new axioms as the target of unification, i.e. mathematical results demonstrable without the new axioms, whose proofs with their aid would be simpler, more perspicuous, and brought together in a unified way. The focus on new theorems going beyond the old axioms is characteristic of Friedman's program, but Gödel also explicitly recognized the "fruitfulness" of extensions of set theory by large cardinals, as contrasted with their negations, in that new theorems about the integers become provable (Gödel [1947], p. 483), and he even wrote, of large cardinal axioms due to Mahlo, that they "increase the number of decidable propositions even in the field of Diophantine equations" (Gödel [1947], p. 477). Thus, Friedman's focus on new theorems of the character of "ordinary mathematics" is an entirely natural outgrowth of Gödel's ideas.

hold, especially if one is skeptical about plato. Remaining at the level of formal consistency claims thus appears, from this perspective, analogous to remaining at the level of empirical generalizations in science. It is usually a safer course than adopting a highly theoretical, explanatory hypothesis concerning "unobservables." But imagine where physics would be today if the safer course had generally been followed!

Now it must be conceded that we are only at the early stages of probing Gödel's doctrine and strategy for justifying transfinite set theory in the manner of Friedman's program. Not surprisingly, we simply do not yet have a rich body of results exemplifying unification of mathematics (old and new) by means of higher set-theoretic axioms, especially if the mathematics is required to be combinatorial in character. (The Paris–Harrington result dates only from the late 1970s, after all. That's *less* than 30 years ago.) So the jury is still out, one may say. But the tide certainly seems to be in the direction one might have suspected, namely that predicativity, for all of its proof-theoretic interest and epistemological merit, marks no natural boundary on set-theoretic strength required for genuine mathematics at the level of combinatorial or ordinary mathematical problems. Indeed, there may be no stopping point at all, not even in Cantor's heaven.

Acknowledgement: A near ancestor of this paper was presented at a symposium, "Predicativity: Problems and Prospects," held at the joint meetings of the American Philosophical Association and the Association for Symbolic Logic, in Seattle, March 28, 2002. I am grateful to cosymposiasts Solomon Feferman and Jeremy Avigad and to Linda Wetzel for their comments and also to Harvey Friedman for helpful discussions and correspondence.

References

Drake, F. R. [1974] *Set Theory: An Introduction to Large Cardinals* (Amsterdam: North Holland).

Feferman, S. [1988] "Weyl vindicated: *Das Kontinuum* 70 years later," reprinted in *In the Light of Logic* (New York: Oxford University Press, 1998), Chapter 13.

Feferman, S. [1998a] *In the Light of Logic* (New York: Oxford University Press).

Feferman, S. [1998b] "What rests on what? Proof-theoretic analysis of mathematics," in *In the Light of Logic* (New York: Oxford University Press).

Feferman, S. and Hellman, G. [1995] "Predicative foundations of arithmetic," *Journal of Philosophical Logic* **24**: 1–17.

Ferreira, F. [1999] "A note on finiteness in the predicative foundations of arithmetic," *Journal of Philosophical Logic* **28**: 165–174.

Friedman, H., Robertson, N., and Seymour, P. [1987] "The metamathematics of the graph minor theorem", in Simpson, S. G., ed., *Logic and Combinatorics, Contemporary Mathematics* (Providence, RI: American Mathematical Society), pp. 229–261.

Gödel, K. [1931] "On formally undecidable propositions of *Principia Mathematica* and related systems," reprinted in van Heijenoort, J., ed., *From Frege to Gödel: A Source Book in Mathematical Logic, 1879–1931* (Cambridge, MA: Harvard University Press, 1967), pp. 596–616.

Gödel, K. [1947] "What is Cantor's continuum problem?," reprinted in Benacerraf, P. and Putnam, H., eds., *Philosophy of Mathematics: Selected Readings*, 2nd edn. (Cambridge: Cambridge University Press,1983), pp. 470–485.

Hellman, G. [1989] *Mathematics without Numbers: Towards a Modal-Structural Interpretation* (Oxford: Oxford University Press).

Hellman, G. [1994] "Real analysis without classes," *Philosophia Mathematica* **2**(3): 228–250.

Hofweber, T. [2000] "Proof-theoretic reduction as a philosopher's tool," *Erkenntnis* **53**: 127–146.

Simpson, S. [1985] "Nonprovability of certain properties of finite trees," in Harrington, L. A., Morley, M., Scedrov, A., and Simpson, S. G., eds., *Harvey Friedman's Research on the Foundations of Mathematics* (Amsterdam: North-Holland), pp. 87–117.

Simpson, S. G. [1999] *Subsystems of Second Order Arithmetic* (Berlin: Springer).

Smorynski, C. [1982] "The varieties of arboreal experience," *Mathematical Intelligencer* **4**: 182–189.

Weyl, H. [1918] *Das Kontinuum. Kritische Untersuchungen uber die Grundlagen der Analysis* (Leipzig: Veit).

10   On the Gödel–Friedman Program

10.1   Introduction

As is well known, Gödel [1931] showed, in principle, given any "good" mathematical theory T extending elementary arithmetic, how to construct sentences in the language of T which are undecidable in T, where "T is *good*" means: (1) T is formally consistent, (2) T is formal, in the sense that the set of (Gödel codes of) its theorems is recursively enumerable, and (3) T numeralwise represents the recursive functions of natural numbers. As an immediate corollary, it follows that no such T can prove all and only the true sentences of (even first-order) arithmetic (i.e. true in the standard model). The first example of an undecidable sentence constructed in the proof of Gödel's first incompleteness theorem is known as a Gödel sentence, G, which, standardly interpreted, says of its own code that it is not the code of a theorem of T (where that is spelled out in terms of an arithmetized formalization of the syntactic proof relation for T, along with an ingeniously constructed substitution function). This G is, of course, a very arcane sentence from the standpoint of ordinary working mathematics. It is properly described as "metamathematical" rather than "(ordinary) mathematical."[1] Somewhat closer to home is the example proved to exist by Gödel's second incompleteness theorem of the same great 1931 paper, viz. the statement 'Con(T)' of the proof-theoretic *consistency* of T, formulable in the language of T. But this too is metamathematical rather than (ordinary) mathematical in character, despite the fact that mathematicians would readily acknowledge the importance of consistency of formalized counterparts of their informal mathematical theories.

---

[1] Sentences that express syntactic (e.g. proof-theoretic) or semantic properties and relations are paradigmatically "metamathematical," although they count as "mathematical" as well if the syntactic or semantic properties and relations can be expressed via a coding in terms of numbers or sets. We use the phrase "ordinary mathematical" to refer to the sorts of statements that occur in standard texts in pure mathematics, intelligible without recourse to codings of syntax.

Of course, these unprovable sentences of any given good theory T can be added to T as further axioms; but then Gödel's metatheorems apply to any such extension T' of T to generate similar undecidable sentences for T'.

Some philosophers[2] took the metamathematical character of Gödel's undecidable sentences to suggest that the phenomenon should not be seen as undermining the foundational program of logicism precisely because the phenomenon did not really affect ordinary mathematics, for example number theory or analysis.

In the 1960s and 1970s, however, this situation started to change. A major result was the demonstration, by Paul Cohen,[3] that Cantor's famous "continuum hypothesis," CH, could not be proved in the going, very powerful set theory, ZFC, provided it is consistent. CH can be formulated as a statement about sets of real numbers: it states that any infinite set of reals is either countable or is in 1–1 correspondence with all the reals (i.e. there is no infinite cardinality strictly between the countable and the power of the continuum). CH is thus fully "mathematical" as opposed to "metamathematical." Coupled with Gödel's proof in the mid-1930s of the consistency of CH with ZFC, Cohen's proof established the full undecidability of CH from mainstream set theory.

A second important example was discovered by Paris and Harrington [1977], demonstrating an incompleteness in Peano arithmetic, viz. a finite form of Ramsey's partition theorem. Again this was clearly "mathematical" as opposed to "metamathematical" in content. This result was particularly striking as it is at the level of finite sets of natural numbers (codable as numbers), even more elementary than CH.

Since Paris–Harrington, a series of ever-improving independence results has been systematically developed by Harvey Friedman, some of them establishing equivalence between mathematical statements at the level of sets of integers, etc., and statements of consistency of various large cardinals, cardinals whose existence cannot be established in standard mathematics as codified in ZFC. An example will be provided below, in the context of the main questions to be addressed in this paper, concerning two gaps. First, how can we bridge the gap between justifying statements of *consistency* of large cardinal axioms – formulable as statements quantifying over natural numbers – and statements asserting the *existence* of such cardinals? Second, how can we bridge the gap between *actual existence*, asserted by set theorists working within the standard framework of "height-actualism," recognizing

---

[2] For example, Hempel [1945], see n. 6, p. 384 in Benacerraf and Putnam [1983], where the paper by Hempel is reproduced.
[3] Cohen [1963, 1964].

a single all-encompassing universe of "all sets" and "all ordinals," on the one hand, and merely *potential existence*, associated with a "height-potentialist" view articulated with the resources of modal logic, on the other? (Our aim in attempting to bridge both gaps is to do better than the paleontologist, who, according to creationists, manages to fill a gap in the fossil record only by creating two new ones!) Let us now take up these questions.

10.2   On the Gap between Consistency and Existence of Large Cardinals

As a warm-up example, let us look at Kruskal's theorem on embeddability of finite trees, and Friedman's results on this bearing on the "necessary uses of the uncountably infinite."

Define a finite tree T to be (infimum) embeddable into another T_0 just in case there exists a 1–1 mapping $\varphi : T \to T_0$ such that for any $x, y \in T$ $\varphi(x \wedge y) = \varphi(x) \wedge \varphi(y)$. Kruskal's theorem can be stated as: There is no infinite set of pairwise non-embeddable finite trees.

Equivalently, we have the following theorem.

Kruskal's Theorem: *For any sequence $\langle T_i \rangle$ of finite trees, there exist i, j with $i < j$ such that T_i is infimum embeddable into T_j.*

Now this is not a finitary statement as it quantifies over infinite sequences. However, Friedman discovered a finite form of Kruskal's Theorem and proved that it is not predicatively provable. Friedman's finite form of Kruskal's Theorem (*FFKT*) is as follows.

For any positive integer c there exists a positive integer $n = n_c$ so large that if $T_1, T_2, \ldots, T_n$ is any finite sequence of finite trees with $|T_i| \leq c \cdot i$ for all $i \leq n$ then there exist indices i, j such that $i < j \leq n$ and T_i is embeddable into T_j.
(FFFKT)

Next, Friedman proved, over a weak base theory, that *FFFKT* is equivalent to the 1-consistency of the subsystem of classical analysis known as ATR_0 ("arithmetic transfinite recursion," with induction limited to formulas covered by the comprehension axioms). This theory is widely recognized to define a limit to predicative provability, with its proof-theoretic ordinal Γ_0 lying just beyond predicative provability.[4] Indirectly, therefore, the uncountable is implicated: *FFFKT* transcends mathematics restricted to treating only countable sets and functions, as predicative analysis does.

---

[4] Weaver [2018] has argued for a larger ordinal marking the limit to predicative provability.

A convinced predicativist can, however, reply as follows. The statement that ATR_0 is 1-consistent is only a statement at the level of the natural numbers (via well-known Gödelian techniques of arithmetizing syntax); it does not say anything about the uncountably infinite. How can the proponent of full classical analysis and higher set theory answer this reply?

Let us turn to another example of Friedman's, this one implicating large cardinals. It arises in Friedman's [2017] Boolean Relation Theory .

Let $A \cup. B$ denote the union of A and B but implying that they are disjoint. Next define a multivariate function f from $\mathbb{N}^k$ into $\mathbb{N}$ to have *expansive linear growth* just in case there are c, d and $e > 1$ such that for all x in the domain of f, if $|x| > e$, then $c|x| \leq f(x) \leq d|x|$ where $|x|$ is the maximum term of the tuple x. For $A \subseteq \mathbb{N}$ we write fA for the set of all values of f at arguments from A. Now consider the following.

Proposition: *For all multivariate f, g from $\mathbb{N}^k$ into $\mathbb{N}$ of expansive linear growth, there exist infinite sets A, B, C contained in N such that*

$$A \cup. fA \subseteq C \cup. gB$$
$$A \cup. fB \subseteq C \cup. gC$$

Now let MAH+ denote the theory ZFC+ "for all n there exists an n-Mahlo cardinal." Friedman proves the following:

Theorem: *The Proposition is provable in MAH+ but not in ZFC (assuming ZFC is consistent). In fact, the Proposition is not even provable in ZFC+ the scheme "there exists an n-Mahlo cardinal," for each n, as it proves the 1-consistency of the latter (over the subsystem of second-order arithmetic (classical analysis) known as ACA ("arithmetic comprehension axiom").*

Remarkably, if the pattern of letters, A, B, C, is altered at all in the statement of the Proposition, it becomes provable in RCA_0 ("recursive comprehension axiom"), the weakest of the subsystems of analysis studied in reverse mathematics.

Since the first sentence of the Theorem states only a sufficient condition for proving the Proposition, again what is shown indispensable is not the statement of existence of the large cardinals *per se* but rather the 1-consistency of the indicated scheme for them, a statement at the level of natural numbers.

This is symptomatic of a general method, made explicit by Feferman, of "defanging Gödel's doctrine," that is the claim that indefinite extendability of transfinite types is "the true reason" for the incompleteness phenomena. That method is the following: whenever a proof of a sentence A from a large cardinal hypothesis, LC, is given along with a demonstration that A is unprovable from the original theory (if that theory is consistent), it suffices to replace LC with the proof-theoretic reflection principle:

$$\text{Prov}_{LC}(`A') \rightarrow A, \tag{R}$$

for each sentence A of the original language. (In Friedman's program, A is typically of low set-theoretic rank.) This expresses confidence in adding LC as an axiom to our set theory, but (R) itself is not committed to anything of higher type than what is quantified over in A itself. The challenge we face, then, is to bridge the gap between (R) and LC.

It is worth noting that a similar situation arises in considering the link between large cardinals and descriptive set theory. Martin and Steele [1988, 1989] proved that the existence of infinitely many Woodin cardinals implies Projective Determinacy (PD), the statement that all projective sets of reals are determined (in the sense that a player in the relevant Gale–Stewart game has a winning strategy). This result in turn yields a detailed structure theory of projective sets. Moreover, work of Woodin, Martin, and Steele shows that this large cardinal hypothesis is minimal, in the sense that PD and the following statement are equivalent:

$$\text{For each } n, \ \text{Prov}_{ZFC+\text{"There are } n \text{ Woodin cardinals"}}(`A') \rightarrow A,$$

for every sentence A in the language of second-order arithmetic. As Steele sums this up: "PD is the 'instrumentalist trace' of Woodin cardinals in the language of second-order arithmetic."

10.3   Confirmation Short of Proof

Reflecting on the sorts of results just reviewed, it seems unlikely that large cardinal axioms transcending ZFC can and will be proved from intuitively even more compelling new set-theoretic principles. This seems especially so in the cases of the smaller large cardinals, like the inaccessibles and the Mahlo cardinals, whose mathematical existence or logical possibility already seems quite plausible given the principle that any model of set theory can have a proper extension encompassing higher ordinals and sets of higher rank. As a result, it seems that we must lower our sights and seek alternative routes leading to justification of such large cardinal axioms, routes that lead at least to rational credibility rather than the certainty of deductive mathematical proof.

Now, as is well known, Quine proffered a holism of scientific testing that, in principle, extends to mathematics, mainly through its manifold applications in the empirical sciences.[5] As an example, consider the classical Axiom of Infinity ("Infinity" for short), as it occurs in axiomatic set theory, such as Zermelo or Zermelo–Fraenkel set theory. First, note that it *is* an axiom rather than a theorem. It *had* been a theorem in Frege's system of the *Grundgesetze*, where it followed from his fateful abstraction principle, Axiom V, which guaranteed an extension of every well-formed formula with a single free variable, where extensions of formulas are distinct if and only if the formulas are not coextensive. For Frege had shown, in the *Grundlagen*, how to define explicitly "_ is a natural number"; then its extension has to be infinite as guaranteed by the behavior of the successor function. But unfortunately, it turned out that Axiom V led to Russell's Paradox and had to be jettisoned. Furthermore, in Russell's theory of types, there is no type with infinitely many objects unless one builds an infinitude of objects in on the ground floor, or else just stipulates that there is a type which includes the union of all finite types. It thus seemed that the mathematical existence of an infinite totality had to be an axiom. But then how can it be justified as true (or at least possibly true)?

Here Quine and Gödel, as will soon be made clear, come to the rescue, answering "By its fruitfulness, ultimately inductively, by the evidence of its manifold successful applications, especially in classical analysis,[6] along with its uses in mathematical physics." And surely those successes testify to at least the consistency or coherence of Infinity, which for mathematical purposes has often seemed quite sufficient.

In connection with large cardinal hypotheses, however, such a route seems quite implausible, given how much scientifically applicable mathematics can already be done in subsystems of classical analysis, far more limited in their set-theoretic existence postulates than even Zermelo set theory, let alone ZFC or extensions by large cardinals. For example, one must search hard for examples of scientifically applicable results that cannot be comprehended by predicative analysis, which restricts set-existence postulates to the countably infinite. In light of this, we are advised to look elsewhere for routes to confirmation short of proof of large cardinal axioms.

The *locus classicus* describing such a route is Gödel's famous essay, "What is Cantor's continuum problem?"[7] The key passage from our standpoint here is worth quoting nearly in full:

... even disregarding the intrinsic necessity of some new axiom, and even in case it has no intrinsic necessity at all, a probable decision about its truth is possible also in another way, namely, inductively by studying its 'success'. Success here means fruitfulness in

---

[5] See, e.g., Quine [1953].

[6] NB: In mathematical analysis, infinite sequences (of rationals) are needed just to define what a real number is.

[7] Gödel [1947], reprinted in Benacerraf and Putnam [1983], pp. 470–485.

consequences, in particular in 'verifiable' consequences, i.e. consequences demonstrable without the new axiom, whose proofs with the help of the new axiom, however, are simpler and easier to discover, and make it possible to contract into one proof many different proofs ... A much higher degree of verification than that, however, is conceivable. There might exist axioms so abundant in their verifiable consequences, shedding so much light on a whole field, and yielding such powerful methods for solving problems (and even solving them constructively, as far as that is possible) that, no matter whether or not they are intrinsically necessary, they would have to be accepted at least in the same sense as any well-established physical theory. (Benacerraf and Putnam [1983], p. 477)

The key ingredients of the kind of "success" that Gödel is describing as possible for new axioms of set theory, such as for small large cardinals, are *fruitfulness* in the sense of abundance of consequences, especially "verifiable" consequences, and a kind of *unification* brought about in the domain of proofs of consequences, a unification we often associate with the phenomenon of *variety of evidence*. As suggested by Gödel's phrase, "probable decision about ... truth," clearly we are in the domain of *confirmation* of hypotheses by evidence rather than deductive proof. This in turn suggests that the tools of Bayesian confirmation theory may be relevant in fleshing out these ideas, even in the domain of higher mathematics, as well as in empirical science, where Bayesian confirmation has had its main applications. After all, Bayesian tools are valuable to guide rational belief generally, and are not limited to empirical scientific contexts, or so we claim here.

Let us now state Bayes' Theorem and then describe how the Bayesian apparatus helps rationalize the desiderata alluded to by Gödel, that is, fruitfulness, variety of evidence, and unification, both of narrower theories and of bodies of evidence.[8] Letting H stand for a hypothesis or theory, E a body of evidence, and B a given background information, Bayes' Theorem can be stated thus.

Bayes' Theorem: $P(H/E \& B) = \frac{P(H/B) \cdot P(E/H \& B)}{P(E-B)}$, *where*   'P(X/Y)' *denotes the conditional probability (or credence) of* X *given* Y.

The term on the left is called the posterior probability of H given the evidence E and the background information B; the first term of the numerator is called the *prior probability* of H on the background B, the second term in the numerator is called the *likelihood* (of evidence E given H and B), and the denominator is called the *prior* of the evidence E on the background B. This is an elementary result of standard probability theory, independent of any particular interpretation of "probability." In epistemic applications of probability, the theorem instructs how to "update" degrees of rational belief when evidence

---

[8] For sources on these topics, we refer the reader to Howson and Urbach [1989], Earman [1992], and Myrvold [2017], and works cited therein.

for (or against) a hypothesis is gained. Often the hypothesis H together with the background B implies the evidence E, in which case the second term in the numerator is equal to 1 and drops out. Also note that, in assessing the denominator, we are not to use the information – even if we have it – that E actually has been found or established to be true (and so has probability 1), for then no known evidence could boost the posterior of H beyond its prior. (This is known as "the problem of old evidence".[9]) Thus, typically the denominator is taken as strictly < 1. The value of coming to know E is that it "licenses updating," that is, setting $P_{new}(H/B) = P_{old}(H/B \ \& \ E)$.

It should be acknowledged that Bayesian confirmation theory has to deal with the phenomenon of subjectivity of assessments of prior probabilities and the fact that different subjects may assign very different priors to a given hypothesis. This, however, is mitigated to some extent by the phenomenon of "merger of opinion" or "convergence" in the long run. As several mathematical results testify, as more and more evidence accumulates in favor of a given hypothesis, so long as the priors assigned are > 0, posteriors assigned by different subjects move closer to one another, provided updating adheres to Bayes' rule.[10]

Let us now try to connect the Bayesian apparatus with the virtues (of new potential axioms) that we interpret Gödel to have highlighted as providing inductive justification of new axioms, apart from any "intrinsic necessity." The first of these was "fruitfulness," which we interpret as richness in consequences for which we can gain support independently of the particular hypothesis or new axiom, H, that we are assessing. Let us assume that H, possibly together with some background knowledge or assumption, B, implies a rich set of supportable sentences. The hope is then that these can serve in the role of "evidence" for H, relative to the given background, B. Then they serve in the role of $E = E_1 \wedge \ldots \wedge E_n$.[11] Since these occur in the denominator of Bayes' Theorem, obviously the longer a list of the Es continues, the greater potentially is the boost H receives in computing its posterior probability (also called "credibility" in the context of assessing rational belief). We say "potentially," because the degree of incremental support the Es provide is inversely proportional to the prior probabilities of the E_k conditional on the conjunction of the earlier Es, as dictated by the rule for computing the probability of the whole conjunction of the Es. We assume that the *more varied* the evidence statements are, the lower the conditional probabilities of the "next" statement given the conjunction of the previous ones. (Where there is little variety – where all the pieces of evidence are "of a piece," so to speak – such conditional probabilities would be higher.) This is the short and sweet Bayesian

---

[9] For more on this topic, see Glymour [1980], also Howson and Urbach [1989], pp. 270 ff.
[10] For exposition and critical assessment, see Earman [1992], Ch. 6.
[11] Often, we let the evidence statements be ordered temporally, but this is inessential.

story on the virtue of "variety of evidence": the more varied the pieces of evidence, the lower these conditional probabilities among the evidence statements whose conjunction enters in the denominator of Bayes' Theorem.[12]

Note that these virtues of abundance and variety of evidence are accounted for in Bayesian terms by the relevant *inequalities* involved, in the absence of precise information about the prior probabilities of the target hypothesis and the evidence (so long as they are strictly between 0 and 1).

When it comes to the virtue of *unification*, matters are complex, in part because, *prima facie*, there is a large hurdle to get over, as follows. Let H_1 and H_2 be as disunified as you like, let H_1 be supported by evidence E_1 and let H_2 be supported by evidence E_2, and suppose a potential unifier, H_3, along with relevant background knowledge B_3 implies each of H_1 and H_2. Then it follows from the axioms of probability that the posterior of H_3 given E_1 cannot exceed that of H_1 on E_1, and likewise with respect to H_2 and E_2. Moreover, $P(H_3/E_1 \wedge E_2) \leq P(H_1 \wedge H_2/E_1 \wedge E_2)$, and this remains the case even if we suppose that E_1 has no effect on H_2 and E_2 has no effect on H_1. It is true that unifier H_3 would typically be supported by both E_1 and E_2 so that its posterior on those bodies of evidence would rise beyond $P(H_1/E_1 \wedge E_2) = P(H_1/E_1)$ and also beyond $P(H_2/E_1 \wedge E_2) = P(H_2/E_2)$. But we certainly still want a posterior of H_3 to exceed that of the disunified conjunction, $H_1 \wedge H_2$.

Now a resolution of this comes readily to mind if we consider examples of successful unifications of theories in the natural sciences, for example the unification by Newtonian mechanics, NM, of bodies of theory of terrestrial and celestial mechanics taken separately (as they were historically). Even if we suppose that the latter precursors were as good at predicting relevant evidence as NM was, nevertheless, NM over time came to surpass its precursors in evidential support *via new bodies of evidence*, such as predictions of new planets, as well as phenomena outside the solar system (e.g. orbiting binary stars, stability of whole galaxies). In our schematic notation above, we have to take account of much new evidence, E_3, in estimating the posterior probability of H_3. Thus the evidentiary advantages accruing to the unifier, H_3, are accounted for via the mechanisms of fruitfulness and variety of evidence reviewed above. Thus, we do not have to appeal to a special confirmational role of what has been called "common origin inference or explanation," which has proved hard to justify in purely Bayesian terms, despite its functioning as a clear *desideratum* in theory selection in its own right.[13]

---

[12] See Earman [1992], pp. 77–79 for a good discussion. Earman even suggests that we take the smaller measure of the conditional probabilities of components of evidence on "previous" components as our definition of "variety of evidence."

[13] See Janssen [2002] for an account of several important historical episodes. See also Myrvold [2017] for an insightful treatment of unification in Bayesian terms, explicating the evidentiary advantages of unification in terms of increases in mutual information among evidence

The reader now may well ask, however, "What has all this to do with pure mathematics and, specifically, with bridging the gap between consistency of large cardinals on the one hand and mathematical existence of them on the other?"

The idea is to appeal to results such as Friedman's theorem stated above from his Boolean Relation Theory to increase the posterior rational credence of the hypothesis of the existence of Mahlo cardinals. As stated above, we have the following dependencies, where "BR Prop" refers to Friedman's Boolean Relation Proposition, stated above:

$$\text{MAH} + \quad \Rightarrow \quad \text{BR Prop} \quad \Rightarrow \quad \text{1-Cons of Scheme}$$
$$"\exists k \ (k \text{ is n-Mahlo})", \text{ for each } n.$$

For Bayesian-style confirmation of MAH+ to work, we need independent epistemic access to one of its consequences. It would not be the Boolean relation itself, as that presumably is specific to this theorem, not arising in a variety of contexts. On the other hand, the 1-consistency of Mahlo cardinals may well so arise, so that there is a reasonable prospect of confirming it through its own fruitfulness in other applications. Similarly, consistency properties of other cardinals whose existence is derivable from that of the Mahlo cardinals – for example inaccessibles, hyper-inaccessibles, etc. – can further enrich the "evidence" for the Mahlo cardinals, via the Bayesian mechanisms rehearsed above.

Note then a significant difference between Gödel's vision described above and the situation of Friedman's and Woodin–Martin–Steele's (WMS) results: Gödel emphasized "explaining" mathematical facts of low set-theoretic rank, already mathematically proven, by simplifying and unifying their proofs (already known and not requiring large cardinals), whereas Friedman and WMS prove new theorems not derivable without new assumptions, for example of consistency properties of large cardinals or proof-theoretic reflection principles. One obvious reason for Gödel's emphasis emerges on a Bayesian analysis, as sketched above: updating rational beliefs depends on invoking relevant evidence; if you do not have independent access to such evidence, you cannot invoke it!

If we are right, however, not all is lost, as experience with axiom systems, including familiar ones with the addition of consistency statements or proof-theoretic reflection principles, can support rational credibility of those statements and principles. Sometimes a constructive or quasi-constructive model-theoretic argument can even be given. (For instance, the standard intuitionistic argument

statements. Whether common-origins explanatory power can be assigned an additional *evidentiary* role remains problematic.

that the Dedekind–Peano axioms of arithmetic (DPA) are true of our conception of "finitely constructed numbers," combined with Gödel's double-negation translation, does support the claim that *classical* DPA (with second-order induction) is consistent.) Our suggestion here, then, is that independent evidence for consistency and proof-theoretic reflection principles can be gained, even though it falls short of deductive proofs in mathematics of low set-theoretic rank. And then such principles become available in turn as evidence for extraordinary posits such as large cardinals, on the grounds that those posits provide good, unifying explanation(s) of those independently confirmed principles.

It should also be clear that the best judges of the sort of boosts to rational credibility of large cardinals provided by the Bayesian mechanisms are mathematicians (including set theorists) and logicians (especially proof theorists), rather than philosophers. It should also be clear that such success depends on a rich variety of proof-theoretic results. Our guess here is that this is all still a work in progress, even given the impressive results that have been obtained so far.

Finally, to round out this section, let us make two remarks. First, there is not even a whiff of ESP in all this. All of it stands independently of Gödel's controversial remarks (in the same paper from which we quoted above) about faculties of mathematical intuition akin to sense perception.

Second, our main point has been that Bayesian epistemology, though developed primarily with the empirical sciences in mind where ampliative reasoning is the norm, is also applicable in pure mathematics. After all, it is generally applicable in virtually any context where rational belief is at issue. And, despite the fact that mathematical epistemology is rightly concerned with the deductive methods of mathematics, some of its most challenging and intriguing problems concern the identification and justification of basic and newly proposed axioms, as the starting points of deductive proof. As Gödel and Quine have emphasized, inductive methods are very much in play, and so the relevance of Bayesian methods should not be surprising.

10.4   The Second Gap: Between Actual Existence and Logical Possibility

On the received view of axiomatic set theory, its axioms are regarded as true of a fixed, maximal universe of all sets and all ordinals. As explained in detail elsewhere, however, this runs into several problems.[14] For instance, how do we know that this universe really obeys the standard axioms (of ZFC), for example

---

[14] For an overview, with comparisons with other versions of mathematical structuralism, see Hellman [2005] and Hellman and Shapiro [2019].

that the structure is well-founded, that power sets are really full, that the ranks of the structure satisfy the Replacement axioms, and so on? After all, there are other systems of set theory that are perfectly legitimate objects of mathematical investigation; and it seems arbitrary to single out one universe of sets as "the true one," inside which all other universes have to be embedded. Whereas ZFC does a good job in articulating a *structuralist* account of the various branches of ordinary mathematics, including number theory, analysis, algebra, geometry, topology, etc., nevertheless it makes a notable exception of set theory itself, which is not treated structurally, despite the multiplicity of set theories. Furthermore, the idea of a fixed, maximal universe of "all the sets" and "all the ordinals" conflicts with the deep-seated principle that any mathematical totality can be extended to a more comprehensive one.[15] In the case of set theory, the means of such extendability lie within set theory itself, for example by applying operations such as taking singletons, or taking powers sets (or power universes), and so forth.

Such considerations have motivated alternatives to the received view, in particular the alternative view of mathematics as based on modal logic framed in terms of a notion of logico-mathematical possibility.[16] The core idea is to construe a mathematical theory, say classical number theory or analysis, as investigating what would hold in any structure of the appropriate type that there might be; and then to postulate outright that such structures are possible, or possibly exist. The first part of this is called "the hypothetical component" (of the modal-structuralist interpretation, MSI, of the mathematical theory in question). This consists of a translation scheme which sends a sentence S of the theory in question to a modal conditional, for example in the case of the Dedekind–Peano axioms for arithmetic, DPA, of the form,

$$\Box \forall X\ \forall f\ [\wedge \mathrm{DPA} \to \mathrm{S}]^{X}_{s/f},$$

where $\wedge$ DPA stands for the conjunction of the (finitely many) Dedekind–Peano axioms with the second-order axiom of mathematical induction, framed in the language of plural quantifiers, and where the superscript X indicates relativization of all quantifiers to the domain X, and the subscript indicates substitution of the function variable f for the successor function sign s as it occurs in the axioms and the sentence S. Note that this appears to involve second-order quantification

---

[15] See Zermelo [1930] as a major source of this view. Also, see Mac Lane [1986], especially at p. 390.

[16] See especially Putnam [1967], Hellman [1989], and Hellman [1996], reproduced as Chapter 1 in this volume. NB: These interpretations take modal operators as primitive, obeying the S-5 axioms of quantified modal logic. Talk of "possible worlds" is heuristic only: officially, we never quantify over "worlds."

over domains and functions; however, this can be expressed nominalistically using mereology combined with plural quantifiers.[17]

So far, we have what appears to be a kind of second-order modal "if-then" interpretation. But what would it tell us in case there were no possible "models" of DPA or classical analysis, CA?[18] Then, regardless of the consequents of the quantified modal conditionals, all of them would count as true, many vacuously so. To avoid this problem of vacuity, the MSI postulates outright the possible existence of "models" of the appropriate type. In the case of DPA, we thus assert:

$$\Diamond \exists X \exists f[\wedge DPA]^{X}_{s/f}, \tag{Poss Ex}$$

and similarly for CA and other categorical theories, including axioms for initial segments of standard models of ZF^2C, ZFC with a single second-order Replacement axiom (implying second-order Separation). (By "standard model" we mean well-founded, with full power sets, more about which in a moment.) In the case of full ZF^2C, we postulate the possible existence of a "model" on the plan of Poss Ex above:

$$\Diamond \exists X \exists R \, [\wedge ZF^2C]^{X}_{\in / R}.$$

We now highlight some key features of the MSI in connection with set theory.

First, as with the number systems and analysis, one recovers the expressive power of second-order set theories, especially ZF^2C, along with key results of Zermelo [1930].

1. Quasi-categoricity: Given any two "models" with bijectively equivalent urelement bases, one is isomorphic to an initial segment of the other.
2. Small large cardinals: The height of any such "model" is a strongly inaccessible cardinal.

Second, all quantification over individuals behaving as sets is relativized to a "model," more accurately to a given plurally specified domain, X. As in the case of the number systems and analysis, quantification over models is achieved via plural quantification, for example, "There are or could be some things, the X, interrelated thus-and-so. . ." As indicated in the displayed formula above for ZF^2C, set-membership is actually eliminated on the MSI in favor of universal generalization over two-place relations satisfying the axioms. Officially, then, the predicate "set" is also eliminated, just as "number" is.

---

[17] See Hellman [1996] and Burgess, Hazen, and Lewis [1991] in Lewis [1991].
[18] We put "models" in quotes to indicate that we officially reconstrue such in terms of plural quantifiers for domains and the two-place membership relation, following the plan laid out above for "models" of DPA.

Third, proper classes in the usual absolute sense are avoided in favor of unrestricted extendability of "models," expressed modally:

$$\Box \forall M \; \Diamond \exists M' \, [M \prec_e M'], \tag{EP}$$

where M and M' stand for "models" as spelled out in the axiom of possible existence stated above, and where $\prec_e$ indicates that M' is a proper end-extension of M (i.e. sets of the lesser model have the same members in the greater model: mnemonic "no new wine in old bottles"). Thus the Extendability Principle, EP, states that any ZF^2C model there might be would possibly have a proper end-extension. An integral part of the MSI applied to set theory is the relativization of second-order comprehension to the domain of a model; there is no trans-world comprehension. This is naturally motivated by the idea that you can collect "at a world" only what exists at that world. One cannot collect at a world what merely might have existed there. Collections – better pluralities – are world-bound.[19] Thus, it is impossible to have "ultimate totalities" such as a class or plurality of all objects of all possible models. We claim that this restriction on modal comprehension is rooted in our understanding of modality combined with set-theoretic discourse and is by no means an ad hoc restriction.

The fourth and final feature of the MS interpretaton of set theory that should be listed here concerns the ease with which small large cardinals, inaccessibles and Mahlo cardinals are derivable. As noted above, Zermelo [1930] proved that the ordinal height of a standard model of ZF^2C is a strongly inaccessible cardinal. That however involves a perspective outside the model. How do we guarantee that such a cardinal occurs as an element of the domain of a model? Well, that follows immediately from an application of the Extendability Principle, EP, as the ordinal height of a possible model M occurs as a set-ordinal and cardinal inside any proper extension, M', of M. Suppose, then, motivated by the iterability of this process, we add as an axiom to ZF^2C the statement,

$$\forall \alpha \; \exists \beta \, [\beta > \alpha \; \wedge \; Inac(\beta)],$$

where '$Inac(\beta)$' abbreviates the statement that β is strongly inaccessible (i.e. is a regular, strong limit cardinal).[20] Call the resulting theory ZF^2C+. Then the height of any model of this theory is a hyper-inaccessible cardinal, an inaccessible cardinal with inaccessible-many cardinals below it. (That is, the height is

---

[19] See especially Linnebo [2013] for an excellent discussion of this and related principles.

[20] A cardinal λ is *regular* just in case it is not the union of fewer ordinals less than λ; and it is a *strong limit* just in case (i) it is a limit cardinal, that is, of the form $\aleph_\mu$ for limit ordinal μ, and (ii) if $\alpha < \lambda$, then $2^\alpha < \lambda$.

a fixed point of a function f on ordinals α such that $f(\alpha)$ is the least inaccessible $\geq \alpha$.) This then motivates adding a further axiom guaranteeing arbitrarily large hyper-inaccessibles. A fixed point of a suitable *normal* function on ordinals, i.e. a function from ordinals to ordinals that is increasing – if $\alpha < \beta$ then $f(\alpha) < f(\beta)$ – and continuous at limits – if α is a limit ordinal, $f(\alpha) = \bigcup_{\xi < \alpha} f(\xi)$, such as f above, is then a hyper-hyper-inaccessible. Clearly this procedure can be iterated without end. If we now generalize further and adopt as a new axiom the statement, call it F, that *every* normal function has a regular fixed point, we have that the height of a standard model of this extension of ZF^2C is a Mahlo cardinal.[21] Then, passing to the modal interpretation, another application of the Extendability Principle guarantees that a Mahlo cardinal possibly occurs as an "element" of a standard model of ZF^2C+F.

In this manner, we see that these small large cardinals involve very natural extensions of ZFC through iterative processes, invoking the Extendability Principle. The MSI thus provides motivation of such large cardinals "from below." This contrasts markedly from motivation "from above" afforded by Reflection Principles, which implicitly appeal to the "indescribability" of "the universe V of all sets."

This now brings us to the main questions of this section. First, we ask, what happens with respect to confirmation of large cardinal hypotheses (for the small large cardinals considered so far) when *mathematical existence* is construed as "logico-mathematical possibility"? And second, how can mere possibility *explain* relevant consistency and proof-theoretic reflection phenomena?

By way of an answer, consider that *actuality* actually plays no role in purely mathematical explanation, in stark contrast to *causal* explanation, which tends to dominate discussions of explanation in the empirical sciences. This is in part because of the modal force of mathematical truth, reflected in the choice of modal logic, in the case of modal-structural interpretations, the choice of S-5. If sentences Γ are satisfied in a structure, then *necessarily* no contradiction is implied by Γ. It makes no difference if we say, "If sentences Γ *could be* satisfied ... " This is reflected in the fact that the modal-structural construal of mathematical existence as logico-mathematical possibility reflects, *under translation*, the inference from existence to necessary existence. Thus we have, for example,

$$\Diamond \exists X \; [\Phi(X)] \;\Rightarrow\; \Box \Diamond \exists X \; [\Phi(X)],$$

---

[21] See Drake [1974] for details.

which is an instance of the characteristic S-5 modal axiom. Furthermore, the standard platonist inference from possible existence to actual existence is also respected, *under translation*:

$$\Diamond\Diamond\, \exists X\ [\Phi(X)] \;\Rightarrow\; \Diamond\, \exists X\ [\Phi(X)],$$

which is an instance of the characteristic S-4 axiom (which clearly holds also in S-5). Thus, we can say, for instance, that if *possibly* there exists a structure for set theory with Mahlo cardinals, then *necessarily* no contradiction is implied, based on the soundness of the underlying logic. Thus, we have our crucial "explanatory link" between the possibility of small large cardinals and the consistency of statements asserting their mathematical existence, just as desired in the Gödel–Friedman program.

It is also worth pointing out that in the empirical sciences as well, there are important patterns of explanation (of actual phenomena) that invoke possibilities (although those are more constrained than mere logical possibilities). For example, in classical and quantum statistical mechanics, general statements about thermodynamical behavior of complex systems are regularly explained by appealing to the relative "sizes" – in the measure-theoretic sense – of collections of exact microstates compatible with given macroscopic constraints. While the macrostates involved may be actual, only one of the microstates is, all the (infinitely many) others being merely (physically) possible. In effect, possibilities are invoked to explain, say, why heat normally passes from a hotter body to a colder one when they are brought into proximity, not the other way around, despite the fact that, if we could specify the exact microstate of the bodies, and if we could solve the relevant differential equations, then we could in principle explain singular statements about the actual evolution of the complex system in terms of microdynamical laws of motion without making reference to merely possible configuratons of the system. But if we want to explain general regularities concerning thermodynamic behavior, reference to possible microstates is in practice unavoidable.

To sum up, we submit that a modal-structural approach to small large cardinals is not worse off than the more common face-value platonist interpretation when it comes to confirmation via the mechanisms described above in connection with the Gödel–Friedman program.

References

Benacerraf, P. and Putnam, H. (eds.) [1983] *Philosophy of Mathematics: Selected Readings*, 2nd edn. (Cambridge: Cambridge University Press).

Burgess, J., Hazen, A., and Lewis, D. [1991] "Appendix," in Lewis, D., *Parts of Classes* (Oxford: Blackwell), pp. 121–149.

Cohen, P. [1963] "The independence of the continuum hypothesis". *Proceedings of the National Academy of Sciences of the United States of America.* **50**(6): 1143–1148.
[1964] "The independence of the continuum hypothesis, II". *Proceedings of the National Academy of Sciences of the United States of America.* **51**(1): 105–110.

Drake, F. R. [1974] *Set Theory: An Introduction to Large Cardinals* (Amsterdam: North Holland).

Earman, J. [1992] *Bayes or Bust? A Critical Examination of Bayesian Confirmation Theory* (Cambridge, MA: MIT Press).

Friedman, H. [2017] *Harvey's Foundational Adventures: Boolean Relation Theory Book* http://u.osu.edu/friedman.8/foundational-adventures/boolean-relation-theory-book/.

Glymour, C. [1980] *Theory and Evidence* (Princeton, NJ: Princeton University Press).

Gödel, K. [1931] "On formally undecidable propositions of *Principia Mathematica* and related systems," reprinted in van Heijenoort, J., ed., *From Frege to Gödel: A Source Book in Mathematical Logic, 1879–1931* (Cambridge, MA: Harvard University Press, 1967), pp. 596–616.

Gödel, K. [1947] "What is Cantor's continuum problem?," reprinted in Benacerraf, P. and Putnam, H., eds., *Philosophy of Mathematics: Selected Readings*, 2nd edn. (Cambridge: Cambridge University Press,1983), pp. 470–485.

Hellman, G. [1989] *Mathematics without Numbers: Towards a Modal-Structural Interpretation* (Oxford: Oxford University Press).

Hellman, G. [1996] "Structuralism without structures," *Philosophia Mathematica* **4**(2): 100–123.

Hellman, G. [2005] "Structuralism," in Shapiro, S., ed., *Oxford Handbook of Philosophy of Mathematics and Logic* (Oxford: Oxford University Press), pp. 536–562.

Hellman, G. and Shapiro, S. [2019] *Mathematical Structuralism* (Cambridge: Cambridge University Press).

Hempel, C. G. [1945] "On the nature of mathematical truth," reprinted in Benacerraf, P. and Putnam, H., eds., *Philosophy of Mathematics: Selected Readings*, 2nd edn. (Cambridge: Cambridge University Press, 1983), pp. 377–393.

Howson, C. and Urbach, P. [1989] *Scientific Reasoning: The Bayesian Approach* (Chicago, IL: Open Court).

Janssen, M. [2002] "COI stories: explanation and evidence in the history of science," *Perspectives on Science* **10**: 457–522.

Lewis, D. [1991] *Parts of Classes* (Oxford: Blackwell).

Linnebo, Ø. [2013] "The potential hierarchy of sets," *Review of Symbolic Logic* **6**(2): 205–228.

Mac Lane, S. [1986] *Mathematics: Form and Function* (New York: Springer Verlag).

Martin, D. A. and Steele, J. R. [1988] "Projective determinacy," *Proceedings of the National Academy of Sciences USA* **85**: 6582–6586.

Martin, D. A. and Steele, J. R. [1989] "A proof of projective determinacy," *Journal of the American Mathematical Society* **2**: 71–125.

Myrvold, W. S. [2017] "On the evidential import of unification," *Philosophy of Science* **84**: 92–114.

Paris, J. and Harrington, L. [1977] "A mathematical incompleteness in Peano arithmetic," in Barwise, J., ed., *Handbook of Mathematical Logic* (Amsterdam: North
Holland), pp. 1133–1142.

Putnam, H. [1967] "Mathematics without foundations," reprinted in Benacerraf, P. and
Putnam, H., eds., *Philosophy of Mathematics: Selected Readings*, 2nd edn.
(Cambridge: Cambridge University Press,1983), pp. 295–311.

Quine, W. V. [1953] "Two dogmas of empiricism," reprinted in *From a Logical Point of
View* (New York: Harper & Row, 1963), pp. 20–46.

Weaver, N. [2018] "Predicative well-ordering," https://arxiv.org/abs/1811.03543.

Zermelo, E. [1930] "Über Grenzzahlen und Mengenbereiche: Neue Untersuchungen
über die Grundlagen der Mengenlehre," *Fundamenta Mathematicae*, **16**: 29–47;
translated as "On boundary numbers and domains of sets: new investigations in the
foundations of set theory," in Ewald, W., ed., *From Kant to Hilbert: A Source Book
in the Foundations of Mathematics*, Volume 2 (Oxford: Oxford University Press,
1996), pp. 1219–1233.

Part III

Logics of Mathematics

In his influential paper, "Truth by convention,"[1] Quine subjected the linguistic doctrine of logical truth (LD) to a critique that, to many, has seemed devastating. Having granted the conventionalist (what Quine took to be) his/her starting points, Quine caught his opponent in a vicious regress: to proceed from the linguistic stipulations to the (full class of) logical truths requires logical rules themselves in addition to any of the stipulations. What Lewis Carroll's tortoise said to Achilles (on the need to appeal to *modus ponens* to justify any application of it) seemed an arrow in Carnap's heel.

Carnap seems never to have taken the critique very seriously. His reply to Quine's "Carnap and logical truth,"[2] which repeated the upshot of "Truth by convention," is couched in irony. Quine had found LD "empty" and "without experimental meaning"; moreover, he had found it "implying nothing not already implied by" the assertion – which he surely accepted – that logic is obvious. This afforded Carnap the opportunity to point out that LD implied itself and to express his wish that Quine had only *said* that he believed LD true. Further, he said that the emptiness of LD was something he had always accepted, that in Vienna that was what you expected of philosophical truths, and that in so phrasing his finding, Quine was accepting the analytic-synthetic distinction in so many words.[3]

One has the feeling that two of the century's most important thinkers were talking past each other. In order to remedy this unfortunate situation, let us see how, making occasional use of more recent developments, the debate might have continued.

***Carnap**(1):** In my *Logical Syntax* I was perhaps overly skeptical about using our ordinary background language to introduce logical notions.[4] As a clearer alternative, I preferred the method of implicit definition, determining meanings of logical particles by (arbitrarily) laying down postulates and transformation rules. This, I take it, was the source of the procedure you followed in "Truth by

---

[1] Quine [1936], "Truth by convention."
[2] Quine [1963], "Carnap and logical truth".
[3] Carnap [1963], "W. V. Quine on logical truth".
[4] Carnap [1937], see the Foreword, p. xv and p. 18.

convention," where you granted the conventionalist the "starting point" of stipulating sentences of certain forms to be true. This, however, must not obscure that it is central to LD that logical truths are true solely in virtue of the meanings of the logical words. (The logical truths (of a given sufficiently rich and well-regimented language) are, for present purposes, best specified as those truths in which only the logical words occur essentially;[5] the latter we may fix to be 'not', 'and', 'all', and '='. Choice of these (up to classical logical equivalence) is not arbitrary given a general notion of analyticity, but, for reasons that will emerge, I won't press that point here.) As I hinted in my reply,[6] given conventions on the interpretation of words, one is no more free to stipulate truth values for logical sentences than one is for 'the sky is blue'. And the interpretations of the logical words I regard as *capable of being given by stipulation* (as also done in the *Syntax*[7]), for example, by truth-tables (for 'not' and 'and'), by example (for 'all', one can stipulate it to be eliminable in favor of conjunction over finite domains), and by direct specification of extension (for '=', one can stipulate that it holds just of each entity and itself, no matter what it may be).

Quine**(1): Since our last exchange on this topic, I have elaborated on the derivative character of word-meaning from a behavioral standpoint.[8] But those issues need not detain us here. For my procedure in "Truth by convention" was, if anything, favorable to the conventionalist case: by construing the stipulations as rules for assigning truth to sentences of the appropriate forms, I had to do some work to develop the regress toward the end of the paper. If, instead, we regard the stipulations as governing the logical particles directly, along the lines you say you prefer, it becomes even more obvious that those stipulations alone do not suffice to generate the infinitely many logical truths, that in fact they will have to be supplemented by rules of logical inference, which was precisely the outcome of my paper.

Carnap**(2): This point is obvious. We can agree that LD in the following form is false:

LD(I)   Stipulations that particles (of a language) function as the logical words (of our special list) suffice to *generate* the class of logical truths (of that language).

Granting this, however, raises a fundamental problem I have found with your argument. If the task confronting the conventionalist were to *justify* logic, the regress you uncovered would indeed be vicious, for logical rules are indeed

---

[5] See the formulation in Quine [1960a], "Carnap and logical truth," at p. 110 and n.2.
[6] Carnap [1963], "W. V. Quine on logical truth", p. 916.   [7] Carnap [1937], sections 5 and 55.
[8] Quine [1960b], Ch. 2.

required to arrive at (infinitely many of) the logical truths from a finite body of stipulations. But the main thrust of conventionalism is that *logic needs no justification*: as Wittgenstein argued, logical truths are empty, they assert nothing. LD(I) is simply irrelevant to the conventionalist purpose.

Indeed the conventionalist must be able to justify, of any logical truth S, *that S is true by linguistic convention*. Given a *characterization* of the class of logical truths, for example, as containing all sentences of certain forms (axiom schemata) and closed under certain rules (such as *modus ponens*), it can be expected that logical rules will be required in demonstrating, of complex logical truths S, that they are in the class. And if the case is made that, indeed, that class consists of stipulative, empty truths, the demonstration will have achieved its aim.

Given a sentence S (which, it is supposed, has truth-value true or false), we may indeed be uncertain of its status as a logical truth on account of its complexity. In a sense, anyone's assertion *that S* can be said to stand in need of justification. But when we succeed in deriving S from logical truths by sound rules, we show that in fact we were mistaken; we show that in fact S requires no justification precisely because it is a logical truth. The claim *that S requires no justification* is what really required justification; and in making it out we had recourse to logical rules. But that is as legitimate in this quarter as in any. What better way could there be to justify anything?

Quine*(2): The distinction you point to is clearly important. But let us see how it bears on the linguistic doctrine. We have agreed to abandon LD(I). What is to go in its place? From your reply,[9] I gather that you prefer a thesis couched in epistemic terms, such as

LD(II) Knowledge of stipulations (as in LD(I)) suffices to *ascertain* of any logical truth that it is true.

Carnap*(3): That does seem to be a fair reconstruction of the third paragraph of my reply. I meant, of course, that, in contrast with empirical statements, no further knowledge of facts about the world is required.

Quine*(3): As you know, I question the coherence of "facts about the world." But LD(II) well reflects the historical thrust of conventionalism, to account for *a priori* knowledge employing the (potentially) scientific terms of linguistics. Apart from my challenge to this, however, the point that needs making right now is that LD(II) is no better off than LD(I). Surely it is not in general sufficient, to ascertain of a logical truth S that it is true, simply to inspect S and "look up" the stipulations governing the logical words contained in S. To ascertain the truth of the full class of logical truths, logical rules over and above

---

[9] Carnap [1963], "W. V. Quine on Logical Truth," p. 916.

the stipulations must be invoked. This was just the predicament encountered in the last pages of "Truth by convention.'. And if in reply you say "It suffices to justify the rules," you confront the familiar regress: to obtain the conclusion of any rule from its premises, a further rule will be needed, and the regress is under way.

Carnap*(4): If I can foresee at all how this dialogue will develop, I will have another opportunity to respond to this last point. First I should treat some still more fundamental matters.

I entirely agree with what you said on the motivation behind conventionalism. But did you not ignore this when you criticized LD as "empty"?[10] There you pointed out that one attempt to assign experimental content to LD, viz., "Deductively irresoluble disagreement as to a logical truth is evidence of deviation in usage (or meanings) of words," also holds up if 'logical' is replaced with 'obvious'. Similarly, you have argued that certain other connections between logic and language could equally well be accounted for by the recognition that logic is obvious, "without help of a linguistic doctrine of logical truth."[11] Now none of this shows that LD is empty; at most it shows that LD is superfluous. But, I claim, it does not show this either. *For we need an account of **why** logic is obvious.*[12] The sort of account we can give of why "It is raining" is obvious does not work for logic. Here is a major explanatory role for the linguistic doctrine, essentially its traditional role in logical epistemology.

But how are we to frame the linguistic doctrine? Let me now propose an alternative to LD(II) which, I believe, improves upon it and earlier formulations. Roughly, instead of claiming that stipulations on logical words suffice for *ascertaining* logical truth, the conventionalist should claim that those stipulations suffice to *determine uniquely* the class of logical truths and their truth. Let us assume that we are working with a first-order language, L, containing a stock of sentence connectives, quantifier phrases, and equivalence relation signs. (These could be interpreted in various ways, not just as the usual logical particles; for example, a connective could be read as 'because', a quantifier as 'many', etc.) Let us further assume that L is sufficiently well regimented so that all other constructions can be worked into the form of lexicon. Next let us call a *logical structure for L* any set D together with a valuation v assigning (n-tuples of) objects in D to the lexicon of L. Now the key idea is this: once it is fixed which particles are functioning as the logical ones (of our list),[13] there is determined a unique class of sentences, the 'logical sentences', such that any

---

[10]   Quine [1963], "Carnap and logical truth," sections 3 and 4.
[11]   Quine [1963], "Carnap and logical truth," p. 389; cf. Quine [1970], Ch. 7.
[12]   The point is hinted at by Strawson [1971] in his review of Quine's *Philosophy of Logic*, see p. 177.
[13]   Attributions of other truth-functional connectives and of existential quantification may be thought of as eliminated in favor of '∼', '∧', and '∀'.

two logical structures for L agree on the truth values of these, no matter how widely they may otherwise diverge. This is the intended force of

LD(III)   Stipulations (as in LD(I)) suffice to determine uniquely a class of sentences (of the language in question), the truth value of each of which is invariant over the totality of logical structures.

Thus, it follows that stipulations on words suffice to determine a class of sentences, logically valid in the usual sense. This is crucial for the convention-alist view of logic. For it is the logical validity of the logical truths that constitutes their "emptiness": no matter what entities there were, and in no matter what relations they stood, these linguistic forms would still come out true. (Thus, it is really the free version of first-order quantifier logic that is relevant, in which, to take account of the null universe, '$\forall x \varphi x \supset \varphi a$' is replaced by '$\forall x \varphi x \wedge \exists y \, (y = a) \supset \varphi a$'. Identity can be retained as 'logical', but we admit further imperfections in the essential-occurrence test for validity, for example, the not-strictly-valid '$\exists x \, (x = x)$'.)

Notice that LD(III) does not readily generalize so as to give content to the claim that 'analytic sentences' are factually empty. For in order to obtain the analytic sentences from the logical truths, one needs to appeal, not merely to co-extensive terms, but to co-intensional terms. Fixing the intensions of lexical items would fix truth values of a class of analytic sentences (over all relevant logical structures), but how do we 'fix intensions'? Furthermore, LD(III) does presuppose that adequate interpretations of the logical particles can be *stipulated*. (So far we have both been taking that for granted, you perhaps more for the sake of argument than I. Of course, it must be defended independently.[14]) But, as you have argued, and I have been slow to realize, the sense in which meanings of words (in general) can be stipulated is far from clear. Can one stipulate "the meaning" of, say, 'positron' so as to support the LD(III) analogue for analytic truths containing it (and terms used to "define" it)? These are odd questions for *me* to be asking, but philosophy in the 1980s isn't what it used to be. (Sigh!)

Quine*(4): I have no quarrel with LD(III) itself. As you have framed it, I suppose that it is a mathematical fact about models, provided one grants the presupposition on stipulations that you mentioned. My main problem is with your interpretive claim about "factual emptiness". As I have written, it does appear natural to regard logical truths

as the limiting case where the dependence (of the sentence) on traits of the subject matter is nil. Consider, however, the logical truth, 'everything is self-identical', or '$(x)(x = x)$'. We can say that it depends for its truth on traits of language (specifically on the usage of '$=$'), and not on traits of its subject matter; but we can also say, alternatively, that it

---

[14] See below, Carnap*(8).

depends on an obvious trait, viz., self-identity, of its subject matter, viz., everything. The tendency of our present reflections is that there is no difference.[15]

As I see it, you are committed to there being a difference: the first description would be accepted but the second would not. And this seems to undermine the view, for what difference can there be?

***Carnap**(5):** I am glad that you have repeated that argument. If only I could have foreseen the effects it would have on our readers, I would have addressed myself to it in my reply. (I have turned over in my grave many times for not having done so.) First of all, I would not use the word 'obvious' at all. (This I did hint at in my reply.) Even if all logical truths are (potentially) obvious, the converse is so obviously false that it can only beg the question to suggest that there is no other way of phrasing the alternative.

Secondly, the alternative way of phrasing the alternative description is to replace the word 'obvious' with the word 'stipulated'. Self-identity is a paradigm example of what I would call a 'stipulated universal trait': we introduce '=' (or can: if anyone insists that the symbol '=' already has another use, we choose a new symbol, '=*') by stipulating that, in any context whatever, including all modal and counterfactual contexts, '=' ('=*') holds just between any object and itself. In the language of models, '=' ('=*') is assigned the set of ordered pairs of the form (x, x) over any domain of objects whatever. Similarly, a trait such as 'being red or not red' is a stipulated (universal) trait: it is determined to hold of every object whatever in any (putative) domain of objects as a consequence of our stipulations on 'or' and 'not'. (If anyone insists that 'or' and/or 'not' already have different uses, we can introduce new symbols, 'or*' and 'not*', stipulating that they behave in accordance with the truth-tables for (classical) disjunction and negation.) All logically valid open schemata give rise to 'logically valid traits' of this sort, and all are determined to hold of any object(s) in any domain by stipulations on the logical words (in accordance with LD(III)).[16]

Thirdly, once the irrelevant term 'obvious' is replaced by 'stipulated', understood as just described, the conventionalist has *no* objection to saying that the logical truths depend for their truth on stipulated traits of objects (and no others), with one important proviso: so long as the word 'depend' is not used to suggest that, holding the logical part of the language fixed, it could have been otherwise. In the idiom of quantified modal logic, the logically valid traits are necessary traits, precisely because of the stipulations on the logical words. Thus, it is quite

[15] Quine [1963], "Carnap and logical truth," p. 390.
[16] Predicates designating stipulated universal traits would be special cases of Carnap's "Allwörter"; see Carnap [1937], section 76, pp. 292 ff. Note, however, two differences: the predicates of interest here are not relative to any category; and they are not introduced by means of any general notion of analyticity, but rather in terms of explicit stipulations governing logical words.

misleading to suggest (as use of 'depend' does to my ear) that we might somehow vary the trait and, after the fashion of Mill, look for a variation in the truth-value of the logical sentence in question. But if this implication of 'depend' is cancelled, the new "alternative description" is no alternative at all, but just a more longwinded version of the original: logical truths depend for their truth (solely) on stipulations governing the use of the logical words.

Thus, your argument that the conventionalist is committed to an unintelligible difference would seem to collapse, once the alternatives are properly phrased.

Quine*(5): Perhaps your notion of a stipulated (universal) trait can be invoked to give content to your claim that logical truths are "factually empty." (In provisionally granting this, I am granting at most a very limited special case of your analyticity doctrine, as you appear to have recognized already.)

If I understand you, then, the notion of word stipulation is basic to logical metatheory. Using it, but not now your general notion of analyticity, you propose a sufficient condition for "factual emptiness": a sentence S is factually empty if the truth-value of S is unaffected by arbitrary counterfactual assumptions concerning what entities there are and arbitrary such assumptions concerning their traits (including relations), save the stipulated traits. (Our stipulations dictate that we discard or declare automatically true any counterfactual with antecedent such as, 'If Caesar were not self-identical. . .,').

I gather from your nods of approval that my paraphrase is fair. To celebrate our progress in communication, let me mention two important related points on which I believe we actually agree. First, it is no contention of mine that logic "has factual content" for presupposing, for example, that it makes sense to speak of entities and of their satisfying predicates. (Notice that I don't say here, "having properties and standing in relations". In first-order logic we never mention properties or relations; rather we use predicates, as I have said many times. In this regard, logic is ontologically neutral.) Such a contention rests on a confusion, that of a presupposition with an assertion. We agree that logic has certain *preconditions of applicability.* One of them is that any language whose reasoning is being symbolized can be represented with our paradigmatic form of predication, '$F(x)$' (or '$F(x_1 \ldots x_n)$') . If a symbol system (language?) cannot be so represented, our logical schematism does not apply. But that schematism makes no commitment whatever that such applicability conditions actually obtain in any case. The very same can be said with regard to the true/false dichotomy. It is a precondition of applicability, not a commitment. No revision of logic is contemplated in pointing to cases (e.g., imperatives, or grunts for that matter) to which it does not apply. And debates over particular subjects, such as the classical versus intuitionist conceptions of mathematics, are not really debates over (classical) logic at all, but over its applicability to a certain realm of discourse.[17] Logic does not proclaim its own applicability.

Second, even if I should grant the stipulative character of logic, I should not regard logical reasoning as thereby devoid of substantive purpose. For example, the demand for sound rules reflects the purpose: never to pass from truth to falsity, come what may. We agree: *Nothing in the linguistic doctrine diminishes in any way the importance or substantive character of such **purposes** of logic.* Questions concerning them are completely distinct from the question of the empty, assertionless character of logical truths themselves.

But now let us take stock of what has been accomplished. You have agreed to scrap both LD(I) and LD(II). In its place you have proposed a weaker "determination principle," LD(III). Two major problems remain. The first concerns the assumption, thus far unchallenged, that stipulations on the logical particles of the sort you require are really possible. The second concerns the whole epistemological thrust of the linguistic doctrine. Let us take this latter problem first.

The problem is very simple. Even if LD(III), together with your interpretive claim on factual emptiness, can be sustained, how can it be used to solve the problem of logical knowledge? Potentially at least, we know an infinite set of logical truths. Even granting your point that, since empty, they "require no justification," still – as you yourself put it at the outset – infinitely many statements of the form 'S requires no justification' do require justification. But this involves appeal to rules, and these go beyond the word *stipulations*. In the dissolution of one epistemological problem, another has taken its place. The linguistic doctrine LD(III) would seem too weak to solve it. My original regress argument still has its force.

***Carnap**(6):** I said above that I believe this regress argument can be answered. As I see it, the "rules regress" arises due to a far too restrictive view of what it is to justify a (logical) rule. We can agree that what needs justification is that a given rule, let us say *modus ponens* (MP), is sound, i.e., truth-preserving.[18] Now the assertion of soundness *can be given* a conditional form: if the premises (say, A and

---

[17] In Quine [1970], p. 84, I put matters somewhat differently: I described the dissident logician as "rejecting the meaningfulness of classical negation." But this raises the puzzle: if he understands what he is rejecting, how can he reject its meaningfulness? Mere incomprehension on someone's part is of no logical interest. Perhaps the dissident is best viewed as rejecting the classicist's way of representing certain realms of discourse (for bad or better reasons, see pp. 85 ff.). This comes down to questioning a precondition of applicability.

[18] Note that soundness and completeness enter at different levels. It is soundness that is relevant at the "ground level" of justifying particular claims of the form,

S requires no justification,

for, typically, logical rules are used to derive S from axioms whose status (requiring no justification in virtue of emptiness) is evident. In establishing soundness of the rules, we establish in effect that *emptiness is preserved* in their application, and that is all that is required in any particular case. Completeness, on the other hand, enters at a higher level, when we ask whether the method of justification for the particular cases will always work.

A $\supset$ B) are true, the conclusion (B) is (or must be) true. Then it is supposed that, to establish this conditional, one *must proceed* in the manner of natural deduction: assume the antecedent ('A and A $\supset$ B are true') and derive the conclusion. And, of course, such a derivation requires the application of a rule, in this case (the metalanguage version of) MP itself.

But is this the *only* way in which one could come to know that MP is a sound rule? That seems to me absurd. And, before presenting any alternative story, one can see that it *is* absurd, for it would mean that *we really don't (and can't!) know* that MP is a sound rule. I submit that any epistemology committed to this is bad epistemology, for we certainly do know that MP is sound.[19]

> It is important to recognize that establishing completeness is not required as part of the ground level justification procedure, because, in fact, proving completeness requires some relatively sophisticated mathematics whose "definitional" or "conventional" status is far from clear and would be much more difficult to secure than what is involved in soundness. But, because completeness enters only at the "metalevel," it poses no further problem for the conventionalist view of logic. (It may pose a special problem for a conventionalist view of the conventionalist view of logic. In a more traditional formulation, the question would be whether 'logic is analytic' is analytic. This would seem to turn on the status of the mathematics involved in LD(III), a problem that cannot be treated here. What is to be emphasized is that it is a separate problem about which Carnap* and Carnap need not agree.)
>
> It should also be noted that arriving at the soundness of each rule (MP and UG, for simplicity) is, in a significant way, more elementary than arriving at the soundness of the system as a whole. The justification that the logical conventionalist is called upon to supply is, as it were, of the logical truths one by one. That is, Carnap* must be able to show of each logical truth S that it can be known solely from linguistic stipulations. This is weaker than showing that knowledge of system-soundness is a matter of linguistic convention. The difference may be brought out symbolically: writing '$\Diamond$Kn' for 'can be known', the Carnap* position requires an account of
>
> (i) $\forall S: \Diamond \ Kn \ulcorner \vdash S \Rightarrow \vDash S\urcorner$ (S ranges over logical truths),
>
> whereas a stronger position requires accounting for our knowledge of system-soundness, i.e., an account of
>
> (ii) $\Diamond \ Kn \ulcorner \forall S: \vdash S \Rightarrow \vDash S \urcorner$.
>
> Now it is true that (ii) will require appeal to mathematical induction (but see note 22, below), but (i) does not.

[19] Let 'MP_n' denote *modus ponens* for the nth level of the usual Tarski hierarchy of languages, taking L_0 to be a formal language for first-order quantifier logic (as in Church [1956], see note 21) introduced by logicians employing L_1 as their natural language (e.g., Polish, English, etc). Carnap*'s remark here should be understood as saying that it is absurd to think that the only way of acquiring knowledge of the soundness of MP_n is to have acquired knowledge of the soundness of MP_{n+1}. At some stage it must be possible to come to know the soundness of MP_j without invoking a higher level MP. Below, Carnap* sketches a route for the soundness of MP_0 that does not even appear to invoke a higher level MP. Alternatively, Carnap* might argue that, in learning 'if ... then' of L_1, we in fact do learn the soundness of MP_1. As Wittgenstein might have put it, learning that such formal manipulations are correct (preserve acceptability – later truth – of the premises) is inseparable from mastering the 'if ... then' idiom. In this case, MP_1 can be invoked in the course of arriving at the soundness of MP_0 without the threat of regress. A similar line of argument can be developed for the few L_1 logical manipulations (involving 'not', 'and', 'there is') that might – as a matter of psychological fact – be inescapable in the hypothetical process Carnap* proceeds to sketch.

In fact, we are not constrained to the mold (or rut) of a tortoise. We don't *have to* assert soundness in the form of a conditional; nor do we *have to* carry out a logical deduction to see that MP is sound. First, it must be emphasized that the conventionalist is content to confine his claims to idealized, formal languages. It is in these that logical truths may be rigorously expressed, and it is claimed that, so expressed, they are stipulative truths that can be known to be such by rules whose soundness can also be known, given the stipulations. Thus, by 'MP', I am referring to a particular rule governing the truth-functional conditional, which we think of as being introduced into our language. Whether particular uses of 'if . . . then' in English are properly represented by this formal operation is an entirely distinct question. And the details of an alternative justification of soundness of analogues of MP for other conditionals depend on the details of their use or introduction.

This understood, we may assert the soundness of MP in non-conditional form, as follows:

(i)   There is no assignment of truth-values (T, F) that makes both A and A $\supset$ B T and B F.[20]

How can we come to know this? First by coming to know the elementary combinatorial fact that TT, TF, FT, FF are all the possible assignments of T, F to two elements, A, B. Here a story must be told, of course, that does not involve, for instance, deducing this fact in the system of *Principia Mathematica*. (Again, such a story had better be a real possibility, or else it will turn out that we *really don't know* this (or virtually any other!) mathematical fact either.) Second, by consulting our own stipulations on '$\supset$', viz., the truth-table we used in introducing it into our language. By examining each row, we see: (a) it is correctly written (the ink has the right shape, and is sufficiently stable over the time-period of the examination – or else, of course, we can do this in our imaginations as you, the reader, are doing right now), and (b) the only row assigning A T and A$\supset$B T contains T for B. Thus, all that is involved is an elementary finite search; and this need not be construed as a chain of reasoning in a logistic system.

---

[20] Formulation (i) is just a special case of a general metalinguistic stipulation on use of the terms 'sound' and 'rule of inference':

(ii) A rule of inference with premises $P_1 \ldots P_n$ and conclusion C is sound *iff* there is no valuation making (the universal closure of) $P_1 \ldots P_n$ true and (the universal closure of) C false.

The parenthetical condition is needed for the rule of universal generalization which, together with *modus ponens*, suffices to generate all first-order logically valid formulas (from standard axiom schemata). Working with this framework is particularly convenient because it reduces the problem of justification of logical rules to a very simple form.

Since (ii) *is* just a stipulation on some metalinguistic technical vocabulary, no special inference licensing passing from the right side to the left is involved: establishing the right side for a given rule *is* establishing the soundness of the rule. One possible route to regress is hereby blocked.

Quine*(6): I have some misgivings about your strategy concerning the alleged status of (i) as a stipulation introducing a logical particle. But let us see now how you would treat the quantifier rules, which transcend anything so elementary as truth-tables. We can continue working with an axiomatic system of first-order logic with MP and UG as the only rules.[21]

Carnap*(7): The case of UG is quite straightforward. What must be shown is that on the basis of linguistic stipulations, we can know that UG is a sound rule. Since free variables are involved, soundness of UG takes the form:

(iii)   Every valuation (over any (non-empty) domain D) making the universal closure of A true makes the universal closure of $\forall x$A true.

But what problem can there be in knowing this, since the universal closure of A and that of $\forall x$A are the same, up to some permutation of initial universal quantifiers! I suppose it is just here that one must appeal to stipulations governing the interpretation of '$\forall$': implicit in the soundness of UG is the soundness of permuting initial universal quantifiers, and knowing this does rest on conventions such as

(iv)   '$\forall x\varphi(x)$' holds in a domain D (under valuation v) just in case '$\varphi(\bar{d})$' holds in D (under v) for each object d in D as value of (the dummy name) $\bar{d}$.

Given (iv), it is trivial that, for example, '$\forall x\forall y\psi(x, y)$' and '$\forall y\forall x\psi(x, y)$' are logically equivalent, for we have, for any structure (D, v):

(v)   (D, v) $\vDash \forall x\forall y\psi(x, y)$ iff (D, v) $\vDash \forall y\psi(\bar{d}, y)$ for all values d of $\bar{d}$ in D iff
(D, v) $\vDash \psi$ ($\bar{d}, \bar{d}'$) for all values d of $\bar{d}$ in D and all values d of $\bar{d}'$ in D.

But also

(vi)   (D, v) $\vDash \forall y\forall x\psi(x, y)$ iff (D, v) $\vDash \forall x\psi(x, \bar{d}\,')$ for all values d of $\bar{d}\,'$ in D iff
(D, v) $\vDash \psi$ ($\bar{d},\bar{d}'$) for all values d of $\bar{d}\,'$ in D and all values d of $\bar{d}$ in D.

which is identical to (v) given commutativity of 'and', which is also true by stipulation. More generally, one could directly stipulate satisfaction conditions for formulas with arbitrary (finite) strings of initial universal quantifiers:

(vii)   (D, v) $\vDash \forall x_1 \ldots \forall x_n\psi(x_1 \ldots x_n)$ iff (D, v) $\vDash \psi$ ($\bar{d}_1 \ldots \bar{d}_n$) for all n-sequences (with repetitions) from D as values of $\bar{d}_1 \ldots \bar{d}_n$.

---

[21] See, e.g., Church [1956], section 30, p. 172. Here UG can be stated without restrictions: from any wff A to infer $\forall x$A, where x is an individual variable. (This rule, of course, applies only to theorems, not to premises.)

This guarantees arbitrary permuting of initial universal quantifiers since we understand that the order of dummy names doesn't matter: the same set of n-sequences (all of them) is covered in all cases.

In the present context, (vii) is to be preferred as a way of guaranteeing the permutation rule. The more usual route, via (iv), requires a proof by mathematical induction in order to obtain the general result for arbitrary strings of quantifiers. In logic texts that is perfectly in order, but here we wish to rely as little as possible on mathematical reasoning and as much as possible on direct stipulation.[22]

Quine**(7): I am glad to find you adhering to the same strategy in the case of quantifiers that you employed for sentential connectives: the (admittedly clever) strategy of working both a formulation of soundness of a logical rule and a stipulation on a logical particle into one and the same statement (nearly enough). I see no reason to question these formulations in their former capacity. It is the latter that I question. For, as comes through blatantly in (iv)–(vii), satisfaction-conditions for sentences of the form '$\forall x \varphi$' are given by means of 'all'. As I have written:

... one is perhaps tempted to see the satisfaction conditions as explaining negation, conjunction, and .... quantification. However, this view is untenable; it involves a vicious circle. The given satisfaction conditions ... presuppose an understanding of the very signs they would explain, or of others to the same effect.[23]

In short, (iv)–(vii) all presuppose familiarity with the universal quantifier (in one form or another) and are useless as means of introducing it. Similarly, (i) presupposes familiarity with 'and' and 'not' (i.e., with '⊃' in one form or another), and is useless as a means of introducing (first) truth-functional connectives.

Carnap**(8): Does this hark back to the difficulty of "self-presupposition of primitives" that you raised toward the end of "Truth by Convention"?[24] That has long puzzled me, for it seems to conflate two very distinct issues. First, there is the issue of *language learning*, the question of how we can or do acquire our first understanding of logical words. Here you quite rightly point out: not by explicit linguistic stipulations. But second, there is the issue of *language-based truth*, the question of the linguistic character of the satisfaction

---

[22] Even if, at some stage, it had proved necessary to invoke mathematical induction, that by itself would not have constituted a circularity or regress. We do, in fact, know that induction holds of the natural numbers. On my (Carnap*'s) view, this is guaranteed by what we mean by 'the natural numbers'. The fact that logic can be used to make this *rigorous* (as Russell showed) must not obscure the fact that we can, prior to formalization, intend 'the natural numbers' to be understood so as to insure induction. (Author's note: Although this way of phrasing the point, with its soft-pedalling of formalization, might not have appealed to Carnap, its essence emerges quite clearly in Carnap [1937], section 34h, pp. 121–124, where mathematical induction is used in the metalanguage to establish the "analyticity" of object-language induction, and the charge of circularity is explicitly rebutted (see p. 124). Cf. the last paragraph of Carnap*(2), above.)

[23] Quine [1970], p. 21.   [24] Quine [1936], "Truth by convention," p. 104.

principles. Here what is at stake is whether knowledge of language suffices for knowledge of the soundness of logical rules. (Since the rules govern the logician's regimented language, that is the one that is relevant, of course.) It is quite irrelevant that the satisfaction conditions cannot serve the purposes of first language acquisition.

I have already argued for the connection between stipulation of satisfaction conditions (such as (i), (iv), and (vii)) and knowability of the soundness of logical rules. All that remains to be shown is that those conditions really are or can be entirely stipulative. But this is clear, since we *can* use those conditions to introduce new expressions, such as '⊃', '∀', etc., in whose terms logical sentences are to be framed. If these happen to coincide with simple expressions already in use, so much the better; but, as already noted, that is really inessential to the conventionalist case. All that is required is that *some* means be available for expressing the desired stipulations.

Quine*(8): Your new way with the linguistic doctrine, if I understand you, involves detaching it from any general analytic/synthetic dichotomy. Above, I was careful to attribute to you only a sufficient condition for "factual emptiness," one which did not rely on "meaning" or "confirmation." But now I find you invoking the distinction between knowledge of language and knowledge transcending language. But this just is the analytic/synthetic distinction in another form.

Carnap*(9): I have not invoked the distinction in anything like full generality. All that need be granted is the stipulative character of setting up "logical grammar" as you yourself have sketched it.[25]

Thus, I think we do have this general point: *one need not defend a general analytic/synthetic distinction to make the conventionalist case for logic.* One could even grant your point that any line drawn by linguistic science between knowledge of language and knowledge of extra-linguistic fact will be arbitrary in non-trivial ways. But, just as Putnam has detached from this the claim that there are some clear cases of linguistic equivalence,[26] I would insist that there are certain clear cases of linguistic stipulation, constraining any future linguistics. If the introduction of new particles by explicit rules of the sort presented above is not stipulative, what is?

Of course, metalinguistic resources for the stipulations must be available. On this score, I have been presupposing throughout that both you and I are conversant in English, and that the English expressions 'not', 'and', 'all' (or 'each', or 'every', or 'not any ... not ...', etc.) are available for the task, perhaps

---

[25] Quine [1970], pp. 22–25.
[26] Putnam [1962].

supplemented with diagrammatic devices and other "scholia as needed for precision". Do you dispute that?

Quine*(9): Much of my own logical writing seems rather to presuppose it![27] However, in a way I suppose I have disputed it in the case of quantifiers. One may read my ruminations on ontological relativity thus, as casting doubt on anyone's capacity to distinguish certain allegedly deviant interpretations of quantification over infinite sets (corresponding, e.g., to certain of their infinite subsets).[28] Just as I have found unrelativized reference behaviorally inscrutable, I would have to say the same for quantification.

Carnap*(10): This is not the place to descant upon my perplexity over inscrutability of reference.[29] I could cite some recent work that, I believe, tends to undermine its credibility,[30] but let me make these few remarks.

First, even if quantifiers are inscrutable, how does this affect the linguistic doctrine? As I see it, not at all! Its impact is simply that there is systematic ambiguity in the introduction principles (such as (iv) and (vii)). If, for example, D is the power set of the natural numbers, these stipulations would not distinguish among interpretations of $\forall$ that we would (naively?) describe as "the standard interpretation, in which '$\forall$' ranges over all subsets" and "deviant interpretations, in which '$\forall$' ranges over all subsets of a proper elementary submodel guaranteed to exist by the downward Skolem–Löwenheim theorem." But *in any case, the soundness of UG is insured* (even) *by the* (ambiguous) *stipulations, and the rules regress is still blocked.* So long as we don't shift from one interpretation to another in midstream (in the space of a given deduction), nothing can go wrong, since by hypothesis we are dealing with *submodels* of any relevant background theory. Any such shifting that would affect soundness could, I submit, be detected and ruled out in our learning of the quantifier idioms.

---

[27] For example, I have written:

> ... this logical language [of the formal logician] ... has its roots in ordinary language, and these roots are not to be severed ... It is enough that we show how to reduce the logical notations to a few primitive notations (say '~', '.', '∈', and universal quantification) and then explain just these in ordinary language, availing ourselves of ample paraphrases and scholia as needed for precision.

> Quine [1953], "Mr. Strawson on logical theory," p. 150.

[28] Quine [1969], "Ontological relativity," pp. 54 ff.

[29] When I think I understand it, I cannot see how it differs from my own rejection of external ontological questions. But this leaves me thoroughly confused, for (a) you have explicitly rejected the internal/external distinction on which my rejection is based (Quine [1969], pp. 52–53), and (b) as a realist about physics (in a sense which I claim not to grasp), you would seem committed to at least the intelligibility of an "absolutist" ontological position, would you not?

[30] For example, Friedman [1975] and Hellman [1974].

Second, in your own writings, you have consistently dealt with a similar problem concerning *identity* in a way with which I am entirely sympathetic. Problems of individuation of complex objects, such as persons or mountains, are, you have maintained, misconceived as problems of identity; rather they reveal some difficulty concerning the application of the predicates ('person', 'mountain', etc.). Shouldn't any reason you have for this (very sensible) position carry over *mutatis mutandis* to quantifiers? Should we not say: "'All' is the clear notion and retains its clear truth conditions from context to context. Any problem raised by the Skolem–Löwenheim theorem (if one is raised) concerns the particular predicate whose extension is (taken to be) uncountably infinite, for example, 'subset of integers'"?

Finally, it is perhaps an interesting question of psycholinguistics, how in detail we gain our mastery over quantifiers in discourse about the infinite. But is the problem more difficult in principle than that of acquiring dispositions to project an ordinary predicate beyond a finite sample to potentially infinitely many new cases? On the other hand, according to the inscrutability doctrine, a legitimate question has not even been posed since, on that doctrine, it is nonsense to say that we really do learn the standard interpretation of the universal quantifier.

As argued, the linguistic doctrine is in any case safe from inscrutability. As a challenge to linguistics, I find the latter intriguing; but taken literally, it seems to be a form of skepticism undermining the very intelligibility of that challenge. How would you prefer it to be understood?

Quine*(10): We have been at this long enough. Perhaps we can ask Professor Quine himself. I think he has been eavesdropping on us for some time now . . .

Acknowledgement: I owe special thanks to my colleague J. Alberto Coffa for many helpful suggestions and criticisms and for years of patient prodding. For distortions, felicitous and otherwise, in the positions of Carnap and Quine portrayed in the dialogue, I assume full responsibility.

References

Carnap, R. [1937] *The Logical Syntax of Language* (London: Routledge & Kegan Paul).
Carnap, R. [1963] "W. V. Quine on logical truth," in Schilpp, P. A., ed., *The Philosophy of Rudolf Carnap* (La Salle IL: Open Court), pp. 915–922.
Church, A. [1956] *Introduction to Mathematical Logic I* (Princeton, NJ: Princeton University Press).
Friedman, M. [1975] "Physicalism and the indeterminacy of translation," *Noûs* **9**(4): 353–374.
Hellman, G. [1974] "The new riddle of radical translation," *Philosophy of Science* **41** (3): 227–246.

Putnam, H. [1962] "The analytic and the synthetic", in *Mathematics, Matter, and Method: Philosophical Papers*, Volume 2 (Cambridge: Cambridge University Press, 1979), pp. 33–69.

Quine, W. V. [1936] "Truth by convention," reprinted in *The Ways of Paradox and Other Essays*, revised edn. (Cambridge, MA: Harvard University Press, 1976), pp. 77–106.

Quine, W. V. [1953] "Mr. Strawson on logical theory," reprinted in *The Ways of Paradox and Other Essays*, revised edn. (Cambridge, MA: Harvard University Press, 1976), pp. 137–157.

Quine, W. V. [1960a] "Carnap and logical truth," reprinted in *The Ways of Paradox and Other Essays*, revised edn. (Cambridge, MA: Harvard University Press, 1976), pp. 107–132.

Quine, W. V. [1960b] *Word and Object* (Cambridge, MA: MIT Press).

Quine, W. V. [1963] "Carnap and logical truth," in Schilpp, P. A., ed., *The Philosophy of Rudolf Carnap* (La Salle IL: Open Court), pp. 385–406.

Quine, W. V. [1969] "Ontological relativity," in *Ontological Relativity and Other Essays* (New York: Columbia University Press), pp. 26–69.

Quine, W. V. [1970] *Philosophy of Logic* (Englewood Cliffs, NJ: Prentice-Hall).

Strawson, P. F. [1971] "Book review: *Philosophy of Logic* by W. V. Quine," *Journal of Philosophy* **68**(6): 174–178.

12 Never Say "Never"!
On the Communication Problem between Intuitionism and Classicism

12.1 Introduction

It is commonplace that opposing philosophical schools – whatever their subject – have difficulty communicating with one another, and, indeed frequently talk past one another rather than engage in rational debate. At times, this seems to have been Brouwer's own view of the relation between intuitionistic and classical approaches to foundations of mathematics.[1] In such circumstances, it is frequently obscure just wherein genuine disagreement – as opposed to merely apparent or verbal disagreement – actually resides, or even whether there really is at bottom any genuine disagreement at all.

Thus, for example, we all learn in our introductory courses in philosophy of mathematics that "intuitionism rejects the law of the excluded middle and proofs of existence by *reductio ad absurdum*, etc." But later we learn that in fact the intuitionistic "logical connectives" – corresponding in some sense to expressions such as 'or', 'if, then', 'not', 'for every', 'there exist' – are understood very differently from the classical logical connectives (or perhaps we should say that the intuitionistic understanding of these words differs radically from the classical understanding of those very words). We may also learn that even the component mathematical sentences themselves have specifically intuitionistic interpretations. And then it appears that there may be no conflict on the level of logic at all: *everyone* can see shortcomings with, say, p or$_i$ not$_i$-p (where the subscript i indicates an intuitionistic reading of the connectives (and the component sentences)), and they are irrelevant to the classical law, p or$_c$ not$_c$-p (where the subscript c indicates a classical reading). What began as

---

[1] Brouwer ended his inaugural address of 1912, "Intuitionism and formalism," by quoting Poincaré:

Les hommes ne s'entendent pas, parce qu'ils ne parlent pas la même langue et qu'il y a des langues qui ne s'apprennent pas.

Brouwer [1975], pp. 123–138.

192 Logics of Mathematics

a disagreement over basic logical principles seems to dissolve into a disagreement over what language to speak. And here, one might hope that a Carnapian principle of tolerance could be invoked to keep the peace: why should *I* object, after all, if you say that *you* do not understand *my* language and wish to speak your own – so long, that is, as you allow *me* the same privilege and refrain from invidious (audible) asides to the effect that *I* really do not understand what *I* am talking about? Of course, even without the audible asides (not to mention the frontal assaults), it is naturally viewed as something of a scandal that such a situation should develop in connection with the language of *number theory* and *analysis*, the hard core of mathematics, the *lingua franca* of all the exact sciences. If *this* language is not a model of clarity and precision, open to the understanding and employment of any rational intelligence, what is? And so we are motivated to probe more deeply.

The literature, in fact, reveals a wide range of views on the relationship between intuitionistic and classical *mathematics* (as contrasted with the associated philosophies of mathematics). At one extreme, Dummett cites specific intuitionistic analytic theorems as "refuting certain classically valid logical laws," and as showing that "the Contintuity Principle is actually inconsistent with classical logic."[2] Accordingly (taking the author at his word, principles of charity notwithstanding), intuitionistic mathematics is, as a whole, incompatible with classical mathematics. On the other hand, perhaps this should not be taken seriously, as it coheres ill with Dummett's more central (but still extremist) view, namely that classical infinitistic mathematics and its logic, based on the notion of "truth-in-a-structure," are "unintelligible" – that only a logic and mathematics based on "verification-conditions" or "proof-conditions," such as intuitionistic logic and mathematics, are capable of being grasped and communicated by human language users. (I say "coheres ill" for how do you refute meaningless noise?) (Thus, Dummett continues Brouwer's extremist, revisionist stance *vis-à-vis* classical mathematics, but bases it, not on a mystique of the creative subject, but on a verificationist approach to public language.) On the other side of the spectrum. Tait argues that, insofar as intuitionism makes sense, even internally, it must be construed as part of classical mathematics, hence surely compatible therewith.[3] (And, of course, classical mathematics is "intelligible" and hence something with which it is possible for a constructive mathematics to be compatible.) Perhaps somewhere in between is the view of Troelstra that intuitionistic mathematics "describes a part of mathematical experience."[4] I say "perhaps," because this is really quite ambiguous: if all mathematics is somehow understood as "describing mathematical experience" – perhaps on the model of phenomenalist reduction

---

[2] Dummett [1977], pp. 84–85.   [3] Tait [1983].
[4] Troelstra [1977], p. 1 (where the view is also attributed to Kreisel).

programs which take all intelligible descriptions to be "of experience" – then the position endorses one of Tait's conclusions, in that both constructive and classical mathematics are at least on the same footing, the former a part of the latter. (On the other hand, the common footing for Tait is anything but phenomenalist: it is the ordinary classical framework of functions and types.) But if we resist this reconstruing of all mathematics (leaving room for a mathematical subject matter distinct from our experience of it), then we indeed seem to have a distinctive view here, one according to which intuitionistic and classical mathematics would not be immediately comparable: intuitionistic mathematics describes the "rational potential" ("capacities to construct proofs") of "an idealized human mathematician," whereas classical mathematics retains its "objective" content.

Given this wide array of positions, it is comforting to note that there has been significant progress of late, continuing early efforts of Gödel, in the enterprise of giving explicit interpretations of intuitionistic logic and mathematics within classical logico-mathematical frameworks, suitably expanded with new primitives – modal operators in some cases,[5] new predicates in others.[6] Such efforts help realize more precisely and explicitly something that has been implicitly known ever since successful semantical interpretations of intuitionistic systems were devised using a classical metalanguage,[7] namely that from within a classical framework, there is, in a sense, no communication problem at all: intuitionistic mathematics (at least number theory, and probably the theory of choice sequences and extensions as well, although this is less settled, I believe) can be faithfully translated into expanded classical languages, so that not only is it clear that no formal incompatibility arises, but insight into the significance of intuitionistic theorems is also afforded the classicist.[8]

Here I will first call attention to some consequences (which I take to be fairly obvious), drawn from standard explanations of the intuitionistic logical apparatus, concerning the compatibility of intuitionistic and classical mathematics. This will lend support to the program of classical interpretation. Next I wish to approach the communication problem from the other end, as it were, and to ask how much of the classical framework intuitionism can, as it is actually practiced and preached, reasonably be taken to understand? A signficant step along these lines has been taken by Tait, who has argued persuasively that, in practice, intuitionistic mathematics should be understood as working with extensionally identified functions, just as in classical mathematics, the difference being merely in the "principles of construction" recognized as giving rise to such

---

[5] Gödel [1932]. More recent is Stewart Shapiro [1985]. See also other contributions to this volume and references cited therein.

[6] Lifschitz [1985].   [7] Such as the semantics given in terms of Kripke trees, in Kripke [1965].

[8] For example, the connections drawn with Kleene's realizability in Lifschitz [1985] are illuminating.

objects.[9] What I shall argue is that, similarly, a coherent understanding of certain aspects of intuitionistic practice requires that some non-intuitionistic sense be made of infinitistic quantification that is suspiciously like that used in classical mathematics. This suggests that intuitionism is less self-contained than one might have supposed, which in turn has implications for its proper overall assessment.

12.2   The (Intuitionistic?) Explanations of the Intuitionistic Logical Primitives

Fundamental to any comparision of intuitionistic and classical mathematics is an understanding of intuitionistic language. (I base this summary on the efforts of intuitionistics and their sympathizers to explain themselves to the wider scientific community.)[10] Here two interrelated components are crucial: (1) the meaning of a mathematical assertion; and (2) the meanings of the logical constants, which we shall denote $\&_i$, v_i, $\rightarrow_i$, $\exists_i$, and $\forall_i$. (Negation, $\neg_i$, is defined via $\rightarrow_i 0 = 1$.) Concerning (1), meaning is understood as given by (warranted) assertibility conditions, as contrasted with classical truth conditions. One (an idealized human mathematician) is warranted in asserting that p just in case one "can prove" p, where this in turn is understood in terms of having suitable mental constructions or methods for carrying them out. (The precise nature of such constructions and methods remains somewhat moot: it depends on the particular statement p (which may ostensibly be "about" natural numbers, choice sequences of rationals, species, spreads, etc.); moreover, the domain of constructions is thought of as open ended, potentially infinite (this is very important), and not identifiable in advance with the proofs or operations of any formal system. One may read the explanations of the logical constants as in part informing us as to what counts as an "intuitionistically acceptable constructive proof.")

Concerning (2), the explanations which have become standard in the literature (referred to as the BHK explanations, after Brouwer, Heyting, and Kolmogorov) are essentially as follows (in which "proof" always means "intuitionistically acceptable constructive proof").

> A proof of $p \&_i q$ is a pair (c_1, c_2) of constructions such that c_1 proves p and c_2 proves q.
>
> A proof of $p \, v_i \, q$ is a pair (c_1, c_2) such that c_1 proves p or c_2 proves q.
>
> A proof of $p \rightarrow_i q$ is a constructive operation c on constructions such that it can be recognized of c that, when applied to any proof c' of p, it yields a proof $c(c')$ of q.

---

[9] Tait [1983].
[10] Heyting [1966], Kreisel [1962], and Dummett [1977], pp. 12–26, 64–65, *passim*.

A proof of $\exists_i xA$ is a construction c such that c proves $A(\underline{o})$, where $\underline{o}$ denotes some constructed object o from the domain of quantification. (This is relatively unproblematic in the case of the natural numbers, but requires further explanation in the case of "infinitistic" objects, such as real numbers.)

A proof of $\forall_i xA$ is a constructive operation c on constructions of which it can be recognized that, when applied to any construction c' of an object o of the domain of quantification, it yields a proof $c(c')$ of $A(\underline{o})$.

From these clauses, one can gather what the assertibility content of a complex sentence is supposed to be: it is a claim to have a proof of the sentence, or at least to have a method of finding such a proof. Thus, the content of an intuitionistic disjunction is to be contrasted sharply with the truth-conditional content of a classical disjunction: assertion of the former amounts to a claim to have a proof of at least one of the disjuncts. (Notice that the explanation employs the word 'or' or an equivalent. This will be discussed further below.) The contrasts between $\rightarrow_i$ assertions and $\exists_i$ assertions and their classical counterparts are also readily apparent. (Again, notice that the explanation of the latter employs the word 'some', or an equivalent.) The former deserves emphasis in connection with intuitionistic negation: to claim $\neg_i p$ is to claim to have a method of transforming any (putative) proof of p into a proof of an absurdity (say, $0 = 1$). (We may abbreviate this by saying it is a claim to have a refutation of p.) Thus, a strong existential component (with respect to methods or constructions) is involved in the assertion of any intuitionistic negation.

Equally striking – but insufficiently emphasized – is the contrast between $\forall_i$ assertions and their classical counterparts. The former always involve a *strong existential claim*, namely that one is in possession of a method of proving each instance of the statement in question. A fully explicit theory, one would think, would employ a constructive existential quantifier in connection with every universally quantified statement! (Yet again, notice that the explanation of $\forall_i$ employs the word 'any' or an equivalent.)

Now I take it as uncontroversial that, in light of these explanations, the intuitionistic logical constants have radically different interpretations from their classical counterparts, and that the subscripting already employed is needed if we are to avoid elementary fallacies generated from ambiguities. This, obviously, is not to say that such fallacies are inevitable in any context. They are not, for there are many formal analogues between intuitionistically acceptable and classically correct inference patterns (for example, from p & q to infer p, from $p \rightarrow q$ and p to infer q, etc., in which we have purposely deleted subscripts, as the inferences are sound under both readings). However, in many crucial cases, subscripting is required, for example, in "the law of the excluded middle," as noted at the outset.

Moreover, the subscripting should carry over to all logical and mathematical laws themselves. It is highly misleading to speak of "the law of the excluded middle": we have not one, but two "laws," depending on which connectives are employed. And the fact that the component sentences generally have distinct meanings for the intuitionist and the classicist – and the fact that the above explanations of the connectives are intimately bound up (perhaps inseparably so) with the distinctive intuitionistic assertibility content of any mathematical statement – these facts only enhance the necessity of distinguishing formally analogous "laws." As already indicated, for example, *everyone* can see that the intuitionistic "law of the excluded middle" ((LEM)$_i$) is not generally correct for arbitrary propositions p. If p is a problematic statement (say, the Goldbach conjecture) for which we possess neither a constructive proof nor a constructive refutation (nor any method of finding either), then *no one* would want to assert *this* instance of (LEM)$_i$. And, of course, this is entirely irrelevant to the corresponding instance of (LEM)$_c$, which makes no claim whatsoever about anyone having any constructive proofs. Similar remarks can be made about every logical principle, and, indeed, about every mathematical statement. (This, of course, is not to deny that there is a domain of statements – the "decidable" ones – on which no difference emerges between intuitionistic and classical reasoning.) Thus *everyone* can see that you cannot in general *intuitionistically* establish an $\exists_i$ statement (i.e., provide a method for finding a witness) by reducing its negation (either intuitionistic or classical!) to absurdity. (Note that while a classical proof of an $\exists_i$ statement, reducing its classical negation to absurdity, is perfectly cogent (to the classicist) and does establish its truth (that an idealized human mathematician in fact has a method of finding a witness), it may not actually *provide* such a method.) Again, this is irrelevant to the classical inference pattern, which only *classically* establishes (i.e., as true) statements by (possibly non-constructively) reducing classical negations to absurdity.

So intuitionism rejects (LEM)$_i$, but so does classicism. Should we say, however, that intuitionism rejects (LEM)$_c$? In a sense, perhaps, but that can mislead: for intuitionism (in its extreme versions, Brouwerian, Dummettian, etc.) claims not to make sense of (LEM)$_c$, since it makes no sense of the classical notion of "truth-in-a-structure" on which (LEM)$_c$ is based. But in *this* sense, we should equally say that intuitionism (in these extreme versions) rejects *all* classical laws. Better to say it fails to comprehend, or finds unintelligible, classical language *tout court*. (Again, this is not to deny that intuitionism can faithfully interpret in its own peculiar terms a significant portion of classical logic and mathematics. So still better to say intuitionism fails to comprehend, or finds unintelligible, classical language applied to infinite domains, or undecidable cases, or some refinement along these lines.)

To see the importance of these distinctions in mathematical contexts, consider "the Axiom of Choice". It is frequently said that constructivists reject, or are at least skeptical of, "this principle." But, in certain elementary cases, as is well known, if we write it out using intuitionistic quantifiers, and read the whole statement intuitionistically, it is seen to be obviously true! That is (to take the simplest case, in which first-order quantification is over natural numbers and φ is first order),

$$\forall_i x \exists_i y \varphi(x, y) \rightarrow_i \exists_i f \; \forall_i x \varphi(x, f(x)) \qquad (\text{AC})_i$$

clearly holds. For recall what was said about the strong existential content of $\forall_i$ statements: in the antecedent of $(\text{AC})_i$ we have commitment to a method of finding, for each x, a y such that $\varphi(x,y)$; but having such a method is just what is understood by having (or being able to find) a constructive choice function, as claimed in the consequent. (So even the strong requirement imposed by $\rightarrow_i$ is satisfied in this case.) And you do not have to be an intuitionist to see and accept this. (NB: Certain generalizations, e.g., in which $\forall_i x$ ranges over reals, fail due to cases in which any method of the antecedent is multivalued, depending on "how x is presented.")

Should we say that intuitionism rejects $(\text{AC})_c$, the classical axiom? Again, in a sense, but that can mislead. For, in general, intuitionism (in its extreme versions) fails to comprehend either the antecedent or the consequent, since classical infinitistic quantification is involved. But in *this* sense we should say that intuitionism (in its extreme versions) rejects *all* classical mathematical statements involving such quantification. Better to say ...

Incidentally, it is interesting to ask just what it was that the French constructivists (such as Borel, Baire, and Lebesgue) were doubting in "rejecting the Axiom of Choice." Surely not $(\text{AC})_i$. The truth of $(\text{AC})_c$? Or its intelligibility? Probably the latter, but perhaps also the truth of some mixed version in which we move from, say,

$$\forall_c x \exists_{i(\text{or } c)} y \varphi(x, y),$$

to

$$\exists_i f \; \forall_{c(\text{or } i?)} x \varphi(x, f(x)),$$

in which "i" is understood broadly as "constructive." This may be a reasonable interpretation, since the modern set–theoretic notion of "arbitrary function" is rather esoteric, and it seems to have been the traditional formula- or rule-based notion that was implicitly understood in quantification over functions (whereas quantification – especially universal quantification – over, say, points and point sets in a topological or metric space may have been understood classically). As it happened, the French constructivist school's mathematical practice made essential reliance on $(\text{AC})_c$, despite the proclaimed mathematical philosophy.[11] In any case, the need to draw these distinctions should be clear.

[11] See Moore [1982], pp. 93–103. The above suggestion(s) seem to be compatible with Moore's exposition.

12.3   Some Implications and the Problem of Circularity of the Explanations

What has been said with respect to the Axiom of Choice carries over, *mutatis mutandis*, to all mathematical principles. Unless the specific intuitionistic meanings of the logical constants and component sentences are firmly held in mind, apparent conflicts with classical tautologies and theorems will be rife, and the content of intuitionistic mathematics and its relationship to classical mathematics will be thoroughly distorted. For this reason, it is unfortunate indeed that intuitionists themselves have presented, and continue to present, their work in a notational form borrowed from classical logic and mathematics. Our use of subscripts is an improvement, but a really satisfactory notation would presumably involve explicit machinery for referring to the computational states of an idealized human mathematician (henceforth, ihm, for short), and to the methods and constructions of an ihm, and machinery for expressing relationships between an ihm and such methods and constructions, and, perhaps, relationships between these and further "mathematical objects" countenanced by the intuitionist.[12] Explicit translations of intuitionistic mathematics into expanded classical languages (referred to above) represent one type of effort in this direction, but in two respects they fall short. First, they employ classical machinery which intuitionists (of the extreme variety) claim not to understand. That is, a fully explicit presentation of intuitionist mathematics, representing a self-contained intuitionist perspective, should employ only intuitionistically comprehensible primitives. Second, the extant translations are still rather schematic, usually packing a great deal into a single new primitive (e.g., an S-4 modal operator, or a single predicate for "calculable numbers," etc.). In some cases – and in spite of some nice formal results on the faithfulness of the translations – it is clear that some distortion is involved.[13]

In any case, the following phenomenological facts are evident: classicists (such as myself, and, I conjecture, any other willing to take the trouble) are capable of assimilating intuitionistic mathematics as it has been presented: we follow the proofs; we agree that the theorems proved are true (or at least we see no reason not to accept them); and we find no conflict whatsoever between any such result and any classical results with which we are acquainted. This much

---

[12] Indeed, the work of Troelstra [1969] is a step in this direction, although classical-looking abbreviations creep in early on.

[13] See e.g. Shapiro [1985], p. 25, where it is recognized that the proposed translations of some of the intuitionistic logical constants (in particular, the conditional and the universal quantifier) are only approximations and that "an exact translation is not possible."

as least gives us confidence that the program of faithful translation into a consistent classical theory should be fully successful.

To illustrate briefly, consider the "principle of open data," a fundamental axiom in the theory of ("lawless" or "free") choice sequences, resembling the continuity principles (used to "refute classical logical laws," in Dummett's language). Where lower case Greek letters range over such sequences, and '$\alpha \epsilon n$' means that sequence α has initial segment (coded by) natural number n, the axiom scheme of open data is written,

$$A\alpha \rightarrow \exists n \ (\alpha \epsilon n \ \& \ \forall \beta (\beta \epsilon n \rightarrow A\beta)),$$

where A contains no non-lawlike parameter other than α, and in which all connectives are of course intuitionistic. What does this formula say? It says:

An ihm has a method of transforming a proof that $A\alpha$ into a construction of an initial segment of α together with a method of transforming any construction of a sequence β with that same initial segment into a proof that $A\beta$.

More compactly, any (demonstrable) property of lawless sequences is fully determined by an initial segment, in that the property (provably) holds of any sequence with that initial segment. Why should we accept this? Because a proof that $A\alpha$ is achieved on the basis of finite initial data on the development of α (this is part of what it *means* for an ihm to prove $A\alpha$); no further development of α can alter the status of this proof that α meets condition A; therefore, any sequence agreeing with α on that initial segment will be subject to the same proof that established $A\alpha$.

Obviously classicists can reason this way, and obviously there is no conflict between the principle and the classical negation of the (clearly false) classical version of that principle. The very same situation obtains with respect to the various continuity principles of intuitionistic analysis, which merely develop the same general idea in combination with the intuitionistic Axiom of Choice. These in turn lead to the (extended) fan theorem and to the Brouwer continuity theorem (that "every function defined everywhere on the closed interval [0,1] is uniformly continuous on [0,1]," to use unsubscripted, classically shocking language), all of which can be accepted with equanimity.

As a final illustration (more easily presented than the Brouwer theorem itself), consider the following:

Theorem: $\neg_i \forall_i \alpha [\forall_i n \ \alpha(n) = 0 \ v_i \ \neg_i \ \forall_i n \ \alpha(n) = 0]$,
in which $\alpha(n)$ denotes the n'th term in the development of α. (Here α ranges over a more general class of choice sequences, including lawlike sequences,

satisfying certain continuity principles. This is a theorem which Dummett cities as "refuting . . . the law of the excluded middle.")[14]

What does this theorem say? It says: it is refutable that an ihm can have a method of transforming any construction of a choice sequence into a proof that each of its terms is 0 or into a refutation that each of its terms is 0. The reasoning involved in the proof can readily be informally summarized. Suppose there were such a method, then for any α the method applied to α would yield a proof that α is all 0s or it would yield a refutation of this. Consider the first case; by a continuity principle, the proof that any given α is all 0s would have to be based on finite initial data on the development of α, in which case the proof would carry over to any sequence continuing that finite initial segment, which is obviously absurd (since we can continue that segment with 1s if we wish). But α was arbitrary, so this establishes that, for any α, the method yields a refutation that α is all 0s; but this is equally absurd since we may take α to consist of all 0s. QED.

Again, any classicist can follow the reasoning. And, I hope, anyone can see that it has nothing to do with the classical law of the excluded middle.

Thus far I have been taking for granted that we all can work with the BHK explanations of the intuitionistic logical constants sketched above. For classicists (speaking at least for myself now), those explanations serve as an entry visa into the foreign territory of intuitionistic mathematics. But, by simply holding on to them and periodically consulting them (working intuitively, without relying on formal systems or formal semantics), I am able to travel comfortably enough (undaunted even by the jungle of notation). Moreover, never do I experience any inner struggle over the classical texts which – to stave off homesickness – I have packed in my luggage. Heavy though they may be, no tourist attraction ever tempts me to discard them.

But how do I succeed in this? For, as intimated above, there appears to be a circularity in the BHK explanations, analogous to that which arises when one gives Tarskian truth conditions in an effort to "explain" the classical logical constants. (In the metalanguage, one finds oneself employing natural language logical words such as 'or', 'not', 'some', 'every', etc.) At any rate, there would be a circularity if the words in question ('and', 'or', 'any', 'some') as they occur on the right in the BHK explanations could not be independently grasped. And, since the BHK explanations were originally inspired by intuitionists who claimed not to comprehend classical language, we (classicists) may worry that in order to apply for a visa for this journey we must already be in possession of a valid visa!

There are two routes around this impasse, each bearing on different aspects of the communication problem. The first is the most straightforward: it is

[14] Dummett [1977], 84.

simply to read the right sides of the BHK explanations classically. For a classicist, what could be simpler? Moreover – and here I am consciously on the level of phenomenology – it seems to work! At least, I can follow the development of intuitionistic analysis, nodding in assent in the appropriate places. Were I more interested, more gifted, or both, I might even be able to prove new intuitionistically acceptable results. And, as already emphasized, I experience no cognitive dissonance in connection with my classical training. Here I am, of course, assuming that my conscious awareness of what I am doing is accurate: consciously I notice no difference in my reading of the right sides of the BHK explanations and my reading of classical mathematics (which I also regard as quite accurately formalized by classical logistic systems). Perhaps I am mistaken here. I could conceivably even be mistaken about the absence of (overt) conflict. (Obviously, hidden conflicts cannot be ruled out without further argument, involving such things as a suitable formal classical translation and a relative consistency proof.) But I report the phenomenological facts for what they are worth and will leave to psychoanalysis the task of elucidating my unconscious procedures.

This first route (if it works) is open to the classicist, but not to the intuitionist who fails to grasp classical logical language, unless it could be maintained that, in the specific contexts of the BHK explanations, classical logical language is intuitionistically acceptable, or – what may come to the same thing – unless it could be maintained that, in those contexts, there is a primitive sense of the logical words in question which is comprehensible to both the intuitionist and the classicist. The second route seeks to establish this.

In fact, it is sometimes claimed that the BHK explanations are indeed genuine explanations, unlike the Tarskian classical truth conditions, precisely because, on the right sides, the logical particles occur only in restricted contexts, namely in "proof conditions" which are decidable (in the intuitionistically relevant sense of "provable or refutable").[15] If indeed those conditions could all be taken to be of a fixed elementary form, such as "construction c proves statement A," and if this relation could be seen to be decidable in all relevant cases, then we could regard the BHK explanations as having achieved a reduction of all (mathematical) uses of the logical constants to cases of sufficiently low complexity to effect genuine explanations. In particular, the basic, decidable contexts could be regarded as unproblematic: learning to use the logical constants in such contexts can be equated with learning a decision method of the sort that is paradigmatic for constructive mathematics (e.g., like learning to recognize that a symbol is a symbol of a natural number, or perhaps even like learning how to say "I now have a headache").

---

[15] Dummett [1977] claims this, see p. 409.

This matter has been examined in some detail by Weinstein.[16] The upshot of his investigation was this: A case can be made that "c proves A" is decidable in case A is an elementary arithmetic statement (say, equation with primitive recursive terms). Moreover, the clauses for $\&_i$, V_i and $\exists_i$ can be seen to preserve decidability of "c proves A" for A formed by these operators. In the first two cases, all that is required is to recognize whether c is a pair whose coordinates respectively prove each of (in the case of $\&_i$) or at least one of (in the case of V_i) the respective components of A. In the case of $\exists_i$, all that is required is that it be recognized of c whether (to employ only a harmless modification of the BHK explanation as I sketched it above) it is a pair whose first coordinate proves that the second fulfills the quantified condition in A. (Here, the occurrence of "some" ("c is a – i.e., some – pair") is innocuous, involving merely the inspectable form of the given construction c.) The clauses for the conditional and the universal quantifier, however, were found to be more problematic. The reason is that the proof conditions for conditionals and universally quantified statements involve constructive functions applied to arbitrary constructions as arguments. Indeed, Weinstein was able to reduce these proof conditions to those of statements of a very special form, namely either a statement that a decidable property (e.g., being a proof of a certain statement) holds of a given construction or that such a property holds of all constructions. But it is precisely in connection with the latter sort of statement that problems were seen to arise, due essentially to the open-ended nature of intuitionistically acceptable mathematical constructions.

A serious problem was seen to arise in particular for the relation "c proves $\neg_i \forall_i xB$," in which the conditional is combined with universal quantification (recall $\neg_i$ is defined via $\rightarrow_i$ and an absurd statement). The proof conditions for this form of statement require that c operate on arbitrary c' which prove $\forall_i xB$, which in turn operate on arbitrary c'' of objects of the intended domain to yield proofs, $c'(c'')$ of $B(c'')$. Now there is, unfortunately, a disanalogy between the roles of c' and c'' in these conditions. For, if the domain of quantification is canonically generable, as in the case of the natural numbers, the c'' are in turn of a canonically specifiable form. Thus, one could at least argue that the relation, "c' proves $\forall_i xB$," is decidable, even though the proof condition for this involves universal generalization over constructions (of the canonically specified form). However, unlike the natural numbers, intuitionism countenances no canonical mode of generating arbitrary constructions that can count as proofs of a universal statement, i.e., constructions in the role of c' in the proof conditions for $\neg_i \forall_i xB$ above. Even if B is a decidable property of natural numbers, no such mode of generation is envisioned. Indeed, if we had a proof that $\neg_i B(m)$ for (m a numeral of) a particular natural number m, we would thereby have a canonical method of reducing to absurdity any proof of $\forall_i xB$, without having to consider the detailed

---

[16] Weinstein [1977].

nature of such a proof. But intuitionism does not insist that this be the only way of refuting $\forall_i xB$. In fact, if it did, it would accept what is known as Markov's principle,

$$\forall_i x \ (B(x) \ \vee_i \ \neg_i B(x)) \ \rightarrow_i (\ \neg_i \ \forall_i xB \ \rightarrow_i \exists_i x \ \neg_i B).$$

in which the initial antecedent clause expresses the decidability of B. However, this principle is in general *not* intuitionistically acceptable. (This is readily seen by attending to the strong existential component of $\forall_i$ statements: refutability of such a statement does not in general guarantee that a search through the natural numbers will turn up a counterexample, even where B is a decidable property of natural numbers.) Thus, we lack a convincing argument that the proof conditions for negative universals are intuitionistically decidable, and indeed we are led to suspect that they are not. Thus, we lack a convincing case that the BHK explanations are genuinely explanatory by the second route. It thus appears that the first route may be the only route for the classicist trekker.

12.4   Intuitionism as a Self-Sufficient "Never Never Land"?

This last discussion of the status of the intuitionistic proof relation can serve as a point of entry to the problem area now to be explored. Based on the BHK explanations of the intuitionistic logical apparatus, but otherwise naively speaking our native background English, we sketched a difficulty with the idea that the proof relation is in all relevant cases decidable. In the case of negative universals, we argued (following Weinstein) that there is good reason to doubt the decidability, although, NB, we did not *prove* that the relation is indeed undecidable. Our conclusion was that it may well be.

But now consider carrying out the same discussion again but this time employing only intuitionistic logical apparatus. Can it even be done? Consider how we would express that a relation R is (or may be) undecidable. What we need to get across is that there are (or may be) cases (possible relata of R, say a construction c and a sentence A, where R stands for the proof relation) such that it cannot be proved that $R(c,A)$, nor can it be refuted that $R(c,A)$. And here, "cannot be proved or refuted" means that an ihm could *never* prove or refute $R(c,A)$. For, in entertaining the (absolute) undecidability of a question, we are not merely considering a particular state of information (ignorance) in which the question happens to be unanswered; rather we are considering the possibility that it may remain unanswered no matter how far a search for an answer might be carried. Thus, if we attempt to formalize – using the intuitionistic logical constants – the assertion that (possibly) $R(c,A)$ is (absolutely) undecidable, we obtain,

(possibly) $\forall_i c'[\neg_i (c' \text{ proves } R(c,A)) \ \&_i \ \neg_i (c' \text{ proves}(R(c,A) \rightarrow_i 0 = 1))].$

$$(*)$$

Here the variable c' may be thought of as ranging over the intuitionist universe of constructions. (This is canonical, but more graphically it could be thought of as ranging over states of information of an ihm, or over moments of time in the mental life of an ihm, etc., in which case the predicates used in the following clauses would have to be suitably adjusted.) But this says that (possibly) an ihm has a method of transforming any proof c' of $R(c,A)$ into a proof of absurdity and of transforming any refutation c' of $R(c,A)$ into a proof of absurdity. But the first part – transforming any proof c' of $R(c,A)$ into a proof of absurdity – is just what it means to refute $R(c,A)$, which intuitionistically contradicts the second part. (Strictly, this reasoning relies on the intuitionistically acceptable assumption that proofs are "self-certifying," i.e., whenever c proves p there becomes available a c' such that c' proves that c proves p (c' might be taken simply as the method of mentally reflecting that c proves p). Then any method that refutes "c proves p," applicable to any c, also provides a method that refutes p. (Recall the meaning of "refutes.")) Obviously, the use of the combination $\forall_i c' \ \neg_i c' \ \ldots$ to represent "never" completely distorts the intuitive meaning. When we say we may never have a proof of $R(c,A)$ and may never have a refutation, we are envisioning the possibility of perpetual ignorance; but this is diametrically opposed to the standard (BHK) interpretation of (*).

This, of course, is symptomatic of a general situation and does not depend on the particular example. As Dummett points out, intuitionistically there can be no proof of (absolute) undecidability of any statement, for a proof that the statement is (absolutely) unprovable amounts to a refutation of the statement.[17] What the above example brings out, however, is that the problem is worse: it is not merely that undecidability is unknowable, but that undecidability cannot even be entertained (using the intuitionistic logical constants in the most straightforward way) as a possibility, even when this possibility emerges upon reflection on the very meanings of the intuitionistic logical primitives! (Note that merely adding a modal operator for "possibility" does not help; in the above example, we already added it – parenthetically. The difficulty comes in stating what it is that is envisioned as possible.)

It certainly appears that we must go outside the confines of the intuitionistic logical apparatus if we are to express the possibility of never answering a mathematical question. Indeed, we say suppose that, at *any stage*, an ihm can decide the predicate, "I have a proof that A" and hence can also decide its (ordinary) negation, "I lack a proof that A." In such "stage-by-stage" contexts, the implicit quantification over proofs or constructions in these predicates is taken to be innocuous. (We might write them in the form "I am in-a-proof-recognizing-state-with-respect-to A," etc., thereby eliminating the quantification.) However, in the present case, we must achieve the effect of quantifying

---

[17] Dummett [1977], p. 17.

over a potential infinity of arbitrary stages or states of information, and $\forall_i c'$ (or an equivalent) must be brought in explicitly. If (*) does not work, what other intuitionistic formula would?[18]

It might be thought that the problem could be solved by just adding classical negation to the intuitionist language. (If the proof relation is in fact decidable, any distinction between $\neg_i$ and classical negation in (*) would not matter. However, as we have seen (and are at the same time trying to express!), this may not be the case.) But even if that could cogently be done (which is doubtful: what would count as its "proof conditions," and how could the classical conception of truth be resisted at the same time?), it would not suffice, for the result of replacing $\neg_i$ with classical 'not' in (*) still does not express the relevant possibility due to the strong existential import of $\forall_i$, that is, the implication of having a method which shows of any c' that it fails to decide $R(c,A)$, however this latter is expressed. But clearly, in envisioning the unde-cidability of $R(c,A)$, we are not thereby envisioning a capacity to demonstrate this undecidability (which, as just noted, is regarded by the intuitionist as impossible anyway!). Thus, this existential import of $\forall_i c'$ must somehow be stripped away. But it is difficult to see how this can be done without ending up with ordinary classical universal quantification over infinite totalities, and moreover, over such totalities as are not even generable by any effective rule. And, if so, the intuitionist would have embraced the essence of classicism that was supposed to be incomprehensible.

In reply, however, it may be said, "So what? Such speculations as prompted (*) in the first place are not really part of mathematics anyway; we never really need them in our mathematical work. They are just extra-mathematical mus-ings which we are free to carry out in our ordinary background language. Why should we care to express them within a language designed specifically for mathematics?" How seriously can we take this?

Not seriously at all, I think, for the following reasons. To summarize them in advance: first, intuitionism needs to consider the possibility of undecidability in motivating its own internal logic. Not only does it constantly appeal to such considerations in its informal discussions, it even invokes a specific quasi-formal method in the course of its mathematical work – the so-called "method of weak counterexamples" – which rests on the possibility of undecidability. Second, if

---

[18] It should be noted that there are paradoxes that arise from quantificaion over arbitrary construc-tions: see Goodman [1970], section 9; also, for a simple paradox based on "absolute provability," see Weinstein [1983], p. 264. Efforts to circumvent these by imposing stratification on the universe of intuitionistic constructions, such as Goodman's [1970], bring with them problems, such as giving a satisfactory interpretation of the intuitionistic conditional (so as to guarantee *modus ponens*, for example). For a discussion, see Weinstein [1983], pp. 265–266. The problem we are raising persists in the context of stratification. Even a restricted approximation to classical 'never', in the sorts of cases pointed to here, would seem to elude the intuitionist logical apparatus, so long as the (restricted) domain of constructions over which one quantifies is infinite.

intuitionism is to be taken seriously as a philosophy of mathematics, it must (eventually) come to grips with applied mathematics. But it is difficult to see how it can do this without embracing the very sort of language we have encountered and which the above reply seeks to segregate from "mathematical practice proper." These points confront intuitionism with a dilemma: *either* it is genuinely incapable of expanding its own framework to express the sorts of possibilities in question, in which case (as will be argued) it cannot really motivate its own internal logic in a convincing way, it must even renounce some of its own quasi-formal methods, and it cannot do justice to mathematics in a great many real scientific contexts; *or* it *is* capable of such an expansion, in which case we are led to suspect that it must acknowledge itself as part of a broader, essentially classical framework. And in this case, it then becomes a pressing question (one which I do not pretend to answer definitively here) whether intuitionism represents any viable genuine alternative at all to classical mathematics, either applied or pure.

Consider the first point above. Take for example intuitionism's stance on "the law of the excluded middle." How does it motivate its refusal to accept this law (admittedly without having to claim to refute it)? Not by the paltry observation that there are propositions which we now can neither prove nor refute. Granted this is sufficient to withhold present assent to particular instances of $(LEM)_i$. However, should methods of proof or refutation for these be found, those instances would become part of the corpus of an ihm. If, at any stage, no further exceptions remained, $(LEM)_i$ would have become a general law. Now, I take it, what gives real force to intuitionism's stance is that this *may never* happen, that is, (possibly) at any stage, exceptions to $(LEM)_i$ can be found. But if we attempt to express this as,

$$\text{(possibly) } \forall_i s \, \exists_i p \, (\, \neg_i \vdash_s (p \vee_i \neg_i p)),$$

where $\vdash_s$ means "is assertible at stage s," we fail, due to the strong existential import of $\forall_i$. In entertaining the possibility of an inexhaustible supply of unsolved problems, we are not committing ourselves to the possibility of having a method of generating them! Again, we seem to require non-constructive universal quantification over a potentially infinite, non-canonically generable totality.

Similarly, consider the reason intuitionism does not accept Markov's principle. Even for decidable A, "intuitionistically we can only take $[\neg_i \forall_i x \, \neg_i A(x)]$ to mean 'We can derive a contradiction from supposing that we could prove that $A(x)$ failed for every number,' a proposition from which no guarantee can be extracted that, by testing each number in turn, we will eventually find one which satisfies $A(x)$."[19] For all we know, we may *never* find such a number. But this cannot be expressed as

---

[19] Dummett [1977], p. 22.

possibly $\forall_i c$ ($\neg_i c$ proves $\exists_i x A(x)$),

for this is the very different (and more stringent) possibility of being able to refute the finding of such a number. Indeed, this is not ruled out by the informal reason just quoted, but it is surely not the only possibility envisioned.

It may occur to the reader that perhaps intuitionism need not take seriously motivation which involves infinitistic quantification over constructions (or states of information, etc.), since in fact intuitionism can work with Beth trees or finite Kripke trees, and can thereby present counterexamples to the formulas it wishes to avoid in its logic. Space does not permit an adequate treatment of this issue here, but it should be noted that there are serious problems with taking such trees as accurately representing our states of information with respect to, say, elementary arithmetic. In fact, summing up an extended discussion of this, Dummett himself concludes, "Beth trees cannot be seen as providing an actual semantics for an intuitionistic first-order language."[20] It is doubtful that schematic finite countermodels of formal semantics can provide an adequate substitute for intuitive motivation of the familiar sort, which as we have seen, involves the essentially infinitistic concept of "never".

Turning to the "method of weak counterexamples," this is a method whereby a certain formula is shown not to be intuitionistically demonstrable, not by a refutation, but by invoking currently unsolved problems (e.g. whether the sequence "123456789" occurs in the decimal expansion of π) and letting the behavior of certain explicitly defined objects pertaining to the formula in question turn on the answer to the unsolved problem. For example, to show that $x \cdot y = 0 \rightarrow_i x = 0 \text{ v}_i y = 0$ should not be an intuitionistic law of real numbers, let $A(n)$ hold if a "9" in the sequence "123456789" occurs at the nth place in the expansion of π, let $B(x)$ be also decidable but such that it is not known whether $B(k)$ for k the least number, if it exists, such that $A(k)$. Then define real number generators $<r_n>$ and $<s_n>$ via

$$r_n = \begin{cases} 2^{-n} & \text{if } \forall m \leq n \ \neg A(m) \\ 2^{-n} & \text{if } k \leq n \ \& \ A(k) \ \& \ \forall m < k \ \neg A(m) \ \& \ \neg B(k) \\ 2^{-k} & \text{if } k \leq n \ \& \ A(k) \ \& \ \forall m < k \ \neg A(m) \ \& \ B(k) \end{cases}$$

$$s_n = \begin{cases} 2^{-n} & \text{if } \forall m \leq n \ \neg A(m) \\ 2^{-n} & \text{if } k \leq n \ \& \ A(k) \ \& \ \forall m < k \ \neg A(m) \ \& \ B(k) \\ 2^{-k} & \text{if } k \leq n \ \& \ A(k) \ \& \ \forall m < k \ \neg A(m) \ \& \ \neg B(k). \end{cases}$$

---

[20] Dummett [1977] p. 414. These remarks carry over to Kripke trees; earlier (at p. 208) Dummett presents reasons for thinking that "the advantage, in supplying a representation of the intended meanings of the intuitionistic logical constants, lies heavily with the Beth trees as against the Kripke trees."

(All logical symbols are intuitionistic.) Without solving the problem, we cannot assert (what is implied by) the consequent of the law, $<r_n>$ equivalent to 0 $\vee_i$ $<s_n>$ equivalent to 0 (reals being identified with equivalence classes of Cauchy sequences), but obviously we can assert $< r_n \cdot s_n>$ equivalent to 0. Equivalence classes represented by $<r_n>$ and $<s_n>$ respectively will then serve in a counterexample. As Dummett remarks,

[This method] gains its force from the fact that we have a uniform way of constructing similar 'counterexamples' for each unsolved problem of the same form. Since we can be virtually certain that the supply of such unsolved problems will never dry up, we can conclude with equal certainty that the general statement will never be intuitionistically provable. Such a recognition that a universally quantified statement is unprovable does not amount to a proof of its negation ...[21]

Indeed it does not! But were we to formalize the 'never' intuitionistically, that is precisely what it would mean. And the very same problem – irrelevant commitment to a method of generating unsolved problems – just encountered in connection with (LEM) plagues us here.

Thus far, we have been considering serious limitations in the expressive power of the intuitionist logical apparatus that arise internally within constructive mathematics. When we turn to physical applications of mathematics, the problems appear to be even more serious. Others have noted that the intuitionistic constructive meanings of the logical constants are quite inappropriate in many physical applications, and what I have to say here is merely a further reflection of that important fact.[22]

Consider the following commonplace situation: we have a dynamical physical theory T which makes predictions about a certain physical system S that may endure through, or occur repeatedly at, an infinite number of regions in space-time. (Our background cosmology could explicitly allow that the regions be sufficiently separated, if you like.) A typical consequence of our theory T (together with suitable background information, we may suppose) reads,

" At no time (in the future, say) will system S be found in region R of our state space."

We may believe T to a certain degree, but, as rational empiricists (or fallibilists), we do not treat T as bedrock (as we normally treat primitive recursive equations). At best, we are prepared to assert the tentative correctness of T and its consequences such as the above. We certainly do not claim to have any method of reducing to absurdity any instance violating the above prediction. Yet if we read the logical words in it intuitionistically, representing it as of the form,

$$\forall_i t \ \neg_i R(S, t),$$

[21] Dummett [1977], p. 45.
[22] See, for example, Putnam [1975], pp. 60–78, 75.

that is precisely what we would be claiming. And we cannot patch this up simply by reinterpreting $\neg_i$ to mean reducibility to a conflict with T (together with auxiliary hypotheses), for the actual verification and falsification procedures associated with the statement $R(S,t)$ may themselves be (as they often are in real scientific life) quite delicate and a far cry from anything like carrying out a decision method. Thus, as in the mathematical examples already encountered, the constructive meaning of $\forall_i$ also must be altered. Moreover, it cannot simply be weakened to "we possess a method for assessing as more or less probable that ...," for in many cases we may recognize that we may never be in a position to evaluate the prediction even in this weak sense.

In general, our epistemic predicament in physics is radically different from that which normally prevails in constructive mathematics. Lack of proof or refutation is the norm. And this is often intimately bound up with quantification over infinite totalities, such as space-time, regions of phase space, etc. We need to be able to express that (in some cases, no matter how long we were to endure) *we might never know* whether, say, matter fields beyond an event horizon obey certain laws, whether our cosmos is accurately modeled in this way or that, whether certain objects (say, extraterrestrial intelligent beings) will ever be found, whether certain possibilities (say, certain quantum interferences on a macro-level) will ever be actualized, etc., etc. Intuitionistic logic is entirely inappropriate for representing such things.

Note that these points are quite independent of the more technical question of just how much mathematics actually employed in the physical sciences can be carried out constructively (intuitionistically or otherwise). That is, of course, an interesting and important question (and, it would seem, a part of the answer is, "Quite a lot, and without getting into the baroque theory of (free) choice sequences either!"[23]), but the problems raised here arise no matter how rich the constructive substitutes for classical (pure) mathematics may be. For these problems pertain to our perennial epistemic situation in physical applications of entire theories quite independent of their particular mathematical content. As it appears, that situation is systematically misrepresented by intuitionistic logic.

Finally, as the ensuing dilemma sketched above suggests, I do not think the defense, "We are only interested in doing pure, constructive mathematics proper," will wash. No philosophy of mathematics (as opposed to a merely limited formalism for a part of mathematics) can be tenable unless it is able to treat applied mathematics adequately. Moreover, if indeed it is conceded that, in physical contexts, classical quantification over infinite domains *is* after all intelligible and necessary (as Dummett himself in places seems to concede)[24],

---

[23] This is one of the lessons of Bishop [1967].

[24] See, for example, Dummett [1977], pp. 57–58, where it is at least left open that it may be "perfectly intelligible ... to say that there are infinitely many stars [etc.]," without any implication of any infinite process being completed.

then it would seem that the critical thrust of (extremist) intuitionism is undermined. For (a great deal of) classical (pure) mathematics can be modeled within infinite structures recognized in physics (e.g., space-time), and, moreover, it may even be construed as requiring merely the possibility of infinite structures of the appropriate logical type, without any commitment to their actual realization, either as physical, as mental, or as anything else.[25]

12.5   Tentative Conclusion

It would appear that intuitionism has boxed itself in: in its quest for direct subjective meaning (a kind of certainty) in mathematics, having transformed the subject as the study of abstract structures into a kind of idealized psychology of mathematics, it has deprived itself of the means of expressing some of the most important facts of that psychology pertaining to the limits of knowledge, facts of a kind which, moreover, are pervasive in the scientific applications of mathematics. Classicists may be forgiven for declining to embrace intuitionism, for – among other things – that approach appears unable to do justice to the selfsame speculative psychology it has staked out as its own territory without expanding to incorporate key portions of the language and logic of classical mathematics from which it sought to distance itself.

Acknowledgement: I am indebted to Stewart Shapiro for helpful criticism of an earlier version of this paper and to William Hanson and C. Anthony Anderson for helpful discussion.

References

Bishop, E. [1967] *Foundations of Constructive Analysis* (New York: McGraw Hill).

Brouwer, L. E. J. [1975] *Collected Works*, Volume I (Amsterdam: North Holland).

Dummet, M. [1977] *Elements of Intuitionism* (Oxford: Oxford University Press).

Gödel, K. [1932] "Eine Interpretation des Intuitionistichen Aussagenkalkuls," *Ergebnisse eines Mathematischen Kolloquiums* **4**: 39–40.

Goodman, N. [1970] "A theory of constructions equivalent to arithmetic," in Kino, A., Myhill, J., and Vesley, R. E., eds., *Intuitionism and Proof Theory* (Amsterdam: North Holland), pp. 101–120.

Hellman, G. [1989] *Mathematics without Numbers: Towards a Modal-Structural Interpretation* (Oxford: Oxford University Press).

Heyting, A. [1966] *Intuitionism: An Introduction* (Amsterdam: North Holland).

Kreisel, G. [1962] "Foundations of intuitionistic logic," in Nagel, E., Suppes, P., and Tarski, A., eds., *Logic, Methodology, and Philosophy of Science* (Stanford, CA: Stanford University Press), pp. 198–210.

---

[25] See Hellman [1989].

Kripke, S. A. [1965] "Semantical analysis of intuitionistic logic I," in Crossley, J. N. and Dummett, M. A. E., eds., *Formal Systems and Recursive Functions* (Amsterdam: North Holland), pp. 92–130.

Lifschitz, V. [1985] "Calculable natural numbers," in Shapiro, S., ed., *Intensional Mathematics* (Amsterdam: Elsevier), pp. 173–190.

Moore, G. H. [1982] *Zermelo's Axiom of Choice* (New York: Springer).

Putnam, H. [1975] "What is mathematical truth?," reprinted in *Mathematics, Matter, and Method: Philosophical Papers*, Volume I (Cambridge: Cambridge University Press, 1979), pp. 60–78.

Shapiro, S. [1985] "Epistemic and intuitionistic arithmetic," in Shapiro, S., ed., *Intensional Mathematics* (Amsterdam: Elsevier), pp. 11–46.

Tait, W. W. [1983] "Against intuitionism: constructive mathematics is part of classical mathematics," *Journal of Philosophical Logic* **12**: 173–195.

Troelstra, S. A. [1969] *Principles of Intuitionism* (Berlin: Springer).

Troelstra, S. A. [1977] *Choice Sequences* (Oxford: Oxford University Press).

Weinstein, S. [1977] "Some remarks on the philosophical foundations of intuitionistic mathematics," delivered to the American Philosophical Association, Eastern Division, December 29.

Weinstein, S. [1983] "The intended interpretation of intuitionist logic," *Journal of Philosophical Logic* **12**: 261–270.

13   Constructive Mathematics and Quantum
      Mechanics: Unbounded Operators and the
      Spectral Theorem

Introduction

A major outstanding issue in the foundations and philosophy of mathematics
concerns the *indispensability* of classical infinitistic mathematics for the
empirical sciences. Claims of such indispensability form a modern cornerstone
of mathematical platonism and alternative classical realist conceptions as well
(e.g. modal structuralism) (cf. e.g., Quine [1953], Putnam [1967, 1971], and
Hellman [1989a]). This has posed a corresponding challenge to constructivist
views (intuitionistic, Bishop-constructivist [Bishop 1967], and related
approaches). How much of the mathematics actually employed in the empirical
sciences, especially physics, can be carried out constructively (in the various
relevant senses)? It is probably no exaggeration to say that the viability of
a constructivist philosophy of mathematics is here at stake (cf. Burgess [1984]).
From this perspective, the work of Bishop and his followers [Bishop 1967,
Bishop and Cheng 1972, Bridges 1979] can be said to have breathed new life
into constructivism, as it has taken the program of constructivizing scientifically applicable mathematics quite far indeed. No longer, perhaps, can Hilbert's
original analogy (or our bracketed variant) be maintained with confidence (that
depriving the [scientifically oriented] mathematician of classical logic [or
infinitistic functional analysis] is like depriving the boxer of the use of his fists).

Despite these advances, however, serious problems remain, especially in connection with quantum mechanics (QM). As one recent investigation concluded, "It
is clear that a constructive examination of the mathematical foundations of quantum
physics does reveal substantial problems" (Bridges 1981, p. 272). In particular, the
major theorem of Gleason [1957] characterizing the measures on the closed subspaces of Hilbert space of dimension $\geqslant 3$ (and hence the possible probability
measures over quantum events as ordinarily identified) is not constructively demonstrable (in either the sense of intuitionism or that of Bishop-constructivism
[Hellman, 1993]). In the present paper, we examine the question of

constructivizability of central results of functional analysis for QM involving *unbounded* linear Hermitian operators in Hilbert space, especially the Spectral Theorem. It will be argued, quite generally, that *such operators are not even legitimately recognizable as mathematical objects from a thoroughgoing constructivist standpoint*: not only can the Spectral Theorem (in its full generality) not be constructively *proved*, it cannot even be constructively *stated*! As will be seen, there are two independent arguments leading to this conclusion. The first of these (Section 13.2), based on results of Pour-El and Richards [1983], affects a wide variety of constructivist programs including what we may call *liberal* programs: these extol and exhibit the virtues of constructive proofs (and, usually, of constructive objects as well) without necessarily thereby impugning the meaningfulness or the coherence of classical, non-constructive mathematical practice. (The work of Bridges [1979] would appear to qualify as an important example.) The second argument (Section 13.3) applies in the first instance to *radical* constructivist programs: these reject classical, non-constructive, infinitistic mathematics as allegedly *deficient in cognitive significance due to proof-independent truth commitments*. (Brouwer is usually understood in this way; some of Bishop's [1967] remarks suggest that his position belongs in this category; and Dummett [1977] explicitly espouses this view.) To what extent the argument of Section 13.3 affects liberal constructivists as well we do not attempt to settle. Finally, in Section 13.4, we shall return to the question of indispensability of unbounded operators for quantum physics, with an eye towards assessing the viability of constructivist programs in light of the preceding arguments.

13.1   Preliminaries

We review here some relevant definitions and known results that will be appealed to below. (The reader familiar with the theory of linear operators in Hilbert space can skim this section.) Along the way we remark on the constructivity status of standard proofs of some key theorems. We shall assume familiarity with the elements of the theory of Hilbert spaces.

Let $\mathcal{H}$ be a Hilbert space (usually separable and infinitely dimensional, and over the field of complex scalars) and let A be a linear operator acting in $\mathcal{H}$. A sequence ϕ_n of vectors in $\mathcal{H}$ is said to *converge* to ϕ (in $\mathcal{H}$), written $\phi_n \to \phi$, just in case $\|\phi - \phi_n\| \to 0$ as $n \to \infty$, where $\| \ \|$ is the Hilbert space norm (defined by the inner product, $\|\psi\|^2 = (\psi, \psi)$) and $\to$ denotes ordinary scalar convergence. A is *continuous* iff $A\psi_n \to A\psi$ for any sequence ψ_n converging to ψ. A is *bounded* iff there exists positive real b such that $\|A\psi\| \leqslant b\|\psi\|$, for every ψ in $\mathcal{H}$. The least b satisfying this is called the norm of A, $\|A\|$. In fact, A is bounded if and only if A is continuous (see, e.g., Jordan [1969], p. 17).

A set of vectors D is *dense* (in $\mathcal{H}$) just in case for any ψ there exists a sequence ψ_n in D such that $\psi_n \to \psi$. If the domain D_A of an operator A is dense, another operator A^*, the *adjoint* of A, may be introduced: for each ψ there is then at most one ψ^* satisfying $(A\phi,\ \psi) = (\phi,\ \psi^*)$ for every ϕ in D_A; then set $A^*\psi = \psi^*$. If A satisfies $(\phi, A\psi) = (A\phi, \psi)$ for all ϕ, ψ in D_A and D_A is dense in $\mathcal{H}$, then A is said to be *symmetric* (or *Hermitian*). (Then A^* is defined and is an extension of A. When $A^* = A$, A is called *self-adjoint.*) An important theorem (Hellinger–Toeplitz) says that if A is symmetric and is defined on all of $\mathcal{H}$, then necessarily A is bounded (Riesz and Sz.-Nagy [1955], §114). Thus a symmetric unbounded linear operator cannot be defined on all of $\mathcal{H}$ but only on a domain dense in $\mathcal{H}$. In fact, this can be strengthened. A linear operator A is *closed* just in case whenever both $\phi_i \to \phi$ for $\phi_i \in D_A$ and $A\phi_i \to \psi$, then $\phi \in D_A$ and $A\phi = \psi$. Then we have the following theorem.

Theorem *(Closed Graph Theorem). Every closed linear operator defined on all of $\mathcal{H}$ is bounded (Riesz and Nagy [1955], §117; cf. Prugovecki [1981], Ch. 3, Theorem 3.12.)*

We remark that most operators arising in ordinary physical applications are closed, although many garden variety such operators are unbounded. Further, we remark that the proofs cited of this and the preceding theorem rely ultimately on the nested interval property of the continuum, and are therefore non-constructive (cf. Riesz and Nagy [1955], §31).

To state the spectral theorem, we need the notion of a *spectral family* of projection operators. (A bounded linear operator E is a projection operator just in case E is idempotent and self-adjoint, i.e. $E = E^2 = E^*$.) A *spectral family* is a collection of projection operators E_x depending on a real parameter x satisfying the following.

(i) If $x \leqslant y$, then $E_x \leqslant E_y$.

(ii) If ε is positive, then $E_{x+\varepsilon}\psi \to E_x\psi$ as $\varepsilon \to 0$, any ψ and x.

(iii) $E_x\psi \to 0$ as $x \to -\infty$ and $E_x\psi \to \psi$ as $x \to +\infty$, for any vector ψ.

Theorem *(Spectral Theorem for self-adjoint operators). For each self-adjoint operator A (bounded or unbounded) there exists a unique spectral family of projectors E_x such that*

$$(\phi, A\psi) = \int_{-\infty}^{+\infty} x\,\mathrm{d}(\phi,\ E_x\psi)$$

for all vectors ϕ and all ψ in D_A. The latter consists of all vectors ψ for which the integral

$$\int_{-\infty}^{+\infty} x^2 \mathrm{d}\|E_x \psi\|^2$$

converges. (These integrals can be understood as ordinary Riemann–Stieltjes integrals, as $(\phi, E_x \psi)$ is a complex function of x of bounded variation. See Jordan [1969], pp. 39–44, Riesz and Nagy [1955], §§49, 107, 120. For proofs, see the latter; also Stone [1932], Theorem 5.9.)

In the bounded case, constructive versions of the spectral theorem are known. (See especially Bishop [1967], Ch. 9, Theorem 8.) Concerning the unbounded case, however, it should be remarked that the proofs just cited are non-constructive, relying on the classical non-constructive principle that a bounded monotone sequence converges. (See Stone [1932], Theorem 2.40, used to prove Theorem 5.1 and then Theorem 5.9. Cf. Bishop [1967], p. 28, where the principle in question is shown to reduce the halting problem.)

In its generalized form, the Spectral Theorem is very powerful, providing a *functional calculus* for all operators in its scope, that is, a method of defining and calculating a wide class of functions of an operator: given a spectral family E_x for A satisfying the formula of the Spectral Theorem, and a real function, $f(x)$, measurable with respect to $(\phi, E_x \phi)$, each ϕ in $\mathscr{H}$, $f(A)$ is defined by

$$(\phi, f(A)\psi) = \int f(x)\mathrm{d}(\phi, E_x \psi).$$

This induces a correspondence between polynomials in x and polynomials in the operator A. All this makes sense even for unbounded operators lacking eigenvalues and eigenvectors in $\mathscr{H}$. As a salient example, consider the quantum mechanical (one-dimensional) position operator Q acting in $L^2(-\infty, \infty)$ defined by $(Q\psi)(x) = x\psi(x)$. Its spectral family is given by $(E_x \psi)(y) = \psi(y)$ if $y \leqslant x$ and $y = 0$ if $y > x$, for any vector ψ. One then has

$$(\phi, Q\psi) = \int_{-\infty}^{+\infty} x\mathrm{d}(\phi, E_x \psi),$$

and, for functions f measurable and defined almost everywhere with respect to the spectral family, one has

$$(\phi, (fQ)\psi) = \int_{-\infty}^{+\infty} f(x)\mathrm{d}(\phi, E_x \psi).$$

The domain of Q, D_Q, consists of all those ψ for which

$$\int_{-\infty}^{+\infty} x^2 \mathrm{d}\|E_x\psi\|^2$$

converges, a dense proper subdomain of $\mathcal{H}$.

As a second example, consider the operators for linear momentum. In the one-dimensional case we have

$$(P\psi)(x) = -i\frac{\mathrm{d}\psi(x)}{\mathrm{d}x},$$

defined for absolutely continuous differentiable $\psi(x)$ such that $\mathrm{d}\psi/\mathrm{d}x \in L^2(-\infty, \infty)$. (This operator is unbounded. This is readily seen in the case of $L^2(0, 1)$: here, for example, the function $e^{2\pi inx}$ has unit norm whereas its derivative has norm $2\pi n$ which increases without bound for increasing n. Multiplying $e^{2\pi inx}$ by a suitable damping function yields examples for $L^2(-\infty, \infty)$.) The spectral family for P is given by the Fourier transform operator,

$$(F\psi)(k) = (2\pi)^{-3/2}\int_{-\infty}^{+\infty} e^{-ikx}\psi(x)\mathrm{d}x,$$

and the spectral family E_x for Q, via $F^{-1}E_xF$. (F is a *unitary* operator: norm preserving with an inverse (which equals its adjoint).) That is,

$$(\phi, P\psi) = \int_{-\infty}^{+\infty} x\mathrm{d}(\phi, F^{-1}E_xF\psi),$$

for any vector ϕ and any vector ψ in the domain of P.

Given a self-adjoint operator A and its spectral family E_x, a point x not in an interval on which E_x is constant is said to belong to the *spectrum* of A. (A point at which E_x jumps in value is in the *point spectrum* of A; a point about which E_x increases continuously is in the *continuous spectrum* of A (cf. Jordan [1969], p. 44.))

A useful test for boundedness of a self-adjoint operator, provided by the spectral decomposition theorem, is given by the following.

Theorem: *A self-adjoint operator is bounded if and only if its spectrum is bounded. (See, e.g., Jordan [1969], pp. 47–48.)*

A spectral decomposition theorem can also be proved for unitary operators. For such operators, U, the form of the spectral decomposition is given by

$$(\phi,\ U\psi) = \int_0^{2\pi} e^{ix}\mathrm{d}(\phi, E_x\psi).$$

A very important application of this is the following.

Stone's Theorem: *Let U_t be a continuous group of unitary operators para-metrized by t (i.e. $(\phi, U_t\psi)$ varies continuously with t, $U_0 = I$, and $U_t U_{t'} = U_{t+t'}$ all real t and t'); then there exists a unique self-adjoint operator H (not necessarily bounded) such that*

$$U_t = e^{iHt},$$

for all t. ψ is in the domain of H just in case $(1/it) (U_t - I)\psi$ converges to a limit as $t \rightarrow 0$; the limit is then $H\psi$ (Riesz and Nagy [1955], §137).

We remark that Stone's Theorem plays an important role in quantum dynamics, providing an essential step in deriving the form of the time-evolution law (abstract Schrödinger equation) from basic dynamical princi-ples (see, e.g., Jauch [1968], Ch. 10; Jordan [1969], §30).

13.2   Non-Constructivity of Unbounded Operators: Application of a Theorem of Pour-El and Richards

Pour-El and Richards [1983] extended recursive analysis to Banach spaces and went on to characterize the class of non-computable closed linear operators as precisely those that are unbounded. Taking *"computable sequence x_n of elements of a [given] Banach space X"* as a primitive notion, they set out a list of five axioms governing computability of elements and sequences in X. (For this purpose, elements may be identified with one-element sequences.) These were seen as "quite natural – what one would expect any notion of 'computability' on a Banach space to satisfy." (p. 52) The first axiom (Composition) states that if $a\colon \mathbb{N} \rightarrow \mathbb{N}$ is a recursive function and $y_n\colon \mathbb{N} \rightarrow X$ is a computable sequence, then $y_{a(n)}$ is a computable sequence. Second (Insertions), insertion (interpolation) term by term of one computable sequence in another results in a computable sequence. Third (Summation), given a computable sequence x_n, a computable double sequence of real or complex numbers a_{nk}, and a recursive function $d\colon \mathbb{N} \rightarrow \mathbb{N}$, the sums $s_n = \sum_{k=0}^{d(n)} a_{nk} x_k$ form a computable sequence. Fourth (Norms), the norms of items of a computable sequence form a computable sequence of non-negative real numbers. Finally, fifth (Limits), if y_{nk} is a computable double sequence and, as $k \rightarrow \infty, \|y_{nk} - x_n\| \rightarrow 0$ effectively in both n and k, then x_n is a computable sequence (Pour-El and Richards [1983], p. 53). Next, an *effective generating set e_n for X* was defined to be a computable sequence whose linear span is dense in X.[1]

---

[1] As Pour-El and Richards pointed out in a later paper ([1987], p. 7), this condition sufficed to guarantee a second that was included as part of the definition in their [1983] paper, viz. approximation of any computable sequence by e-polynomials. A word is in order about the concept of *effective generating set*: the condition "dense in X" is to be understood classically, i.e. *in fact* dense, not necessarily computably so. That this must be so is clear from the (1987) paper in

We now can state their main result, henceforth referred to as the PE-R Theorem (Pour-El and Richards [1983], p. 54).

Main Theorem *(1983). Let X and Y be Banach spaces with computability theories, and let e_n be an effective generating set for X. Let T be a closed linear operator whose domain includes $\{e_n\}$ and such that Te_n is a computable sequence in Y. Then T maps every computable element of its domain onto a computable element of Y if and only if T is bounded.*

We wish to focus on the "only if" direction and its significance. In this regard, two questions should be distinguished: *first*, what does the theorem *actually show* (*classically*) with respect to constructivist programs more radical than recursive analysis, specifically, Brouwer–Heyting–Dummett intuitionism [Dummett 1977] and Bishop constructivism [Bishop 1967] (which we will henceforth for convenience group together under the rubric "constructivist" unless there is a need to distinguish them), that is, apart from the question of the constructive acceptability of the Pour-El and Richards proof? In effect, we are asking, what can the *classicist* infer from the PE-R Theorem concerning constructivism? *Second*, we may ask what the *constructivist* should consistently infer with regard to the constructivist program in question. Here of course the question of constructive acceptability of the PE-R proof is directly relevant. Let us take these questions up in turn.

Concerning the first question, the answer clearly turns on the meaning of "*T* preserves computability" as it enters into the PE-R Theorem. Since *T* is a closed linear operator on a Banach space, the meaning derives directly from the five axioms. As phrased, these do involve prior concepts from recursive analysis, and so it may seem that a general link between recursiveness and "constructivity" (in the relevant sense, *à la* intuitionism or Bishop-constructivism) must be forged (either by Church's Thesis or its converse or both). For example, the first axiom (Composition) as phrased becomes constructively acceptable if "All recursive functions are constructive" is acceptable (converse Church). Similarly for the fifth axiom (Limits), assuming that "effective convergence" is spelled out in terms of recursive functions. The fourth axiom (Norms) becomes constructively acceptable provided "Any constructive sequence of reals is a computable sequence of reals [in the sense of recursive analysis]" is acceptable (which is implied by Church's Thesis). Alternatively, one can bypass the appeal to such principles at this stage by simply asking what happens if all reference to

---

which unbounded operators *T* may be *effectively determined*, meaning that there exists an effective generating set for *T*, that is, a computable sequence $\{e_n\}$ such that $\{Te_n\}$ is computable and the set of pairs $\langle e_n, Te_n \rangle$ spans a dense subspace of the graph of *T*. This illustrates a general feature of recursive analysis: in contrast to a thoroughgoing constructivism, classical concepts and classical reasoning are legitimately used in obtaining results.

"recursive functions" made explicitly and implicitly in the axioms is replaced with generic reference to "constructive functions of natural numbers." Provided (i) that the axioms thus interpreted can be recognized as indeed fulfilled by the constructivist's conception, and (ii) that the statement resulting from such replacement in the PE-R Theorem – call it the *PE-R Theorem** – is still proved (at least classically) by the PE-R method (possibly replacing "recursive" with "constructive" in the proof as well), then we have our answer: closed linear unbounded operators (meeting what we may call the *conditions** of their theorem, i.e., preserving constructivity of an orthonormal generating sequence) are non-constructive in the sense of failing to preserve constructivity of inputs, in the constructivist's own sense. What can be said on behalf of (i) and (ii)?

Concerning (i), presumably constructivists are in a privileged position to provide an answer, and perhaps we should not second guess them. In any case, we have no reason to doubt Pour-El and Richard's assessment quoted above concerning the naturalness and generality of their axioms. If anything, their assessment seems even less exceptionable with respect to what we may call the "starred version" of their axioms – under consideration here – obtained from their own by systematically replacing explicit and implicit reference to recursive functions with the schematic term, "constructive function." It is indeed hard to find anything objectionable about the resulting closure conditions from the constructivist's standpoint.

Concerning (ii), we can be more definite. An examination of the PE-R proof reveals that, upon substituting "constructive" for "recursive" (and "computable") systematically throughout, it is transformed into a *classically* correct proof (call it the *PE-R proof**) of the PE-R Theorem*. The PE-R proof that if T is unbounded, T fails to preserve computability depends on two lemmas. (For the reader's convenience, we reproduce these and the immediately surrounding relevant portions of the proof of the Main Theorem in an Appendix.) Concerning the proof of Lemma 1, the only possible difficulty is with the last line: one may worry whether "merely waiting" constitutes a constructively acceptable procedure for finding suitable p', unless there is a constructively acceptable proof that some object *will* eventually be found. Indeed, this is a somewhat delicate point, turning on the way we parse "unbounded" and on the constructivist's logic. (We will return to this below.) Under a strong reading of *"unbounded"* (having a method of exceeding any given putative bound), the argument is constructively acceptable. (We note, in passing, that Bishop-constructivism generally prefers strong, positive definitions to weak, negative ones.) However, even on a weaker reading, we should distinguish the question of the constructive acceptability of the proof (our second question), from the question of the conclusion *per se*. Since, classically, we know that an object of the appropriate sort will be found, we can say – from the outside, as it were – that by examining each item in turn, a sequence of the right sort is constructed.

(Since, by hypothesis, the original sequence is generated by a constructive method, and the property of individual members sought is decidable, there is a reasonable sense – still more open-ended than that of recursion theory – in which the desired subsequence is effectively generable, even though it may not be known in advance how long one must wait in order to find p'_n.) In this sense, then, we do obtain a proof of Lemma 1*.[2]

The situation in connection with Lemma 2 is similar. As phrased, in fact, the PE-R argument shows that the element,

$$y = \sum_{k=0}^{\infty} r^{-a(k)} z_k$$

is computable in Y if and only if the set A is *effectively decidable*, in a general constructivist sense. The effective decision procedure for A actually presented is an *intuitively* effective one. (Indeed it shows that A is recursive, provided we invoke Church's Thesis.) Again, one can worry whether the procedure of waiting until a suitable n turns up (i.e. such that $\|y - y_n\| < r^{-a} - [r^{-a}/r - 1]$) is constructively acceptable without a constructive proof that such an n *will* be found. But, as before, the *classicist* knows that such an n will be found, and this justifies the claim that A is decidable. In other words – adhering to the distinction already urged above – we in fact *have* a decision procedure for A, although we may not have a constructively acceptable proof that the decision procedure really works. Let us call such a set *weakly effectively decidable* (w.e.d.).

Beyond the two lemmas, all that is needed to complete the argument is that there be a constructive function $a: \mathbb{N} \to \mathbb{N}$ which *in fact* enumerates a set A such that A is *not* w.e.d. in the sense just explained, i.e. in the sense in which Lemma

---

[2] The point is underscored by remarking that, in passing from Lemma 1 to Lemma 1*, the phrase "there exists" remains classical. Thus, Lemma 1* does not provide a means of *identifying* the desired constructive function $f(n)$ such that $p_n = p'_{f(n)}$, as a fully constructive proof would require. In effect, the classicist is saying to the constructivist: "We know – although *you* may not – that you will effectively generate an appropriate subsequence."

As was brought to the author's attention after completion of this paper, this very distinction has been incorporated in a formal system of Feferman [1984] (there called T_0), designed to permit classical logical investigation of constructively meaningful results (i.e., pertaining to constructively recognizable objects), and thought especially suitable "for representing Bishop-style constructive mathematics" (abstracting from recursiveness as the standard of constructivity). The main result of that paper was to translate and derive the Main Theorem of Pour-El and Richards [1983] in T_0 (Theorem 3.11). That derivation may also be regarded as one natural way of formalizing the argument we have presented here for Proposition 1 (below). Especially in light of the weak consistency strength of T_0 (e.g., a principal subsystem $EM_0\upharpoonright$, sufficient for most purposes, is conservative over Peano arithmetic), this is then further confirmation, not only of Proposition 1, but of the coherence of the enterprise of employing classical methods to investigate constructive mathematical problems. Indeed, Feferman focused on the PE-R Theorem to illustrate precisely this (as it is "an example that cannot be explained adequately in ordinary constructive terms" (Feferman [1984], p. 143; cf. Bridges [1991], p. 225), although he was not concerned there, as we are here, to explore limitations of constructive mathematics for physical applications.

2 derives effective decidability of A from the hypothesis that the element y introduced is computable. Call such a constructive function a a *non-w.e.d. generator.* (Note that, in this definition, it is not required that the range A of a non-w.e.d. generator be recognized by the constructivist as a mathematical object, but merely that the generating function be so recognized. It is sufficient that the range A be classically recognized.) Given this, A will behave just like a recursively enumerable non-recursive set for the purposes of the PE-R proof. We thus have argued for the following.

Proposition 1: *The PE-R proof* classically establishes their Theorem*, with respect to any framework of constructivity recognizing non-w.e.d. generators. In particular, if T is an unbounded (closed linear) operator on a Banach space meeting the conditions* of the PE-R Theorem, then T fails to preserve constructivity of inputs.*

Two remarks are in order concerning the scope of this proposition. First, it should be stressed that so long as a generator a as described is recognized within a constructivist framework, Lemma 2 applies to show that a relevant Banach space element y is "non-constructive" in any sense of "constructive" which is narrower than PE-R's "computable," i.e. such that for any sequence s of Banach space elements,

> *If s is non-computable, then s is non-constructive.*   (*)

Note that this rests on the *contrapositive* of Church's Thesis, constructively weaker than that thesis. Moreover, the constructively problematic steps in the proofs of Lemmas 1 and 2 hinge, not on (*), but on the converse thesis that computability is sufficient for constructivity. Indeed, this is problematic because of the intentional character of "constructive" ($\cong$ "knowably constructive"). This, however, is not a reason to reject (*). And, if (*) *is* accepted, the PE-R Theorem* applies *a fortiori*.

Secondly, apart from Church's Thesis, the constructivist program will at least recognize (indefinitely many) generators of effectively enumerable sets A such that it will (currently) in fact *lack* even a weak decision procedure for A (e.g., the set of codes of theorems of first-order predicate calculus, etc.). Such A can be used in the PE-R reasoning behind Proposition 1 to establish, not quite the final conclusion of that proposition, but something in practice just as limiting from the constructivist's standpoint: *any unbounded (closed linear) operator T (meeting the PE-R conditions*) cannot be constructively known to preserve constructivity.* (To see this, simply substitute reference to such a set A for the A in the penultimate paragraph of the PE-R proof of their Main Theorem and observe that Lemma 2 applies to show that $y = Tx$ is *not known to be constructive.* This is because – as the classicist knows – were y known constructive, A would in fact be weakly effectively decidable, even if the constructivist did not knowingly have a decision method for A. In this sense, *such unbounded operators cannot be recognized as constructive operators by the type of constructivist program in question.* (Assuming an inexhaustible supply of such sets A, this constitutes a general argument by counterexample for the non-constructivity of such operators. Even should a particular A become decidable, another could be invoked in its place, so that the conclusion has a permanent validity.) This completes our discussion in the first question raised above on the classically discernible implications of the PE-R Theorem.

We turn now to the second question on the implications of the PE-R Theorem from the *constructivist's* standpoint. Here we must consider the constructivity (or failure thereof) of the PE-R proof. Problems arise at the points already indicated above in connection with Lemmas 1 and 2. In the first case, *do* we have a constructively acceptable method of selecting a subsequence p_n of p'_n with $\|Tp_n\| > 10^n \|p_n\|$? As already suggested, this depends on just what the constructivist requires. If the demand is made (as it is by, e.g., Kreisel [1962] and Dummett [1977]) that a constructive method of finding an object of a given sort carry with it a proof that certifies that the method indeed works, then "merely waiting" until an appropriate object turns up is insufficient unless a *constructive* proof is provided that indeed such an object will be found. Is such a proof provided? This depends in turn on how we articulate the constructivized version of the hypotheses of the lemma, in particular the hypothesis that "T is unbounded." Using *constructivist logical connectives* (*including quantifiers*) interpreted, say, in the manner of Kreisel [1962]; (cf. Dummett [1977]), this can be formalized in (at least) two distinct ways, either as

$$\neg\, \exists b \forall x \|Tx\| \leqslant b\|x\| \quad (\text{weak unboundedness of } T),$$

or as

$$\forall b \exists x \|Tx\| > b\|x\| \quad (\text{strong unboundedness of } T).$$

Clearly, if the second is understood, the method given in the PE-R proof of Lemma 1 is constructively acceptable as it stands: the constructive existence of suitable p'_n is already guaranteed by hypothesis of strong unboundedness of T. (Again, we remark that this would presumably be the Bishop-constructivist's preferred reading.) If, on the other hand, merely weak unboundedness is hypothesized, then we may be stuck. We may indeed infer,

$$\forall b \neg \forall x \|Tx\| \leqslant b\|x\|,$$

based on the usual (Brouwer–Heyting–Kreisel) explanations of the constructive (intuitionistic) quantifiers. To pass, however, from this to strong unboundedness requires (at least) an application of Markov's principle,

$$\forall x (Bx \vee \neg Bx) \;\rightarrow\; (\neg \forall x B \rightarrow \exists x \neg B),$$

which is not, in general, intuitionistically acceptable. (Having a refutation that every x is B does not automatically yield a method of finding an x as counterexample to B, even when B is decidable and quantification is over the natural numbers. This stems from the strong existential import of intuitionistic '$\forall$' – that *some* method of proving Bx for each x be *available*. Cf., e.g., Dummett [1977], p. 22, also Hellman [1989b], pp. 51, 58–59, reproduced as Chapter 12 in this volume.) In addition, the inference would appeal to *trichotomy*, which is not constructively valid. Thus, if "*T is unbounded*" is understood as "weakly unbounded," then the PE-R proof of Lemma 1 is not fully constructive as it stands. (Note that these distinctions were not needed in stating the (classical) PE-R Theorem*, since the classical meaning of "unbounded" is unambiguous.)

A similar conclusion is reached concerning the proof of Lemma 2. In order to test for membership in the given set A, we wait to find an n such that the norm $\|y - y_n\|$ is sufficiently small. Classically, of course, we know that eventually such an n will indeed be found since we know, classically, that the series of partial sums y_n of the series for y converges in norm to y. But do we know this constructively? Not until we know that $\|y - y_n\| \to 0$ *effectively* (i.e. *constructively), for that is the only kind of convergence the constructivist recognizes*! But (as Pour-El and Richards note) this is just what has to be proved. Thus, as a fully constructive proof of decidability of A, the proof as it stands would be circular. Finally, moreover, the last step in the proof of the Main Theorem – that $Tx = y$ since T is closed – is non-constructive, since convergence of the series for y is non-effective.

This of course does not answer the question whether a fully constructive proof of the PE-R Theorem is possible. (The closest constructivist analogue of the theorem would assert that from the hypothesis that T is unbounded, a contradiction could be derived, since a constructive T *ipso facto* preserves constructivity; cf. Bridges [1991]). It merely indicates that, as it stands, the proof does not fulfill every constructivist demand. This conclusion, however, in no way affects the truth of the Proposition 1. Rather, we have here a non-trivial case in which classical reasoning enables us to see limitations on constructivity which cannot (at least not yet) be seen by the constructivist. One cannot help being reminded of Plato's parable of the cave . . .

13.3   The Problem of Domains

As indicated in the Introduction, there is an independent line of reasoning leading to much the same conclusion reached in Proposition 1, but affecting in the first instance *radical* constructivism. Once again, we employ classical results to answer questions about limitations of constructive methods.

The reasoning rests on the Hellinger–Toeplitz Theorem, reviewed in Section 13.1, that unbounded symmetric operators A are defined, not on the entirety of the given Hilbert space $\mathcal{H}$, but on a domain D_A dense in $\mathcal{H}$. Thus, given an arbitrary vector ψ in $\mathcal{H}$, it is a relevant mathematical question whether $\psi \in D_A$. From the Spectral Theorem, this is equivalent to the question whether the integral,

$$\int_{-\infty}^{+\infty} x^2 \, d\|E_x \psi\|^2$$

converges, where E_x is the spectral family for A. One certainly suspects that there cannot be a constructive method (even depending on A) for answering this question. One can, in fact, be more definite: any such method, for *any* A, is impossible for it could be used to solve the halting problem. To see this, consider one of the simplest examples. Let $\mathcal{H}$ be L^2, let ϕ_n be a complete orthonormal system, and define,

$$A\phi_n \equiv n\phi_n.$$

The domain of A, $D_A = \{\psi = \sum_{n=1}^{\infty} x_n \phi_n \text{ such that } \sum_{n=1}^{\infty} n^2 |x_n|^2 < \infty\}$, a dense linear submanifold of $\mathcal{H}$. A is unbounded: $\|A\phi_n\| = (A\phi_n, A\phi_n)^{1/2} = (n\phi_n, n\phi_n)^{1/2} = n$ which increases without bound. (A, as expected, is not continuous: $f_n \equiv \frac{1}{n}\phi_n \to 0$ as $n \to \infty$ whereas $Af_n = \phi_n$ does not converge to 0. That D_A is a proper subset of $\mathcal{H}$ is illustrated by letting f have components $x_n = 1/n$ (in the basis ϕ_n): then $f \notin D_A$ since $\sum_{n=1}^{\infty} n^2 |\frac{1}{n}|^2 = \infty$.

That a constructive method of answering the question,

$$\psi \in D_A?$$

would effectively solve the halting problem can then be seen by constructively specifying a vector ξ as follows: its components in the ϕ_n basis are given by a rule such as

$$(\xi, \phi_n) = \begin{cases} \dfrac{1}{n} & \text{if } 2j = \text{ the sum of two primes, all } j \leqslant n; \\ \\ 0 & \text{otherwise.} \end{cases}$$

Then it follows (constructively) that $\xi \notin D_A$ if and only if no counterexample to the Goldbach conjecture can be found. Thus, this reasoning provides a counterexample to any constructive method of deciding membership in D_A. (This should be unsurprising: in general how could there be a constructive method – one that treats infinite sequences by finite approximations – for deciding membership in a dense subset of a topological space?)

In fact, this example can readily be generalized. (This too is unsurprising, but a general theorem serves as a useful guidepost in assessing the status of unbounded operators in QM, as in Section 13.4.) First let us define a separable Hilbert space $\mathcal{H}$ to be (*minimally*) *constructivized* if and only if (i) $\mathcal{H}$ is presented with a constructive orthonormal sequence $\{e_n\}$ and (ii) $\{e_n\}$ constructively spans $\mathcal{H}$ in the sense that any $f \in \mathcal{H}$ can be identified with a constructively convergent sum, $f = \sum_{i=1}^{\infty} c_i e_i$ where the coefficients are constructive (complex) scalars (possibly $= 0$). Next define a set $X \subseteq \mathcal{H}$ to be (*weakly*) *complemented* just in case a witness $w \in \mathcal{H} - X$ can be constructively identified if in fact $\mathcal{H} - X$ is non-empty. Then we may prove the following:

Theorem: *Let T be an unbounded linear symmetric operator defined on a constructive orthonormal basis, $\{e_n\}$, of a constructivized Hilbert space, $\mathcal{H}$. Then, if D_T is complemented, a decision method for membership in the domain D_T of T effectively reduces the halting problem.*

Proof (classical). By the Closed Graph (or Hellinger–Toeplitz) Theorem, D_T is a dense proper subset of $\mathcal{H}$, and by hypothesis some $\psi \notin D_T$ can be constructively identified, as $\psi = \sum_{i=1}^{\infty} c_i e_i$. By linearity and the hypothesis that $\{e_n\} \subset D_T$, $T(\sum_{i=1}^{k} c_i e_i) = \sum_{i=1}^{k} c_i T e_i$ is well defined, and the partial sums on the left, $\psi_k \to \psi$ constructively. Thus, we may constructively specify a vector ξ by a rule such as the following: the components of ξ in the e_n basis are given by

$$(\xi, e_n) = \begin{cases} c_n & \text{if } 2j = \text{ the sum of two primes, all } j \leqslant n; \\ 0 & \text{otherwise.} \end{cases}$$

Thus, we have (constructively) that if $\xi \notin D_T$, $\xi \neq \psi_k$ for any k, and there is no counterexample to the Goldbach conjecture; and if $\xi \in D_T$, $\xi \neq \psi$ and it is absurd that no counterexample to Goldbach can be found. Thus, constructive decidability of D_T has the same effect as constructive solvability of the halting problem. (Clearly, the latter could have been employed directly – though less graphically – in introducing ξ.)

It follows that, *if* constructivism is held to the standard of constructively answering the question, "Is ψ in the domain of operator A?," for any operator A that it recognizes as a mathematical object, it cannot consistently recognize unbounded (linear symmetric) operators (defined on a constructive orthonormal basis and with complemented domain). Here, by "recognize as a mathematical object" we mean recognize as a value of a quantifiable variable over objects of that same type. There is nothing to prevent the case-by-case *use* of, say, a differential operator, to perform particular calculations. But the *theory* of such operators cannot be carried out without quantification over them, and that is what is in question here. We remark that, typically, an unbounded operator – say a differential operator, T – leads outside of the Hilbert space (say an L^2 space), and this can happen even when the input, ψ, is constructive and the output, $T\psi$, is constructive as well, the problem being that $\int |T\psi|^2 dr$ diverges. Such an operator is not regarded as specified unless its domain of definition is specified (cf., e.g., Jordan [1969], p. 130, Problem 3.) To what standard should the radical constructivist be held regarding the question "Is ψ in the domain of A?"?

To answer this, we must recall the philosophical basis for the radical constructivist critique of non-constructive classical mathematics. Following Dummett [1977], we emphasize the following verifiability criterion of cognitive significance:

A mathematical concept is not meaningfully applicable apart from an idealized mathematician's having a constructive method that shows that it applies. In sum, no proof-independent mathematical facts are countenanced.[3]

(PCCM)

(The acronym is for "proof-conditional criterion of meaningfulness.") Now, if the PCCM is consistently adhered to, it seems clear that the above question, "Is ψ in the domain of A?," must be answerable constructively, if an operator A is to be regarded as constructively specified. This standard is merely an extension to the question of definedness (in this case in the (constructivized) Hilbert space) of the radical-constructivist standard for recognizing a constructive function in the first place, namely, that there should be provided a constructive method for answering the meaningful mathematical question, "What is the value of $f(x)$?" As the PCCM expresses, essential to the radical-constructivist conception of a function is the doctrine that there be no separation between the *truth* of a statement of the form $f(x) = y$, and possession (by an idealized mathematician) of a method for computing y from x (and thus of deciding whether $f(x) = z$, for any z). But if both f and x are constructive objects, then a constructivism consistently adhering to the PCCM should *also* provide a constructive method for deciding whether f is *defined* at x. (NB: We are not saying that a *uniform* method be provided, independent of f, or even independent of x, although uniformity in x may be inevitable on a proper constructivist interpretation.) The "slippery slope argument" for this runs thus: After all, the concept of a function's being *defined* – especially in the case of linear operators as just indicated – is certainly one that the constructivist uses and needs to use all the time. As announced in the PCCM, like other mathematical concepts, its conditions of applicability must be fully grounded in methods of construction: *it*

---

[3] As Dummett writes:

"From an intuitionistic standpoint . . . an understanding of a mathematical statement consists in the capacity to recognize a proof of it . . . and the truth of such a statement can consist only in the existence of such a proof." (1977, p. 6)

Moreover,

"[Mathematical objects] exist only in virtue of our mathematical activity, which consists in mental operations, and have only those properties which they can be recognized by us as having." (1977, p. 7)

*should not be allowed meaningfully to apply independently of an idealized math-
ematician's having a construction that shows that it applies.* Otherwise a realm of
proof-independent mathematical fact is implicitly being recognized (whether
f really is or is not defined at x), which is the essence of the mathematical realism
the constructivist seeks to avoid. If such proof-independent facts are recognized in
one quarter (definedness of functions), why not in another (values of functions)?
Thus the proposition of this section applies at least to radical constructivism.

Proposition 2: *Let C be a variety of constructivism recognizing only operators
for which it provides constructive definedness conditions as well as constructive methods of computing values from arguments. Then C cannot recognize
unbounded linear symmetric operators (defined on a constructive orthonormal
basis and with complemented domain) in Hilbert space as constructive
objects.*

To conclude this section, let us briefly examine the resistance to this
proposition implicit in the practice of some leading constructive mathematicians. The standard invoked – decidability of domains of constructive
functions – may not be accepted for the same reasons that lead to recognition of non-effectively decidable sets generally. (Given such a set X, one
may stipulate that, e.g., a "trivially calculable" function, say $f(x) = x$, be
defined on X.) A favorite illustration – and the focus of a recent controversy
on constructive analysis[4] – is the set $F = \{0\} \cup \{x \in \mathbb{R}: x = 1$ &
Fermat's Conjecture fails$\}$. Lacking at present a method of deciding whether
1 belongs to F, why does the constructivist now already recognize F as
a constructive object? The reason given (e.g., by Richman [1987]) is that
F turns out to be the range of a perfectly computable function of natural
numbers, namely $f(n) = 1$ exactly when there are positive integers x, y, z, m all
less than n such that $x^{m+2} + y^{m+2} = z^{m+2}$, $x^{m+2} + y^{m+2} = 0$ otherwise. Therefore, it
is argued, "the set F won't go away" (Richman [1987], p. 24). More generally,
the only restriction on sets recognized by leading constructive analysts (e.g.,
Bridges [1979]) seems to be that they be specified by a (presumably predicative) formula (predicative comprehension principle). What are we to make of
this?

Without attempting to settle the matter definitively here, we would like to
point out that the above appeal to comprehension,

(i) "Every function has a range,"

---

[4] See the exchanges between Ian Stewart [1986, 1987] and Fred Richman [1987] in *The
Mathematical Intelligencer.*

can hardly be regarded as conclusive by the *radical* constructivist, who after all is prepared to sacrifice (at least, not accept) even classical logical laws in the interests of a philosophical principle such as the PCCM. Moreover – and this concerns liberals as well – (i) competes with another classical principle,

(ii) "A set exists iff its characteristic function does,"

which even has its classical computable analogue,

(ii') "A set is recursive iff its characteristic function is,"

unlike (i), whose computable analogue (i') (the result of modifying "function" and "range" by "recursive") fails. So why should not (ii), rather than (i), be preserved in constructive mathematics, i.e., why not end up recognizing as constructively existing only those sets with constructive characteristic functions? (In terms of the "fishy Fermat" example, if we can make its characteristic function "go away," why cannot we make F itself "go away" too?) Whatever the radical's answer, it cannot be that that would cripple constructive mathematics (if indeed it would), for how could such utilitarian considerations matter in the face of a philosophical principle of cognitive significance?

13.4   Are Unbounded Operators Required to Sustain Life in the Quantum World?

Perhaps many forms of life can survive in the quantum world, lower as well as higher. And perhaps we need not regard the lower as having evolved earlier; perhaps we should even recognize parasitic lower forms evolving only after the higher have established their niche. We shall not attempt to settle these questions definitively here, but will content ourselves with fleshing out the metaphor.

As already indicated in Section 13.1, *unbounded operators abound in QM* (both ordinary and relativistic). (As Prugovecki [1981] tells us, "most of the operators of interest in quantum physics are unbounded" (p. 180).) The operators for position, linear momentum, and very important functions of these, especially Hamiltonian operators representing total energy, and the so-called creation and annihilation operators as well, are unbounded.[5] Even the most elementary textbook problems (e.g. the harmonic oscillator) cannot be treated without such operators (see e.g., Jauch [1968], Ch. 12, §6). Moreover, typically

---

[5] The Hamiltonian operators, H, and the annihilation and creation operators, A and A^*, respectively, are expressible as functions of the position and momentum operators, Q and P, in a general form as follows:

such operators are defined on an orthonormal basis (e.g. eigenfunctions of the Hamiltonian) which, at least in standard problems, presumably are amenable to a constructive treatment. (For example, eigenfunctions of the Hamiltonian for the quantum mechanical harmonic oscillators take the form

$$\frac{1}{\sqrt{2^n n! \sqrt{\pi}}} H_n(x) e^{-x^2/2},$$

where the $H_n(x)$ are the Hermite polynomials,

$$H_n(x) = \sum_{k=0}^{[n/2]} \frac{(-1)^k n}{k!(n-2k)!}(2x)^{n-2k},$$

where $[n/2]$ denotes the greatest integer $\leqslant n/2$.) Thus, the condition of definedness of the operators on a constructive orthonormal basis, entering into both the PE-R Theorem and our own of Section 13.3, is not difficult to meet. To make plausible the claim that the proof of the latter is applicable to rule out the constructivity of these operators, we need to argue that a suitable ψ can indeed be found, i.e., such that ψ is in the constructivized Hilbert space $\mathcal{H}$ but not in the domain of definition of the operator in question. Consider the position operator Q (for simplicity, in one dimension), and suppose we have a particle confined to values of $x \geqslant 1$; we can construct $\psi(x)$ as $\psi(x) = 1/x$ for $x \geqslant 1$. Then

$$\|\psi(x)\|^2 = \int_{+1}^{+\infty} \frac{1}{x^2}\,dx = 1,$$

whereas

$$\|Q\psi(x)\|^2 = \int_{+1}^{+\infty} x^2 \frac{1}{x^2}\,dx = \infty,$$

so that Q cannot be defined at ψ. (Note that for this argument, it is not necessary that Q be constructively recognized as a mathematical object, but merely that ψ be and that the integral on the right be known constructively to diverge.)

$$H = \tfrac{1}{2\mu} P^2 + V(Q),$$

where μ is a constant (satisfying $m/\mu = h$) and V is a potential function representing the effect of external forces,

$$A = \tfrac{1}{\sqrt{2}}(Q + iP), \text{ and } A^* = \tfrac{1}{\sqrt{2}}(Q - iP).$$

(see, e.g., Jauch [1968], Ch. 12, §§5, 6). For a wide class of potential functions, V, the domain of the corresponding Hamiltonian, H (for a many-particle quantum system), is a set of infinitely many-times differentiable functions with compact support which is dense in the L^2 Hilbert space (see, e.g., Prugovecki [1981], Ch. 2, Theorems 5.5 and 5.6.)

Finally, the constructivist can evaluate the components of ψ in an energy eigenbasis $\eta_k(x)$ by evaluating the inner products (η_k, ψ), which in standard cases presents no special problems. In fact, we may suppose – without loss of generality – that inner products with an energy eigenbasis are also computable, otherwise the constructivist could not answer elementary questions arising in QM for such state functions (namely, "What is the probability in state ψ of finding eigenvalue E_k of the Hamiltonian H?").

The situation in connection with momentum operators is similar. For example, in the L^2 space of functions $\psi(x)$ with $0 \leqslant x \leqslant 2\pi$, the differential operator T given by $T\psi(x) = -i(d/dx)\psi(x)$ is Hermitian provided its domain is restricted by choosing some fixed real b and requiring that $\psi(2\pi) = e^{i2\pi b}(\psi(0))$. Absolutely continuous functions in this L^2 space possessing square-integrable derivatives but not satisfying the restriction at the extremities 0 and 2π will lie outside the domain of T (but, in fact, in the domain of its adjoint, T^*) (cf. Riesz and Nagy [1955], §119). Presumably, the constructivist has no difficulty in recognizing such functions. Similar remarks carry over in the case of the square of momentum, kinetic energy, and a wide class of Hamiltonian operators (cf., e.g., Jordan [1969], p. 131, Problem 4). Thus, a great many of the standard operators of QM fall within the purview of the Theorem of Section 13.3.

Two questions now arise concerning the role of such operators in quantum physics. First, we may ask whether their *use* at the empirical level is genuinely necessary. And second, in any case, we must ask whether, in their theoretical role, it is necessary to quantify over them as mathematical objects.

The first question arises as follows. In any real physical situation (or applicable hypothetical one)[6] – the argument might run – we need only evaluate probabilities of finding values of physical quantities within an interval $\Delta = (x, y]$ of the real line. But to calculate answers to such questions, it suffices to employ only *bounded* operators, of the form E_Δ, a projector $= E_y - E_x$, these belonging to the spectral family for the original observable A. (Although the spectral theorem for A is unavailable, we may imagine the constructivist coming up with the spectral family independently.) Charges of parasitism apart, however, this strategy cannot do justice to real scientific practice. For how are questions concerning expectation values in ensembles to be framed? Here we must confront pairs of incompatible observables (corresponding to non-commuting operators), such as position and momentum. As is well known from Fourier

---

[6] We are intentionally side-stepping the point that many or even perhaps all of the physical applications of quantum theory involve idealization, and may rest on assumptions known to be literally false (e.g. Galilean as opposed to Lorentz invariance). In the present context, this is appropriate, for, so long as the models invoked are coherent possibilities (and especially if such models are actually useful), a mathematical framework should accommodate them. It must not turn out that some species of constructivism survive only by a lucky accident (despite the prevailing metaphor of this section)!

analysis, if one of these quantities is confined in a state to a bounded interval of the real line, the other can take on arbitrarily large (positive and negative) values with positive measure, so that calculation of the expectation value of that quantity in that state will in general require an unbounded operator.[7]

Not only would unbounded operators seem to be thus empirically indispensable. Theoretically, their role is absolutely central: the very structure of quantum mechanics and its relationship to classical mechanics are bound up with unbounded operators and their interrelations, treated as objects in their own right. Under this heading we may list at least the following, all of which involve consequences of the Spectral Theorem and/or Stone's Theorem. (i) *The canonical commutation relations* between position and displacement operators (the Weyl relations), which invokes Stone's Theorem (quantifying over operators), forging an important link between physical spatial properties (localizability of particles and homogeneity of space) and the commutation rules for position and momentum operators (as generators of the unitary groups invoked in expressing those properties) (see, e.g., Jauch [1968], Ch. 12). (ii) As noted in Section 13.1, *the dynamical structure of the theory*, including the role of unbounded Hermitian operators in the abstract time-dependent Schrödinger equation, is derived from a fundamental symmetry, time-translation invariance, again by invoking Stone's Theorem. Further considerations of Galilei invariance enable us to identify these unbounded operators as Hamiltonian operators. Stone's Theorem enters again in this derivation (see Jauch [1968], Ch. 12, §5). (ii) *The Correspondence Principle*, expressing essential and deep ties between quantum and classical mechanics: that if a classical physical quantity is represented as a (suitably well-behaved) function of position and momentum, $f(q, p)$, then in QM a corresponding observable is represented by $f(Q, P)$ (Q the QM position operator, P the QM momentum operator), probably the most important application of which is in forming the QM Hamiltonian itself, $H = (1/2m)P^2 + V(Q)$. (This, of course, depends on the functional calculus provided by the Spectral Theorem.) Finally, (iv) we may mention the *classical limit theorems* and the Ehrenfest equations, showing that the expectations of the QM Q and P operators obey the classical Newtonian equations governing the time-derivatives of the corresponding classical functions. Quantification over unbounded operators enters into the standard derivation, via the important QM theorem,

---

[7] Indeed, the exact value could be approximated with a bounded operator, but how are we to know in general and in advance how close is "close enough"?

For a related discussion of the suggestion that unbounded operators be dispensed with in favor of their spectral projections (motivated, however, by problems of *physical* interpretation quite distinct from those pertaining to mathematical computability addressed here), see Heathcote [1990].

$$\frac{\mathrm{d}}{\mathrm{d}t} \langle A \rangle_t = \frac{i}{\hbar} (\psi_t, [HA - AH]\psi_t),$$

i.e., the time-derivative of the expectation of A in the state ψ_t at t is given by taking the inner product of ψ_t with the result of operating with the commutator of the Hamiltonian H and A. This is a theorem universal in both H and A. It allows us to derive the Ehrenfest equations without knowing a particular Hamiltonian for the system in question, as would be necessary were we allowed only to work with particular unbounded operators.

From this brief list alone, it seems clear that the theory of unbounded linear operators – not just their individual use – enters crucially in developing a large part of the core content of quantum mechanics. If we are right, then, the quantum world is indeed a hostile environment for many species of constructivist mathematics. Whether any of them can survive at all – and if so, in what (possibly mutated) forms – are questions that such considerations as those addressed here may help to settle.

Acknowledgement: This work has been carried out under support of the National Science Foundation, Award DIR-8922435, which is gratefully acknowledged. I am also indebted to David Malament for helpful comments on an earlier draft, and to a referee for a number of clarifications and for calling my attention to the interesting exchange of n. 4.

References

Bishop, E. [1967] *Foundations of Constructive Analysis* (New York: McGraw Hill).

Bishop, E. and Cheng, H. [1972] *Constructive Measure Theory* (Providence, RI: American Mathematical Society).

Bridges, D. [1979] *Constructive Functional Analysis* (London: Pitman).

Bridges, D. [1981] "Towards a constructive foundation for quantum mechanics," in Richman, F., ed., *Constructive Mathematics*, Springer Lecture Notes in Mathematics, No. 873 (Berlin: Springer), pp. 260–273.

Bridges, D. [1991] Review of Pour-El and Richards, *Computability in Analysis and Physics* (Springer-Verlag), *Bulletin of the American Mathematical Society* 24(1): 216–228.

Burgess, J. [1984] "Dummett's case for intuitionism," *History and Philosophy of Logic* 5: 177–194.

Dummett, M. [1977] *Elements of Intuitionism* (Oxford: Oxford University Press).

Feferman, S. [1984] "Between constructive and classical mathematics," in *Computation and Proof Theory*, Lecture Notes in Mathematics 1104 (Berlin: Springer), pp. 143–162.

Gleason, A. [1957] "Measures on the closed subspaces of a Hilbert space," *Journal of Mathematics and Mechanics* 6: 885–893.

Heathcote, A. [1990] "Unbounded operators and the incompleteness of quantum mechanics," *Philosophy of Science* **57**(3): 523–534.

Hellman, G. [1989a] *Mathematics without Numbers: Towards a Modal-Structural Interpretation* (Oxford: Oxford University Press).

Hellman, G. [1989b] "Never say 'never'!: On the communication problem between intuitionism and classicism," *Philosophical Topics* **17**(2): 47–67.

Hellman, G. [1993] "Gleason's theorem is not constructively provable," *Journal of Philosophical Logic* **22**: 193–203.

Jauch, J. M. [1968] *Foundations of Quantum Mechanics* (Reading, MA: Addison-Wesley).

Jordan, T. F. [1969] *Linear Operators for Quantum Mechanics* (New York: Wiley).

Kreisel, G. [1962] "Foundations of intuitionistic logic," in Nagel, E., Suppes, P., and Tarski, A., eds., *Logic, Methodology, and Philosophy of Science* (Stanford, CA: Stanford University Press), pp. 198–210.

Pour-El, M. and Richards, I. [1983] "Computability in analysis and physics: a complete determination of the class of noncomputable linear operators," *Advances in Mathematics* **48**(1): 44–74.

Pour-El, M. and Richards, I. [1987] "The eigenvalues of an effectively determined self-adjoint operator are computable, but the sequence of eigenvalues is not," *Advances in Mathematics* **63**(1): 1–41.

Prugovecki, E. [1981] *Quantum Mechanics in Hilbert Space* (New York: Academic Press).

Putnam, H. [1967] "Mathematics without foundations," reprinted in Benacerraf, P. and Putnam, H., eds. *Philosophy of Mathematics: Selected Readings*, 2nd edn. (Cambridge: Cambridge University Press, 1983), pp. 295–311.

Putnam, H. [1971] *Philosophy of Logic* (New York: Harper).

Quine, W. V. [1953] "Two dogmas of empiricism," reprinted in *From a Logical Point of View* (New York: Harper & Row, 1963), pp. 20–46.

Richman, F. [1987] "The frog replies" and "The last croak," *Mathematical Intelligencer* **9**(3): 22–24, 25–26.

Riesz, F. and Sz.-Nagy, B. [1955] *Functional Analysis* (New York: Ungar).

Stewart, I. [1986] "Frog and mouse revisited: A review of . . . *Constructive Analysis* by Errett Bishop and Douglas Bridges (Springer: 1985) and *An Introduction to Nonstandard Real Analysis* by A. E. Hurd and P. A. Loeb (Academic Press: 1985)," *Mathematical Intelligencer* **8**(4): 78–82.

Stewart, I. [1987] "Is there a mouse in the house?" and "A final squeak," *Mathematical Intelligencer* **9**(3): 24–25, 26.

Stone, M. H. [1932] *Linear Transformations in Hilbert Space* (New York: American Mathematical Society).

Appendix

For the reader's convenience, we reproduce here relevant portions of Pour-El and Richards' [1983] proof of their Main Theorem.

Now suppose that T is not bounded. We need to find a computable element $x \in \text{domain}(T)$ such that Tx is not computable in Y. The construction of x is based on two lemmas.

Lemma 1: *Take the assumptions of the Main Theorem, with T unbounded. Then there exists a computable sequence of e-polynomials p_n such that Tp_n is computable in Y and*

$$\|Tp_n\| > 10^n \|p_n\| \text{ for all } n.$$

Proof of Lemma 1. By definition of e_n, the linear span of $\{e_n\}$ is dense in X. Since the operator T is closed, T cannot be bounded on the span of $\{e_n\}$; else T would be bounded on X. Now we sweep out the set of all finite (rational/ complex rational) linear combinations of the e_n in an effective way: this is easily done using any of the standard recursive enumerations of all finite sequences of integers. Thus we arrive at a computable sequence of e-polynomials $p'_n \in X$ which runs through all finite (rational/complex rational) linear combinations of the e_n.

By hypothesis, the sequence Te_n is computable in Y. Since Tp'_n is an effectively generated sequence of linear combinations of the terms Te_n, it follows from the summation axiom that Tp'_n is computable in Y.

We now construct a computable subsequence p_n of p'_n such that $\|Tp_n\| > 10^n \|p_n\|$ for all n. By the composition axiom, any recursive process for selecting a subsequence of indices n also produces computable sequences p_n and Tp_n in the Banach spaces X and Y, respectively.

As we have seen, the set of ratios $\{\|Tp'_n\| / \|p'_n\|, \ p'_n \neq 0\}$ is unbounded. On the other hand, by the norm axiom, the sequences $\|Tp'_n\|$ and $\|p'_n\|$ are computable. So we can effectively select a subsequence p_n of p'_n with $\|Tp_n\| > 10^n \|p_n\|$, merely by waiting, for each n, until a suitable p'_n turns up.

Lemma 2: *Let $r > 2$ be a computable real. Let z_n be a computable sequence in Y with $\|z_n\| = 1$ for all n. Let $a: \mathbb{N} \to \mathbb{N}$ be a one-to-one recursive function which enumerates a set $A \subseteq \mathbb{N}$. (The set A is recursively enumerable, and may or may not be recursive.) Then the element*

$$y = \sum_{k=0}^{\infty} r^{-a(k)} z_k$$

is computable in Y if and only if the set A is recursive.

Proof of Lemma 2. The "if" part is trivial. If A is recursive, then the series for y converges effectively, and we apply the limit axiom.

For the "only if" part: Assume that y is computable, and let y_n denote the nth partial sum of the above series. By the summation and norm axioms, the sequence of norms $\|y - y_n\|$ is computable. These norms converge to zero, and we wish to prove that the convergence is effective. Since the sequence $\|y - y_n\|$ is not necessarily monotone, this is not automatic; it depends on the assumption that $r > 2$.

Since $r > 2$, each term in the series $\sum_{a=0}^{\infty} r^{-a}$ is strictly larger than the sum of all the following terms: the ath term is r^{-a}, whereas $\sum_{b=a+1}^{\infty} r^{-b} = r^{-a}/(r - 1)$.

This leads to an effective decision procedure for the set A, proving that A is recursive (and incidentally proving that $\|y - y_n\| \to 0$ effectively). To test whether an integer a belongs to the set A, we merely wait until we have found an n for which

$$\|y - y_n\| < r^{-a} - [r^{-a}/(r - 1)].$$

If a has occurred as some value $a(k)$, $0 \leqslant k \leqslant n$, then $a \in A$; otherwise $a \notin A$. To prove this, we argue as follows.

Suppose $a(k)$, $k > n$, takes some value $\leqslant a$, and let $c = a(m)$ be the *least* such value. Then r^{-c} exceeds the sum of all other terms $r^{-a(k)}$, $k > n$, by at least $r^{-c}[1 - (r - 1)^{-1}] \geqslant r^{-a}[1 - (r - 1)^{-1}]$. Now consider the corresponding series of Banach space elements,

$$y - y_n = \sum_{k > n} r^{-a(k)} z_k.$$

The term $r^{-c} z_m$ has norm r^{-c} (since $\|z_m\| = 1$). However, by the triangle inequality, the sum of the other terms has norm

$$\leqslant \sum_{a > c} r^{-a(k)} = r^{-c}/(r - 1).$$

Hence,

$$\|y - y_n\| \geqslant r^{-c} - [r^{-c}/(r - 1)],$$

and since $c \leqslant a$, this contradicts the previous inequality on $\|y - y_n\|$.

Proof of the Main Theorem concluded. Following Lemma 1, we take a computable sequence p_n in X such that Tp_n is computable in Y and $\|Tp_n\| > 10^n \|p_n\|$. Then the sequence of norms $\|Tp_n\|$ is also computable, and we define the computable sequences

$$z_n = Tp_n / \|Tp_n\| \text{ in } Y,$$

$$u_n = p_n \,/\, \|Tp_n\| \text{ in } X,$$

with $z_n = Tu_n$, $\|u_n\| < 10^{-n}$, and $\|z_n\| = 1$.

Let $a \colon \mathbb{N} \to \mathbb{N}$ be a recursive function which enumerates a recursively enumerable non-recursive set A in a one-to-one manner. Define x and y by

$$x = \sum_{k=0}^{\infty} 10^{-a(k)} u_k,$$

$$y = \sum_{k=0}^{\infty} 10^{-a(k)} z_k.$$

Then by Lemma 2 the element y is not computable in Y. On the other hand, since $\|u_k\| < 10^{-k}$, the series for x converges effectively, and by the limit axiom we see that x is computable in X.

Finally, $Tu_n = z_n$, and the series for x and y converge (although not necessarily effectively). Since the operator T is closed, it follows that x belongs to the domain of T, and $Tx = y$.

14   If "If-Then" Then What?

14.1   Background

The roots of the view of mathematics known as "if-thenism" or "deductivism," like its near cousins, logicism and formalism, are to be found in the nineteenth and early twentieth centuries' "new birth" of mathematics. As summarized nicely by Maddy [to appear], this period witnessed a multifaceted transformation from mathematics as investigating formal properties of aspects of material reality, such as space, time, and motion, to its being an exploration of abstract concepts and structures in their own right quite apart from any material applications that they might have.

The main components of this transformation include the rise of non-Euclidean geometries, the work of Dedekind, Cantor, et al. in the foundations of number systems and analysis, the emergence of transfinite set theory due to Cantor and then Zermelo, and the great advances in mathematical logic due principally to Frege but also to the contributions of Russell and Whitehead and Hilbert and Ackermann. The case of multiple geometries is strikingly clear, as it led to a reconceptualization of geometry as having become the study of a variety of purely mathematical spaces or structures, free of claims to describe actual physical space or space-time. In the same period, the work of Dedekind, Cantor, et al. in the foundations of number systems and analysis, along with Hilbert's work in foundations of geometry, led to the articulation of systems of axioms describing relevant structures in ways akin to the workings of abstract algebra. The new freedom to explore abstract structures not realized in the material world reached a high point (or many high points – I acquiesce in the pun) with Cantor's far-reaching discoveries in transfinite set theory, transcending anything that could be conceived as describing material reality.

All these factors, combined with the discovery of the set-theoretic paradoxes, set the stage for the development of the "big three" early twentieth century foundational programs of logicism, intuitionism, and formalism, with intuitionism being a critical reaction against transfinite set theory, and formalism, as articulated with the help of Hilbert's proof theory, being a quest for a finitistic justification of transfinite set theory via a proof of formal consistency of relevant axiom systems.

On the more philosophical side, there were diametrically opposing ways of viewing the new, abstract-structures-oriented mathematics. One of these, of course, was a platonism or objects-realism, with especially strong ontological commitments to accommodate Cantorian set theory. As already indicated, intuitionism was a severe reaction against this classical platonism, but it was such at the cost of rejecting non-constructive mathematics *überhaupt*, not merely that pertaining to transfinite set theory, but also involving significant parts of classical analysis. Quite naturally, even before the advent of intuitionism, there was a search for what amounted to less destructive alternatives, and it was in this setting that "if-thenism" or "deductivism" emerged. Through its lens, mathematics was conceived as the systematic study of "what follows logically from what." Axioms were regarded as merely the starting points of proofs of theorems; the axioms need not be regarded as true or descriptive of reality, abstract or concrete; indeed axioms never really needed even to be *asserted*. According to if-thenism, they occur at best merely as antecedents of *conditional* assertions, assertions whose truth is a matter of deductive logic, free of ontological commitments, entirely so in the case of (free) first-order logic, and at least largely so in the case of second-order logic (although this very distinction was not clearly available when Russell enunciated his "if-thenism" in 1903, only being drawn clearly by implication in the late 1920s by Hilbert and Ackermann).[1]

Now, although some leading thinkers, including for instance Russell, Hempel, and Putnam (if only briefly), supported if-thenism,[2] it was subject to a number of objections[3]; and, while many of us read about the view and objections to it in studying philosophy of mathematics, it never gained the kind of following enjoyed by its cousins, logicism and formalism. *Whether some version of if-thenism can be forged into a viable philosophy of mathematics, improving on logicism and formalism, is one way of framing the overarching question to be addressed in this paper.* The occasion for our raising it is recent work of Maddy [to appear], which significantly improves upon earlier articulations of if-thenism and systematically addresses the main objections or challenges that have been raised against those earlier formulations. Maddy calls her version "Enhanced If-Thenism" (abbreviated EIT), intended as an

---

[1] See Russell [1903] and Hilbert and Ackermann [1928]. The latter, I believe for the first time, articulated a system of first-order logic. Formal, axiomatic second-order logic came still later, e.g. with Church [1956].

  The realization that second-order logic would suffice to encompass virtually all of modern mathematics came even later in the twentieth century. See e.g. Shapiro [1991] and sources cited therein.

[2] In addition to Russell [1903], see Hempel [1945] and Putnam [1967a].

[3] See especially Quine [1936], Resnik [1980], and Maddy [to appear].

improvement over "Simple If-Thenism" (SIT), and she makes a strong case that it overcomes the objections she considers. In subsequent sections we will first describe the main objections and challenges to early formulations of if-thenism, and then describe Maddy's EIT. Finally, we will examine how EIT fares with regard to those challenges, as well as two further challenges that we claim need to be overcome.

14.2   Challenges to Simple If-Thenism

Following Maddy [to appear], the first three challenges to SIT are as follows.
1. (From history of mathematics) While SIT may be applicable to mathematics since the mid- to late-nineteenth century, how does it apply to earlier mathematics, antedating the rise of the axiomatic method? In short, it seems that in many cases there is nothing to go into the "If" clauses (with Euclidean geometry standing as a notable exception).
2. (From scientific applications of mathematics) When mathematics is applied in natural scientific contexts, it seems that one needs to assert the "Then" clauses – the theorems – of the relevant conditionals, for example the inter-mediate value theorem, the extreme value theorem, etc. Even so elementary an application as the Newtonian law of universal gravitation seems to require commitment to the existence of numbers and functions (the gravitational constant, the mass and distance functions involved). This goes well beyond merely asserting conditionals whose antecedents would be relevant axioms pertaining to numbers, functions, and sets. (See e.g. Putnam [1975] and the literature on indispensability arguments for mathematical realism.)
3. (From arbitrariness of axioms and loss of distinctive content) According to SIT, mathematical assertions of axioms and theorems are replaced by relevant conditional assertions, which describe purely logical implication or deducibility of the consequents from the antecedents. As Quine [1936] pointed out, the SIT method could be applied to any subject whatever, for example sociology or even mythology, if one could state suitable axioms for those subjects, an easy way of "logicizing away" any characteristic content. If mathematics is to retain its standing as a discipline with distinctive content, an account must be given of its axioms (the contents of the "if" clauses), an account that accounts at least for their non-arbitrariness.

   To these should be added two further challenges.
4. (Problem of vacuity) Suppose the conjunction of the axioms entering into the antecedent of a conditional is false or, worse, inconsistent. Then all conditionals with that antecedent count as true vacuously, with non-theorems as well as theorems as consequents. This is an intolerable

situation, even for an ardent if-thenist (if one can be ardent about such a hedged position!). To deal with this, it seems that it must be asserted, at a minimum, that the axioms are *consistent*. But wait: we have known since Gödel [1931], with the arithmetization of syntax, that consistency of axioms of a formal system is statable as a sentence in the language of arithmetic. Such a sentence says that no number is the code of a proof from the specified axioms with a "bad sentence," such as '0 = 1', as the last line of the proof. But this is an infinitistic, quantified statement about numbers, and so a categorical mathematical claim, not merely a conditional one, contrary to the whole thrust of the if-thenist enterprise.[4]

Finally, we raise what seems to be the most fundamental of all the challenges.

5. (Change of subject, loss of *structural* content) With their focus on what is provable from what, both SIT and EIT are in danger of transforming the subject of mathematical problems from questions about what is true in mathematical structures of interest to questions about what can be proved from some starting points or others. Consider, as an example, a question Weierstrass addressed and famously answered (in the affirmative): Is there a continuous function of reals that is nowhere differentiable? This is simply a very different question from the question of whether the existence of such a function can be proved from these or those starting points. It is not that the latter sort of question is not also part of the investigation, as obviously it is. Both kinds of questions are important, but they are *different* questions, and both matter to the mathematician. The first one is a precise mathematical question (thanks to the work of great nineteenth century mathematicians on precise definitions of continuity, real numbers, etc.), whereas the second one is a metamathematical one and a rather *imprecise* one at that, due to uncertainty early in the investigative process as to just what may be needed and/or sufficient to derive the answer to the properly mathematical question.[5] Indeed, the second question could be replaced by the obviously imprecise question "What will it take to informally prove . . . ?" The distinction remains as relevant today as it was in the nineteenth century. Consider a still open problem in number theory, such as the twin prime conjecture or the Goldbach conjecture. The

---

[4] See Resnik [1980], Ch. 3. Also, note that it is to avoid this vacuity objection that Hellman's [1989] modal-structural interpretation includes an axiom to the effect that possibly the axioms in question are simultaneously satisfied.

[5] Here it is important to distinguish the question "What axioms are necessary in order to prove S?" from questions of the form "Is S provable from such-and-such axioms?" The former can be made precise, as is done in the Reverse Mathematics program of Friedman and (see Simpson [1999]) of modern proof theory. But it is the latter type of questions that we are claiming could not be properly posed in many specific cases historically, and that cannot even be formulated in many contemporary cases of open problems.

properly mathematical questions – questions which we all understand – are whether there are infinitely many twin primes, and whether every even number is the sum of two primes. What it will take to settle these questions is far from clear. It is even possible that the axioms of ZFC do not suffice. We simply do not have a precise metamathematical question to pose that singles out the relevant axioms. Furthermore, even if we were to frame a precise metamathematical question or hypothesis about formal provability of a mathematical statement from specified formal axioms, this would be mathematically equivalent to a question or hypothesis of formal number theory, but this is out of bounds for if-thenism, which seeks to avoid commitments to truth of unbounded quantified statements of formal number theory. Thus, if-thenism faces a dilemma: *either* it rests with imprecise questions and statements of informal provability from informally described starting points, *or* it implicitly recognizes facts of formal number theory in the form of unbounded quantified statements about formal provability.

Let us next turn to Maddy's analysis and proposal for dealing with these objections and challenges, developing EIT as an improvement on SIT.

14.3   Maddy's Enhanced If-Thenism

Before exposing EIT, it is appropriate to describe Maddy's way of responding to the first two challenges just listed, as it is in response to the third that EIT is formulated.

In considering the first challenge, from the history of mathematics prior to the rise of the axiomatic method, Maddy prescribes a dose of modesty: we can and should lower our sights and develop an account of *modern* mathematics deploying the axiomatic method, recognizing that mathematics prior to that can be understood differently. If-thenism is tailored to axioms-based mathematics, and it should be evaluated in terms of its account of that much mathematics. Historically, mathematics has evolved substantially, and it is perfectly in order that different accounts apply to different stages of its development. Once we have described EIT and seen how it deals with the other challenges, we can return to this historical challenge in the context of comparing EIT with an alternative, structuralist conception.

The second challenge, from applications of mathematics in the sciences, calls for a re-examination of the indispensability arguments. *Prima facie* it certainly seems that, especially in the science of physics, the very *formulation* of laws and regularities requires substantial mathematics. We have seen how reference to numbers and functions arises already in Newtonian gravitation theory. When it comes to twentieth century physics much more is involved: quantum mechanics (QM) requires substantial portions of functional analysis (the whole Hilbert-space formalism and a generalized probability theory), while general relativity (GR) requires several chapters of differential geometry just for its proper formulation. When we describe for example the hydrogen atom in terms of appropriate rays and operators in the appropriate Hilbert space, we are committed not merely to a conditional with the axioms of Hilbert-space theory as antecedent, and our description of the atom in the consequent! It seems that in recognizing the atoms and particles of modern physics, we are also willy-nilly committed to the mathematical existence of Hilbert spaces and relevant operators thereon.

Maddy's response to all this consists in a fictionalist approach. At the outset, she notes that mathematical models involved in physics frequently are highly idealized, often making reference to objects that do not occur in the actual world and are known not to occur, such as perfect spheres, regions of completely empty space, even point-masses, etc. We no more take such models as directly referring to or describing physical reality than we would take King Lear as an actual person when describing an analogy between aspects of Shakespeare's play and the dynamics of an actual aging patriarch and his family. In short, models can be useful without their satisfied sentences' being taken to be literally true.

All well and good, you may say; but what about the models themselves, however devoid of actual objects or true descriptions they might be? Is one not at least committed to the mathematical existence of models as abstract objects in their own right? Models may be "fictional" in the sense of not being constituted of or accurately describing actual objects and relations; but they seem to merit recognition as *abstracta* nevertheless. This would clearly be an unwelcome conclusion to an if-thenist who seeks to avoid platonist commitments associated with mathematical axioms taken as true assertions. And true to form, Maddy refuses to draw such a conclusion. Instead, in a move reminiscent of Nelson Goodman's nominalism, she focuses on "model-descriptions," which are bits of natural and/or formal languages, themselves taken pretty much at face value. Of course, a thoroughgoing nominalist does owe an account of languages, sentences, words, etc., an account that is clearly fully "nominalistic" and adequate for all legitimate purposes. This is indeed a fairly tall order, as we learned from Goodman and Quine's efforts in their important paper, "Steps toward a constructive nominalism,"[6] which sought a fully nominalist account of a formal system of set theory, deploying myriad devices to make sense even of *variables* and *punctuation symbols*, as well as *formulas* and *proofs*. Of course, everything needed to be described in terms of tokens, rather than types; moreover, all infinities were to be avoided. And, of course, any talk of *possibilities* of constructions was out of bounds. Whether

---

[6] Goodman and Quine [1947]. For a recent discussion, see Hellman [2009].

Maddy would wish to adhere to all of these strictures remains to be determined.[7]

Turning to the third challenge, the need to motivate the choice of axioms along with the need to address the loss of content due to the failure to assert axioms of the "If" clauses of the if-thenist account, here Maddy recognizes that SIT needs to be enhanced. She agrees with Resnik's [1980] analysis that, as it stands, SIT indeed leaves mathematics devoid of characteristic content, construing it as an elaborate exercise in deductive logic. One does not need to embrace a platonist perspective to recognize this as a potentially fatal objection to if-thenism.

The enhancement of SIT that Maddy proposes is to supplement it with characteristically mathematical motivation for articulating certain axiom systems that come to characterize the main subject areas of pure mathematics. One can find such motivation in the history of mathematics, at least from the mid-nineteenth century to the present, when the axiomatic method came to the fore. As an initial example, Maddy considers group theory. Here she cites Kline [1972] who describes the goal of unifying a variety of subject areas with sufficiently abstract algebraic axioms to admit of diverse applications in both pure and applied mathematics.

As a second example (ours, not Maddy's), consider the axiom systems for non-Euclidean geometries arising from modifying the Euclidean Parallels Postulate (EPP). Here, again, the motivation is clear and distinctive, and seemingly amenable to an if-thenist treatment: one wanted to explore the consequences of satisfying alternatives to the EPP, proving theorems deriving from those alternatives, viz. allowing infinitely many parallels through a point not on a given line (giving rise to hyperbolic geometry), or, alternatively, allowing no parallels (spherical or, more generally, elliptical geometry).

As a final example, Maddy points to the axioms of Zermelo–Fraenkel set theory (with the Axiom of Choice, ZFC). These were chosen with the aim of providing a framework for systematizing and justifying known classical mathematics.[8] Systematization was achieved by providing full-fledged (Nagelian) reductions of theories, like those of the number systems and

---

[7] In correspondence, Maddy suggests separating such questions concerning the ontological status of descriptions, etc., from philosophy of mathematics proper. Fair enough; but at some point those questions do need to be addressed, and the answers may have implications for philosophy of mathematics.

[8] For this purpose, set theories far weaker than ZFC would suffice. In particular, the Axiom of Replacement is not needed for recovering classical analysis. Presumably, however, Maddy could cite the goal of framing a good theory of ordinal arithmetic within set theory itself, which provides ample motivation for Replacement. (Without it, for instance, one is not even assured of the ordinal $\omega + \omega$!)

Furthermore, as Maddy [2017] lays out in detail, what we have summarized here is just a part of the story of intrinsically mathematical motivation for ZFC and extensions by large cardinals. In particular, those axioms aim to provide a "generous arena" for pursuing set-theoretic

analysis, defining their primitives set-theoretically, and using the set-theoretic axioms to derive the axioms of those theories. And justification was achieved by demonstrating that those axiom systems are consistent (relative to the consistency of ZFC), by proving the existence of set-theoretic models (relative interpretations). This, as is well known, has become the "gold standard" of mathematical justification. That it has proved so effective serves as ample reason for focusing on the ZFC axioms.[9]

Thus, EIT cites mathematical motivation for adopting certain axiom systems rather than others, thereby rebutting the objection that the if-thenist reduces mathematics to exercises in deductive logical reasoning from arbitrary assumptions.

Let us now turn to how EIT fares with respect to the last two challenges listed above, the Problem of Vacuity and the question of Change of Subject.

14.4   The Problems of Vacuity and Change of Subject

Recall that the problem of vacuity is that, if the "If" clause of a conditional asserted by EIT is inconsistent, then every conditional with that antecedent is true, regardless of the consequent. Thus it appears that EIT must accept the formal consistency of the axioms entering into the antecedents of its conditionals, whereas such consistency is equivalent to an unbounded quantified statement of formal number theory, commitment to which EIT (as well as SIT) is designed to avoid. How does EIT propose to deal with this?

Maddy's strategy is, first, to distinguish between formal mathematical theories, on the one hand, and informal theorem-proving of ordinary mathematical practice, on the other. Much as idealized models in physics are useful guides to predicting and explaining the behavior of actual physical systems, formal axiom systems of pure mathematics are useful guides to mathematical practice. Next, Maddy observes, just as statements true in models in physics need not be taken as true of actual physical systems (e.g. that they contain point-masses or perfect spheres, subject to only such-and-such forces, etc.), so statements about formal mathematical systems, including statements of formal consistency, need not be asserted as literally true of informal theorem-proving. Rather, based largely on experience, if we regard some formal mathematics as providing a good model guiding practice, and if the formal axioms of such a model are

---

mathematics, going well beyond founding classical analysis. There is even the prospect of genuine progress on the Continuum Hypothesis.

[9] To be sure, this does not rule out developing and exploring other axiom systems of set theory, either by adding to the axioms of ZFC (e.g. with large cardinal or determinacy axioms), or by replacing axioms of ZFC with alternatives, e.g. to allow for non-well-founded sets. Also, more can be said by way of motivating the axioms of ZFC, namely by appealing to an informal iterative conception, spelled out in terms of "stages." See Boolos [1971] and Chapter 4 in this volume.

consistent *in the model*, then we should expect that our theorem-proving practice will not lead us into a contradiction. But, although this expectation is supposed to abstract from contingent limitations of time and space – we are claiming more of practice than that no contradiction will be encountered simply because proofs will not get too complex, as a matter of contingent fact – still we are not committed to the truth of a number-theoretic consistency claim. That applies only *in the model*, not in actual mathematical practice.

Two questions can be raised about this strategy. The first is whether the imprecise claim of informal consistency is good enough to counter the vacuity objection. As indicated, that claim is not to be understood as an ordinary prediction, that informal practice will not, as a matter of contingent fact, encounter a contradiction from axioms of, say, Peano Arithmetic (PA), or axioms of Real Analysis (RA), etc. Surely, we want to claim more, to the effect that "in principle," a contradiction *cannot* arise from correct deductions from these axioms. Thus, if we aim for mathematical-logical precision, we will appeal to formal logical rules applied to the axioms. We will thus be appealing at least to a potential infinity of correct proofs; but then it seems we are back to asserting a syntactic equivalent of an infinitistic number-theoretic statement, just as we do in our formal model. It seems that EIT can only get around this by refusing to state consistency claims with logical or mathematical precision. (We will return to this matter below in connection with Hilbert's program and the Gödel incompleteness theorems.)

The second question concerns the status of formal models themselves, the question broached above in describing the second challenge, from scientific applications of mathematics: Is not EIT committed to the existence of idealized models as abstract mathematical objects, not only in the area of scientific applications, but also now in connection with proof-theoretic models of theorem-proving practice? We noted above that Maddy's strategy here is to replace apparent reference to models with reference to "model-descriptions." We then pressed the point that this seems to involve commitment to linguistic types, just as abstract as models themselves, as the committed nominalist should recognize. Here Maddy suggests[10] that we simply rest with our ordinary understanding of "descriptions," and not worry about a genuine nominalistic replacement in terms of tokens. Again, we are moved to ask whether this is good enough. After all, a genuine nominalist substitute would very likely involve the use of modal operators, enabling talk of possibilities of sufficiently many tokens to do the work of "model-descriptions" (which may involve, for example, reference to infinitely many sentences satisfied by a model). Embracing modality, however, would seem out-of-bounds for EIT, as it would enable a modal or modal-structural account of mathematics not subject

---

[10] In correspondence.

to the challenges to EIT described here, especially the fifth challenge concerning "change of subject" and "loss of structural content."

Turning now to that fifth challenge, the concern is that the characteristic content of mathematics, inherent in its substantial theorems, is lost when we restrict ourselves to asserting conditionals even with suitable, well-motivated axioms as antecedents and those theorems as consequents. Rather than asserting, for instance, that there are infinitely many primes, we merely assert that *if* the Dedekind–Peano (DP) axioms of arithmetic hold, *then* there are infinitely many primes. Now that difference would vanish if we were to go on and assert that those axioms do hold or are true (or at least possibly true), as we would then be able to detach the consequent by *modus ponens*. But EIT steadfastly avoids asserting axioms as true (or even possibly true), so we end up with merely the statement that there being infinitely many primes is deducible from the DP axioms, or equivalently that the relevant conditional is a truth of first-order logic.[11] Thus EIT in effect replaces mathematical theorems, as ordinarily understood, with claims of deducibility in first-order logic. In many cases, a theorem of great mathematical interest is replaced with a much less interesting statement of a logical fact, as already illustrated with the example of the Weierstrass theorem on the existence of continuous, nowhere differentiable functions. Moreover, the logical facts concerning provability recognized by EIT must be stated *informally*. If stated with respect to formal systems, they would be equivalent to infinitistic, number-theoretic statements, out of bounds for EIT, much as formal statements of consistency are.

The same goes for the most momentous metatheorems of metamathematics, to wit the Gödel incompleteness theorems. Call a theory T "good" just in case it is logically consistent, formal (in the sense that the Gödel codes of its theorems form a recursively enumerable set), and strong enough to numeralwise represent the recursive functions of natural numbers. The first Gödel theorem (Gödel I) then says that no good theory is negation-complete; every such theory admits undecidable sentences; so for no good theory, T, can all and only the arithmetical truths be provable in T. And the second incompleteness theorem (Gödel II) says that no good theory, T, can prove the number-theoretic statement of T's consistency, effectively dooming Hilbert's program in its original form, at least given the assumption that any "finitistic" consistency proof of a mathematical theory such as DPA or classical analysis or even transfinite set theory could be formalized in a good theory, T, of elementary arithmetic. What becomes of these remarkable, deep theorem-assertions in the hands of EIT? Here the proponent of EIT looks for some informal axioms from which the Gödel

---

[11] NB: EIT cannot frame this in terms of second-order logic, or higher-order logic, generally, since, as is well known, there is no effective proof procedure for higher-order logics. As a result, statements about higher-order logical implication have mathematical content beyond formal deducibility, as they involve quantifying over models satisfying premises and conclusions, and thus go well beyond if-thenism.

theorems themselves can be derived, and then it rests with the corresponding conditionals, or the mere assertions that those axioms suffice to derive the metatheorems. Maddy identifies an informal proof theory, PT, with assumptions about mathematical syntax (governing formulas and proofs), along with informal number theory, PA; then we can assert the conditionals "If PT + PA, then Gödel I and Gödel II," or corresponding metastatements about first-order logical derivability. But the remarkable, deep content of Gödel I and II themselves remains out of reach.

It should also be noted that even the statement "There are good theories," in the above sense of "good," cannot be asserted categorically by EIT, as that would involve quantifying over an infinity of theorems or proofs involved in the conditions of formal consistency, formality, and representability of recursive functions. Presumably, then, "There are good theories," or even "Robinson Arithmetic is a good theory," etc., can only be asserted conditionally upon PT + PA. Even the (quite remarkable) statement "Robinson Arithmetic numeralwise represents the recursive functions" is not assertible; what can be asserted is merely that this follows from PT + PA or some similar informal assumptions!

What then becomes of the momentous inference normally drawn from Gödel II that Hilbert's program in its original form is doomed? EIT views the metatheorem as part of an idealized model; it need not be asserted as true. Instead it guides informal practice. In the present case, the lesson is that we should not seek an informal, finitistic consistency proof of any extension Q^+ of Robinson Arithmetic, Q, as it qualifies as a "good theory"; in our sense. So, Maddy asks, why does this not spell doom for Hilbert's program? Well, in a practical sense perhaps it does, as we infer, in her terms, that "we should not bother to seek an informal consistency proof of Q (say from PT + PA)." But this is not nearly as precise nor as decisive as the standard inference from Gödel II that it is *mathematically impossible* for there to be such a consistency proof: no such proof mathematically exists! It was *this* result, we submit, that Gödel sought and achieved, and that the logic and mathematics community understood – but it is beyond the reach of EIT.

How does the proponent of EIT deal with this challenge? If I understand Maddy's position, it regards this challenge not as highlighting a defect but as recognizing a feature. The "loss of content" we have described is the price EIT pays to avoid what Maddy elsewhere calls "robust realism,"[12] which includes any form of objects-platonism and any form of truth-values realism (such as modal and modal-structural interpretations). Thus understood, EIT seems committed to what Shapiro has called a "philosophy first" view, that the content of mathematical theorems as normally understood is to be set aside for philosophical reasons. One cannot help recalling here David Lewis' "Credo," where he writes:

---

[12] See Maddy [2011].

"I'm moved to laughter at the thought of how *presumptuous* it would be to reject [any] mathematics for philosophical reasons."[13]

Although directed toward a nominalist who rejects classes, the rest of the passage, suitably adjusted, seems applicable in the present context, where what is rejected is the truth (or possible truth) of core axioms and theorems derived from them. It reads as follows:

How would *you* like the job of telling the mathematicians that they must change their ways, and abjure countless errors, now that *philosophy* has discovered that there are no classes [or that axioms and theorems are not to be understood as true in relevant structures, or even possibly true]? Can you tell them with a straight face to follow philosophical argument wherever it may lead? If they challenge your credentials, will you boast of philosophy's other great discoveries: that motion is impossible, that a Being than which no greater can be conceived cannot be conceived not to exist, that it is unthinkable that anything exists outside the mind, that time is unreal, that no theory has ever been made at all probable by evidence (but on the other hand that an empirically ideal theory cannot possibly be false), that it is a wide-open scientific question whether anyone has ever believed anything, and so on, and on, *ad nauseam*?

Not me![14]

Now, in fairness to Maddy and EIT, that view need not be taken to imply that mathematics needs to change its ways. It is compatible with EIT as a philosophical view – indeed an integral part of the view – that *whatever* helps mathematicians get on with their work is fine and not being questioned on that score. Rather EIT aims to offer a reflective view of the nature of the subject, one that avoids or mitigates philosophical problems of metaphysics and epistemology that have plagued other approaches. Mathematics, after all, is a human phenomenon, accessible in varying degrees to lay people as well as experts, and philosophy of mathematics should be free to work on how best to interpret mathematical results from a broad perspective that includes, but is not limited to, furthering the success of mathematical practice.

This response is analogous to my own response to Burgess and Rosen's critique of nominalist reconstruction programs (programs mainly due to Chihara, Field, and Hellman),[15] which critique confronts those programs with a *dichotomy* of *either* subscribing, preposterously, to "hermeneutical nominalism," *or* advocating, recklessly, a "revolutionary nominalism," the former purporting to give an accurate account of the real "deep-structural meaning" of mathematical propositions, the latter proposing a radical transformation of mathematical practice that might well block mathematical

[13] Lewis [1991], p. 59. The bracketed text is my own.
[14] Lewis [1991], p. 59. The bracketed clause is supplied by me.
[15] See Burgess and Rosen [1997] and my critical study thereof, Hellman [1998], reproduced as Chapter 6 in this volume; and, for the main nominalistic reconstruction programs, see Chihara [1990], Field [1980], and Hellman [1989].

progress. My response was that this is a false dichotomy, that the nominalist reconstruction programs are properly viewed, neither as hermeneutical nor as revolutionary, but rather as *rational reconstructions* aimed at overcoming or mitigating epistemological and metaphysical problems while leaving mathematical practice entirely intact.

Thus, in the end, assessing EIT should be on its merits, whether it provides a satisfactory understanding of modern mathematics, and whether it does justice to modern mathematical practice, pure and applied. Also, we should reassess EIT's way with the history of mathematics and ask whether an alternative perspective on modern mathematics might lead to a more unified and accurate view of the history of mathematics. Let us now turn to these matters in our final section.

14.5   The Balance Sheet

Let us first return to EIT's response to the first challenge, that if-thenism does not make good sense applied to mathematics prior to the rise of the axiomatic method. Recall that Maddy's response was to grant the point but then simply restrict the target of EIT to what we have been calling "modern mathematics," mathematics pursued independently of direct applicability to the natural world, deploying the tools of modern logic, including the axiomatic method.

As we have acknowledged, this is a perfectly cogent response, and it effectively focuses attention on mathematics to which EIT is designed to apply. There is, however, a nagging worry: take the case of classical analysis.[16] This of course had its roots in the seventeenth century discovery of the calculus, well prior to the great transformation that gave rise to modern abstract mathematics. Should not a philosophy of mathematics recognize some significant continuity between the discoveries by Newton and Leibniz of the calculus and the nineteenth century advances due to Bolzano, Cauchy, Weierstrass, Dedekind, Cantor, et al.? They were, after all, describing very similar if not exactly the same operations on a common core of (continuous) functions of real variables, later extended to functions of complex variables, etc.

Without offering a detailed positive account of this sort of continuity, it does seem that a broadly *structuralist* conception of mathematics provides a good setting for such an account. While it would be anachronistic in the extreme to attribute to Newton and Leibniz any of the modern forms of mathematical structuralism – for example based on set theory, or category theory, or Shapiro's *ante rem* structuralism, or my modal-structuralism – one can describe

---

[16] I hasten to say here that I am not even an amateur historian (of any kind, let alone of mathematics). I cheerfully defer to experts to assess what I say in this section as far as historical accuracy is concerned.

the early forms of calculus as pertaining, at least in part, to curves in a (3 + 1 dimensional) space-time manifold, representing for example trajectories of particles in motion. And such an account, it certainly seems, provides for greater continuity in the history of analysis than any framework, like EIT, that focuses primarily on logical deduction and "what follows from what," deploying the logical tools provided by Frege and his successors.

Regarding the second challenge, from the use of idealized models in scientific applications of mathematics, we saw above that Maddy avoids having EIT committed to the truth of mathematical sentences satisfied in the models, opting instead for a fictionalist account. (Recall the King Lear analogy.) Nor is EIT committed to the mathematical existence of models as *abstracta*. Instead, "model-descriptions" given in actual scientific languages are supposed to suffice for the purposes of applications. Whether this is sufficient to avoid all platonist commitments turns on whether an adequate account of linguistic descriptions can be given in terms of tokens rather than types. That in turn turns on whether a potential infinity of such descriptions must be accommodated. Since presumably only finitely many actual description-tokens exist or ever will exist, recourse would be needed to speak of *possibilities* of tokens,[17] but this goes beyond the limits of a strict nominalism. If one needs to provide for an infinitude of descriptions, for example of logical consequences of assumptions built into scientific models, then this would be a real problem for EIT. But perhaps all legitimate purposes can be served by only finitely many descriptions. Further work is needed to settle this.

Regarding the third challenge on selection of axioms, what Maddy has to say about this seems fine as far as it goes, and it provides a genuine improvement over SIT. I do think there is an important aspect of axioms that is missing from Maddy's treatment, but this is best discussed below under the fifth challenge concerning "change of subject/loss of structural content."

Turning to the fourth challenge, the apparent need to assert the consistency of relevant mathematical axioms, we saw that Maddy only acknowledges this for informal mathematical practice, treating formal consistency claims and their numerical equivalents as belonging to models of informal practice, thereby bypassing any need to assert such claims as actually true. Our main question here is whether this is good enough. We submit that it is not: we should be prepared to assert – even if we recognize that it is possible we are wrong – that derivations of a contradiction from formal mathematical axioms are *mathematically impossible*, that no correct proof of a contradiction *mathematically exists*, not merely the informal and imprecise statement that a contradiction "will not emerge from ordinary practice." In short, we submit that Maddy's EIT here is not

---

[17] As in Chihara's [1990] semantic version of a simple theory of types (for pure mathematics).

meeting a high enough standard of precision and rigor, expected of modern mathematics and logic.

Finally, regarding the fifth challenge, change of subject/loss of structural content, from our discussion of this above, it seems clear that this is the most significant and far-reaching of all the challenges we have considered. As we also noted, however, it is also a challenge that EIT does not seek to overcome, beyond giving its account concerning selection of axioms, but rather lets stand as a necessary price worth paying for advantages in avoiding metaphysical and epistemological concerns that beset other philosophical approaches to mathematics.

In closing, let us draw out a connection between the fifth challenge and Maddy's handling of the third regarding choice of axioms, which constitutes the main difference between EIT and its predecessor, SIT. The connection is this: Maddy points to various choices of axioms as geared to achieving certain mathematical aims, such as, in the case of group theory, providing for a unifying treatment of a host of interesting applications in both pure and applied mathematics. This is surely a valid point. And so is the one she makes about the motivation for axiomatic set theory, to provide a unifying foundational framework for classical mathematics, one free from the paradoxes that confronted Frege's logicism and Cantor's informal set theory. However, a major motivation behind the choice of axioms in modern mathematics – one that Maddy does not recognize – is the *description of relevant structures* fulfilling the requirements of the branches of mathematics in question. Consider, as a paradigm case, the choice of the so-called "Peano Postulates" for arithmetic. As Peano acknowledged,[18] these were due to Dedekind in his great essay, *Was sind und was sollen die Zahlen?*[19] The example is very instructive, because in the section (71) where Dedekind introduces these "axioms," they serve, not as axioms in the traditional Euclidean–Fregean sense of "true assertions" taken as starting points in proofs of theorems, but rather as *defining conditions* of a type of structure, what Dedekind explicitly called a "simply infinite system" (later called a "progression" by Russell, and known as an "ω-sequence" in set theory). Indeed, Dedekind went on to prove his celebrated isomorphism theorem showing that these defining conditions (which include the second-order statement of mathematical induction) characterize such structures uniquely, up to isomorphism (i.e. the "axioms" are categorical).[20]

---

[18] In Peano [1891], p. 93.

[19] Dedekind [1888].

[20] The second-order statement of mathematical induction is crucial to this result, which rules out non-standard models of arithmetic, which satisfy the axioms with the first-order induction scheme. This distinction was not available to Dedekind, but he worked in an informal set theory, enabling the statement of a set-theoretic version of induction, essentially equivalent to a second-order logical formulation.

Much the same can be said about "axioms" characterizing other number systems, such as the reals and the complexes. Axioms, as defining conditions for a complete, separable, totally ordered continuum (with axioms on the operations of a field, as usually presented), characterize the real number system up to isomorphism. (This requires either second-order logic or some set theory to formulate the completeness axiom, much as Dedekind's isomorphism theorem for the naturals requires the second-order statement of mathematical induction, as just reviewed.) Similarly, in abstract algebra, the "axioms" for algebraic structures, such as groups, rings, fields, etc., serve as defining conditions (although not categorical in these cases, which proves advantageous for the sorts of purposes that Maddy alludes to). Of course, in all these cases, the axioms do serve as starting points of proofs – but they also serve the equally important function of *defining* relevant structures of mathematical and scientific interest. These dual purposes are also served by axioms defining the various types of spaces of functional analysis, geometry, topology, category theory, etc.

What about set theory? Here it may seem that the axioms of ZFC, the system of primary interest in contemporary set theory, are of the Euclidean–Fregean type, asserted as truths, as well as serving as starting points of proofs. And here, EIT claims, contrary to platonist conceptions, that the latter function is all that mathematics really requires, that the axioms do not need to be asserted as truths. There is, however, an alternative conception, one articulated (at least in part) by Zermelo in his great [1930] paper. There the axioms are viewed as describing structures of interest, and not a unique "real world of all the sets." Indeed, as the title of that paper indicated, multiple "domains of sets" are recognized, each being a full, standard model of the axioms[21] (which then included the Replacement axiom, intended as a second-order statement rather than a first-order scheme). And while, as Zermelo recognized, these domains are not all mutually isomorphic – the axioms are not categorical – they are, as he proved, determined up to isomorphism by just two parameters: the cardinality of the urelement basis, and the ordinal height of the domain (measuring how far the power set operation is iterated), what Zermelo called a "boundary number" (*"Grenzzahl"*), which as he proved is always a strongly inaccessible cardinal (whose existence cannot be proved in ZFC, provided *it* is consistent, on pain of conflicting with Gödel's second incompleteness theorem). Thus, Zermelo [1930] established the "quasi-categoricity" of the ZF^2C axioms: any two models

---

[21] Here "full" means that power sets are full, containing *all* subsets of the given set. And "standard" means well-founded. As is well known, due to the compactness of first-order logic, the first-order ZFC axioms have non-standard models in this sense, despite the axiom of Regularity, intended to rule out infinitely descending $\in$-chains.

with equinumerous urelement bases are such that one is isomorphic to an end-extension of the other.[22] Furthermore, Zermelo formulated and adopted an *Extendability Principle*, that every model of ZF^{2}C has a proper end-extension. Thus, on Zermelo's view, set theory, properly understood, deals not with a single maximal world of sets, but with a multiverse of models of ever increasing height. This picture has been taken further by Putnam [1967b] and Hellman [1989], who have used modal logic to formulate the extendability principle to stipulate that, necessarily, any full, standard model possibly has a proper end-extension.[23] Thus, on a Zermelian multiverse view, set-theoretic axioms too should be seen as defining structures of interest, not just as starting points of proofs.

To sum up, our most significant criticism of EIT is the fifth challenge, that the subject of mathematics as the investigation of key mathematical structures has been changed to the mathematically-logically imprecise, informal investigation of "what follows from what," and that this involves an appreciable loss of content. This is reinforced by our last point above, that EIT's view of axioms is incomplete: while of course they serve as starting points of proofs, and while of course mathematics is largely about proving interesting theorems from axioms, axioms also serve the purpose of defining structures that mathematicians and scientists deem worthy of study. Thus, in our view, at best, EIT tells a significant part of the story, but it needs to be supplemented with some version of mathematical structuralism to present a more complete picture.[24]

---

[22] The superscript 2 in ZF^{2}C indicates that the Replacement axiom is the second-order statement, not the first-order scheme. This is necessary for quasi-categoricity.
  A model $\mathcal{M}'$ is defined to be an "end-extension" of a model $\mathcal{M}$ just in case every set of (the domain of) $\mathcal{M}$ is also a set of (the domain of) $\mathcal{M}'$, and no new members of such sets are added to them in $\mathcal{M}'$.

[23] See Putnam [1967b], Hellman [1989], and Hellman and Shapiro [2019], Ch. 5. Also see Chapter 4 in this volume.
  Modal logic is especially useful in avoiding commitment to a maximal model, the existence of whose domain is implied by a second-order logical comprehension axiom (guaranteeing a class consisting in all members of all models of a transfinite sequence like the V_k for k strongly inaccessible). Zermelo's non-modal formulation is subject to this unwanted derivation of a maximal model, although Zermelo could not have foreseen this, lacking an axiomatic formalism for second-order logic. On a modal interpretation, there is every reason not to recognize a class of "all possible sets of domains of models," as classes (or pluralities) are conceived as "world-bound": they can contain only what exists at a given world (although talk of "worlds" is heuristic only – officially the modal-structural interpretation does not quantify over worlds).

[24] I am grateful to Penelope Maddy and Michael Resnik for very helpful correspondence on the subject matter of this paper.

References

Benacerraf, P. and Putnam, H. (eds.) [1983] *Philosophy of Mathematics: Selected Readings*, 2nd edn. (Cambridge: Cambridge University Press).

Boolos, G. [1971] "The iterative conception of set," reprinted in *Logic, Logic, and Logic*, Jeffrey, R., ed. (Cambridge, MA: Harvard University Press, 1998), pp. 13–29.

Burgess, J. P. and Rosen, G. [1997] *A Subject with No Object: Strategies for Nominalistic Interpretation of Mathematics* (Oxford: Oxford University Press).

Chihara, C. S. [1990] *Constructibility and Mathematical Existence* (Oxford: Oxford University Press).

Church, A. [1956] *Introduction to Mathematical Logic I* (Princeton, NJ: Princeton University Press).

Dedekind, R. [1888] "The nature and meaning of numbers," reprinted in Beman, W. W., ed., *Essays on the Theory of Numbers* (New York: Dover, 1963), pp. 31–115, translated from the German original, *Was sind und was sollen die Zahlen?* (Brunswick: Vieweg, 1888).

Field, H. [1980] *Science without Numbers*, 1st edn. (Princeton, NJ: Princeton University Press).

Gödel, K. [1931] "On formally undecidable propositions of *Principia Mathematica* and related systems," reprinted in van Heijenoort, J., ed., *From Frege to Gödel: A Source Book in Mathematical Logic, 1879–1931* (Cambridge, MA: Harvard University Press, 1967), pp. 596–616.

Goodman, N. and Quine, W. V. [1947] "Steps toward a constructive nominalism," *Journal of Symbolic Logic* **12**: 105–122.

Hellman, G. [1989] *Mathematics without Numbers: Towards a Modal-Structural Interpretation* (Oxford: Oxford University Press).

Hellman, G. [1998] "Maoist mathematics? Critical study of John Burgess and Gideon Rosen, *A Subject with No Object: Strategies for Nominalist Interpretation of Mathematics* (Oxford, 1997)," *Philosophia Mathematica* **6**(3): 357–368.

Hellman, G. [2009] "Mereology in philosophy of mathematics," in Burkhardt, H., Seibt, J., Imaguire, G., and Gerogiorgakis, S., eds., *Handbook of Mereology* (Munich: Philosophia Verlag), pp. 412–424.

Hellman, G. and Shapiro, S. [2019] *Mathematical Structuralism* (Cambridge: Cambridge University Press).

Hempel, C. G. [1945] "On the nature of mathematical truth," reprinted in Benacerraf, P. and Putnam, H., eds., *Philosophy of Mathematics: Selected Readings*, 2nd edn. (Cambridge: Cambridge University Press, 1983), pp. 377–393.

Hilbert, D. and Ackermann, W. [1928] *Principles of Mathematical Logic* (Berlin: Springer).

Kline, M. [1972] *Mathematical Thought from Ancient to Modern Times* (Oxford University Press).

Lewis, D. [1991] *Parts of Classes* (Oxford: Blackwell).

Maddy, P. [2011] *Defending the Axioms: On the Philosophical Foundations of Set Theory* (Oxford: Oxford University Press).

Maddy, P. [2017] "Set-theoretic foundations," *Contemporary Mathematics* **690**: 289–322.

Maddy, P. [to appear] "Enhanced if-thenism," in *A Plea for Natural Philosophy and Other Essays* (Oxford: Oxford University Press).

Peano, G. [1891] "Sul concetto di numero," *Rivista di Matematica* **I**: 87–102.

Putnam, H. [1967a] "The thesis that mathematics is logic," reprinted in *Mathematics, Matter, and Method: Philosophical Papers*, Volume I (Cambridge: Cambridge University Press, 1979), pp. 12–42.

Putnam, H. [1967b] "Mathematics without foundations," reprinted in *Mathematics, Matter, and Method: Philosophical Papers*, Volume I (Cambridge: Cambridge University Press, 1979), pp. 43–59.

Putnam, H. [1975] "What is mathematical truth?," reprinted in *Mathematics, Matter, and Method: Philosophical Papers*, Volume I (Cambridge: Cambridge University Press, 1979), pp. 60–78.

Putnam, H. [1979] *Mathematics, Matter, and Method: Philosophical Papers*, Volume I (Cambridge: Cambridge University Press).

Quine, W. V. [1936] "Truth by convention," reprinted in in Benacerraf, P. and Putnam, H., eds., *Philosophy of Mathematics: Selected Readings*, 2nd edn. (Cambridge: Cambridge University Press, 1983), pp. 329–354.

Resnik, M. [1980] *Frege and the Philosophy of Mathematics* (Ithaca, NY: Cornell University Press).

Russell, B. [1903] *The Principles of Mathematics* (New York: Norton).

Shapiro, S. [1991] *Foundations without Foundationalism: A Case for Second-Order Logic* (Oxford: Oxford University Press).

Simpson, S. G. [1999] *Subsystems of Second Order Arithmetic* (Berlin: Springer).

Zermelo, E. [1930] "Über Grenzzahlen und Mengenbereiche: Neue Untersuchungen über die Grundlagen der Mengenlehre," *Fundamenta Mathematicae*, **16**: 29–47; translated as "On boundary numbers and domains of sets: new investigations in the foundations of set theory," in Ewald, W., ed., *From Kant to Hilbert: A Source Book in the Foundations of Mathematics*, Volume 2 (Oxford: Oxford University Press, 1996), pp. 1219–1233.

15 Mathematical Pluralism: The Case of Smooth Infinitesimal Analysis

15.1 Introduction

A remarkable twentieth century development in mathematics has been the construction of smooth inifintesimal analysis (SIA) and its extension, synthetic differential geometry (SDG), realizing a non-punctiform conception of continua in contrast with the dominant classical (set-theoretic) conception.[1] Here one is concerned with "smooth worlds" in which all functions (on or between spaces) are continuous and have continuous derivatives of all orders. In this setting, the once discredited notion of "infinitesimal quantity" is admitted and placed on a rigorous footing, reviving intuitive and effective methods in analysis prior to the nineteenth century development of the limit method. The infinitesimals introduced, however – unlike the invertible ones of Robinsonian non-standard analysis – are nilsquare and nilpotent. While they themselves are not provably identical to zero, their squares or higher powers are set equal to 0. As will be explained, contradictions are avoided by retreating to intuitionistic logic, giving up the law of excluded middle (LEM) in its full generality (whereas the logic underlying non-standard analysis is fully classical). That this weakening of the logic suffices for consistency (at least relative to classical analysis (CA)) is demonstrated in the framework of category and topos theory which have been deployed to define models of these theories. Toposes themselves are categories fulfilling generalizations of rich universes of sets, for example, closure under products and "exponentiation"; in particular, the machinery for defining "subobjects" allows for "multiple truth values" beyond just "true" and "false" and gives rise to an internal logic which in general is intuitionistic, just as SIA and SDG require.

Despite the internal, formal coherence of all this, on reflection these developments present an interesting challenge to philosophy of mathematics and logic. What precisely is the relationship between SIA and CA? Are they even

---

[1] These theories originate in the work of Lawvere [1979].

mutually consistent? *Prima facie*, one would say not, as SIA derives theorems which formally are negations of classical ones, notoriously:

$$\neg\forall x[x = 0 \vee \neg(x = 0)]. \tag{1}$$

In this respect, the situation is *formally* analogous to that arising from Brouwerian intuitionistic analysis, which, in contrast to the (logically) conservative constructive analysis of Bishop, also derives theorems which look like contradictories of classical laws (indeed, for example, the very one displayed, with the variable ranging over infinitely proceeding sequences of rationals). Now, there is a well-known reconciliation between intuitionistic analysis and classical analysis based on simply keeping track of radically distinct meanings given to the logical and mathematical symbols involved.[2] When this is done, for example, the intuitionistic statement of (propositional) LEM really asserts that a method is available for either proving or refuting any mathematical statement, something no one would want built into one's logic. Similarly, the intuitionistic version of the universal formula negated in the display above asserts that a method is available for proving, of any presented real number generator, either that it converges to 0 or that that leads to absurdity, something that, not surprisingly, is itself refutable in the intuitionistic framework. Again, the classicist cheerfully accepts all of this. The question then naturally arises, whether anything along these lines applies in relation to SIA and CA. If not (as will be argued and as seems to be recognized by those well acquainted with SIA), what alternative reconciliation might there be? And, more pressing, what reasons *can* be given for renouncing LEM if "meaning change" is not an option? As Bell brings out, one can construct arguments for limiting LEM from various smoothness requirements, and these will be examined below (and found wanting). Closely related and more generally, how should we interpret SIA? While topos theory provides models of SIA, these are highly abstract, whereas SIA has its roots in geometry and intuitive ideas about continuity. Can SIA be interpreted on its own, independently of topos models? How is its subject matter related to that of CA, identical or distinct, and if the latter, how? If '=' really is logical identity, how can there be objects, nilsquares, whose identity or distinctness vis-à-vis 0 is metaphysically (as opposed to epistemically) indeterminate? As will be explained, in SIA, it cannot be proved that 0 is the *only* nilsquare, but neither can it be proved that nilsquares *other than* ($\neq$) 0 do exist. One speaks of these things as having "only a potential existence" (cf. Bell [1998], 7). What sense can be made of this? As will be brought out below, there is an interesting connection with recent philosophical literature on *vagueness*, in particular the question whether the notion of "vague objects" – whose status as "vague" is not traceable to use of language – is coherent. Finally, can a structuralist interpretation of SIA be given, and are any of

---

[2] See, e.g., Hellman [1989b], reproduced here as Chapter 12.

the ideas of modal-structuralism relevant? The "points" of classical punctiform analysis and geometry are treated as "merely possible" according to this view; how then would nilsquare or nilpotent infinitesimals differ in their "merely potential" status?

Let us make clear that these questions are raised in searching for a substantive interpretation of SIA, one which respects its geometric roots and regards it as having a subject matter ("smooth worlds"), even if only in a hypothetical, idealized sense (as we would argue is also the case with respect to classical analysis and set theory). We are explicitly *not* suggesting that, in the absence of such an interpretation, somehow SIA would be illegitimate *as mathematics*. On the contrary, we are in sympathy with Carnap's remark that "in logic [and, we would add, in mathematics] there are no morals" – meaning that these subjects must be free to be developed without *a priori* constraints beyond minimal requirements of clarity and rigor.[3] The results can then be judged according to their fruitfulness along a variety of dimensions, i.e., "no morals" certainly does not mean "no standards," especially *after* new developments! In the case of SIA and SDG, it is already clear that these theories are presented with as much clarity and rigor as one finds in standard mathematics, and at least they present sufficient conditions for a self-consistent use of infinitesimals in place of the standard limit method in analysis, and that is already historically and philosophically interesting. (Whether these theories will help solve new problems in analysis or differential geometry remains, I believe, to be determined.) Even if it should prove impossible to provide a realist or structuralist interpretation, there remains a fallback "formalist" or "instrumentalist" stance. These theories develop various formal methods for solving problems, in many cases much more simply than their classical counterparts (replacing limit computations with straightforward algebra). Of course, one would like more. Our overarching question can be put: Can one obtain it?

In Section 15.2, we present a brief overview of the basics of SIA (following Bell's [1998] *Primer*), concentrating on those aspects most relevant to the problems of interpretation highlighted here, in particular the essential role of intuitionistic logic. In subsequent sections we take up in turn the question of justification for this restriction and the problem of interpretation, the relevance of work on vagueness, and the bearing of structuralism.

15.2   Leading Ideas of Smooth Infinitesimal Analysis

Here we concentrate on just the basic system of analysis on the real line as presented in Bell [1998]. The basic motivations are geometric and go back to

[3] This famous remark of Carnap's occurs in his book, *The Logical Syntax of Language* [1937], Section 17, entitled "The principle of tolerance in syntax," p. 52. Carnap's comment appended to this section makes it clear that he thinks the attitude of tolerance is applicable to mathematics and is "tacitly shared by the majority of mathematicians" (p. 53).

ancient times. Two traditional notions of "infinitesimal" are brought together in the notion of *nilsquare infinitesimal*. One is the idea of linear infinitesimal "segments" of a curve, along which the length between two given points is thought to be measured as a sum of "microsegments" which are part of the curve but actually straight. The second notion relates to the familiar Archimedean method of defining, for example, the area under a curve (bounded say by the x-axis and two vertical lines) as a sum of infinitesimally thin (along x) rectangles. The first notion embodies the following (informal) principle.

Principle of Microstraightness: *Given any smooth curve C and any point P on it, there is a non-degenerate microsegment about P which is straight.*

This leads to the central (formal) axiom of SIA (called the *Principle of Microaffineness*, below), and it turns out to be sufficient for the Archimedean method. Given a point P on a curve C, the Principle of Microstraightness implies that an infinitesimal segment about P along C is straight, and then the "area defect," ∇, between that of a true (micro) rectangle, $\square$, under the curve about P and the true area is proportional to ε^2, where ε is the base of the rectangle along x (see Figure 15.1). For the method to work, this latter length must not be identified with 0 whereas the area defect should be equal to 0, i.e., we should have $\varepsilon^2 = 0$. That is, we require that *the set of nilsquares*, $\Delta =^{df}$ *the set of magnitudes ε such that $\varepsilon^2 = 0$, not be* = $\{0\}$. But this too is guaranteed by the Principle of Microstraightness: just apply it to the curve $y = x^2$ at the origin (Figure 15.2).

The formal development of SIA proceeds with algebraic ordered field axioms governing the smooth or affine line R, familiar from standard axioms

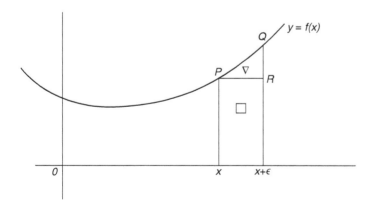

Figure 15.1 Motivating $\varepsilon^2 = 0$

or theorems for the classical real line, $\mathbb{R}$, except that, along with LEM, the principle of trichotomy, "$x < y \lor x = y \lor x > y$," is avoided, as is decidability of identity, "$x = y \lor x \neq y$." Instead, SIA gets by with the axioms "$0 < x \lor x < 1$," and "$x \neq y \rightarrow x < y \lor y < x$." Also, the axiom governing multiplicative inverses, "$x \neq 0 \rightarrow x/x = 1$," exempts nilsquares from invertibility since, as will emerge, the antecedent in such a case cannot be demonstrated. Functions from R (or a closed interval) to R are introduced and associated with curves as graphs in $R \times R$ in the usual way. Informally, one now assumes that the Principle of Microstraightness applies to all curves (i.e., in a "smooth world") determined by functions $f : R \rightarrow R$. This means that all functions behave locally like polynomials $f(x) = a_0 + a_1 x + a_2 x^2 + \cdots + a_n x^n$, in which case $f(\varepsilon) = a_0 + a_1 \varepsilon$, for ε in $\triangle$ (squares and higher powers of which vanish). This means that the graph of f restricted to $\triangle$ is a unique microstraight segment about $(0, f(0))$ on the curve and tangent to it at that point, i.e., that the restriction g of f to $\triangle$ is *affine on* $\triangle$. This is summarized formally in the following.

Principle of Microaffineness: *For any map $g : \triangle \rightarrow R$, there exists a unique b in R such that for all ε in $\triangle$,*

$$g(\varepsilon) = g(0) + b\varepsilon.$$

That is, the graph of g satisfies the equation of a straight line with slope b passing through $(0, g(0))$. This is the central axiom of SIA, also known as the

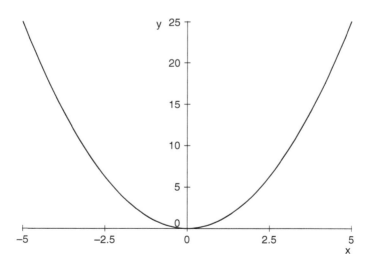

Figure 15.2 $y = x^2$

Kock–Lawvere Axiom (KL, Kock [1981]). It says, in effect (as Bell puts it), that the microneighborhood $\triangle$ of 0 can be translated and rotated but it cannot be bent. It behaves as an infinitesimal "rigid rod" or generic tangent vector to curves in R^2.

The unique existence of the slope b stipulated by the KL Axiom yields three immediate and fundamental consequences.

1. The nilsquares-microneighborhood, $\triangle$, of 0 does not reduce to $\{0\}$. (Apply the axiom to the identity function, $g(\varepsilon) = \varepsilon$. If ε could only take the value 0, then the displayed equation would be satisfied with $b = 0$ or 1, contrary to uniqueness ($1 \neq 0$ is axiomatic).) Thus, it is proved that

$$\neg \forall \varepsilon \text{ in } \triangle(\varepsilon = 0).$$

2. **Principle of Microcancellation**: *For any x, y in R, if $\varepsilon x = \varepsilon y$ for all ε in $\triangle$, then x = y.* (Simply apply the axiom to the function $g(\varepsilon) = \varepsilon x$.)

3. As expected, the slope b of the axiom allows a simple definition of the derivative of a function. Given $f: R \to R$, fix x in R and define $g_x : \triangle \to R$ by $g_x(\varepsilon) = f(x + \varepsilon)$. By KL, $f(x + \varepsilon) = g_x(\varepsilon) = g_x(0) + b_x\varepsilon = f(x) + b_x\varepsilon$, where b_x is unique. This allows introduction of $f'(x)$ defined by

$$f(x + \varepsilon) = f(x) + \varepsilon f'(x).$$

Concerning the first point, the restriction to intuitionistic logic is essential. We *cannot* go on to infer

$$\exists \varepsilon \text{ in } \triangle(\varepsilon \neq 0)$$

for, by the second of the order axioms noted above, this would allow a nilsquare to be ordered with respect to 0, i.e., we would have an ε such that $\varepsilon < 0$ or $\varepsilon > 0$, and either alternative implies $\varepsilon^2 > 0$, contradicting the definition of $\triangle$. With LEM, the KL Axiom immediately leads to inconsistency. Thus, SIA actually *refutes* the universal "decidability of identity," i.e., we have

$$\neg \forall x(x = 0 \lor x \neq 0),$$

as claimed above.

Concerning the second and third points, these are the key tools in developing the differential calculus bypassing limit arguments. The essential idea can be illustrated in about the simplest possible case, the polynomial rule in the case of $f = cx^2$. To find f', substitute in the defining equation:

$$c \cdot (x + \varepsilon)^2 = cx^2 + \varepsilon(cx^2)';$$

then

$$c \cdot (x^2 + 2x\varepsilon + \varepsilon^2) = cx^2 + c2x\varepsilon;$$

equating the right sides, simplifying, and microcancelling then yield

$$(cx^2)' = c2x,$$

the familiar result. This is typical. In particular, note that we reason with ε standing for an *arbitrary* microquantity, which then drops out on microcancellation. We never single out a particular element of $\triangle$ (with the exception of 0, i.e., '0' still behaves as a name). This will be important below.

The only other axiom needed for carrying out differential and integral calculus is the following.

Integration Principle: *Given any (smooth)* $f: [0,1] \to R$, *there exists a unique (smooth)* $g: [0,1] \to R$, *such that* $g' = f$ *and* $g(0) = 0$.[4]

This suffices for the introduction of definite integrals, for the first and second fundamental theorems of calculus, and, with standard machinery for dealing with Cartesian products, the development of multivariate calculus and complex analysis, with standard applications to geometry (areas, volumes, arc lengths, etc.) and physics. (For details, see Bell [1998]).

We close this section by calling attention to one further highly non-classical consequence of SIA, the *indecomposability of the smooth line R*. Call a part J of R "microstable" if it is closed under addition of nilsquares. From the Integration Principle (and other basic results, e.g. the sum rule for differentiation), one derives the following.

Constancy Principle: *Let* $f: J \to R$, *J microstable, such that* $f' = 0$ *identically. Then f is constant.*

Now call a part (subset) U of R "*detachable*" just in case, for all x in R, $x \in U \vee x \notin U$. Then it follows from the combination of the KL Axiom and the Constancy Principle that *the only detachable parts of R are the empty part and R itself.* This is thought to express a fundamental aspect of smoothness. Note that, although the result depends on limiting the LEM, and that a similar sounding result is a feature of intuitionistic analysis, the proof here turns, not on an analysis of "constructive real number," but on the completely different concept of nilsquare infinitesimal and the notion of derivative it provides, ideas as foreign to intuitionism as they are to classical analysis.

15.3   The Problem of Interpretation

As we have seen, it is essential that the Law of Excluded Middle be restricted if SIA is to avoid outright self-contradiction. But this in itself can hardly qualify

---

[4] Here there appears to be a trade-off vis-à-vis classical analysis, where the Integration Principle is derived using the method of limits.

as a sufficient justification for restricting or "giving up" a logical rule, at least if we are trying to conceive of SIA as more than a calculational device (going beyond the "fallback formalist" position). We do not normally have the option of saving a theory from inconsistency merely by declaring some logical principle inapplicable. How does one answer the critic who says, "SIA really *is* self-contradictory. Its practitioners delude themselves into thinking otherwise simply by not themselves drawing the conclusions implicit in their assumptions. That, however, does not in any way alter the logical facts, i.e., that contradictions *are* implied by the assumptions. Merely declaring that only the rules of intuitionistic logic are to be used in this case is like sticking one's head in the sand!"

As indicated in Section 15.1, the defender of intuitionistic mathematics (à la Brouwer, Heyting, Dummett, et al.) has a ready answer to such a critic. The logical connectives and quantifiers in that context have distinctive constructivist meanings, quite different from the classical ones, and these constructivist meanings both support the logical axioms and rules that are accepted and disqualify those that are not accepted, LEM and related principles, such as the law of double negation and proof of existence by *reductio ad absurdum*. For instance, if asserting a "disjunction" (symbolized with $\vee$) is taken to imply availability (to an idealized human mathematician) of a method of determining which disjunct holds, and if asserting a "negation" ($\neg$) is understood as implying availability of a method of refuting what is negated, then *of course* one should not build LEM ($p \vee \neg p$) into one's logic. (Classicists agree completely with intuitionists about *this*!) And if the formula

$$\forall \sigma (\sigma = 0 \vee \sigma \neq 0) \tag{2}$$

is understood as asserting the availability of a method of proving of any presented choice sequence (of rationals) that it constructively converges to 0 or that this is refutable, then all parties should agree that this ought not to be a theorem and that it might well itself be refutable (as indeed it is in intuitionistic mathematics)![5] Thus, a fully explicit formalism should use different symbols from the classical ones to indicate the different operative meanings; then it would be clear that it is not strictly accurate to speak, for example, of "*the* law of excluded middle," but rather that we have to deal with different statements of similar formal shape, a classical one (LEM_{Cl}, say) and an intuitionistic one (LEM_{Int}). The former is guaranteed by classical truth tables, whereas the latter, as just observed, is not generally correct for the intended meanings.[6]

---

[5] Cf., e.g., Dummett [1977]), p. 84, where the statement is even more elementary, viz. that σ either consists of all zeros or not.

[6] A presupposition of particular *applications* of classical logic (not of the logic itself!) is that the reasoning in question can be represented as involving bivalent statements or propositions, i.e.,

Unfortunately, as we have already suggested, no such neat resolution is available in the case of SIA in comparison with CA. Not only do the inventors and expositors of SIA not offer any such justification for restricting the logic to be intuitionistic;[7] it is reasonably evident that such a justification is not in the cards. For starters, consider the status of the nilsquares, members of $\triangle$. None other than 0 is explicitly recognized let alone "constructed" in any mathematical sense. Formally, as already noted above, the "existence of non-zero nilsquares" is actually *refutable*, i.e.,

$$\neg \exists \varepsilon \in \Delta (\varepsilon \neq 0)$$

but in the background one assumes their "possibility" in making sense of the whole theory. So clearly this possibility could not rest on any claim of "constructibility," which would lead directly to contradiction! Rather this "possibility" is simply *assumed* in the background as an expression of a geometric intuition (as in the Principle of Microstraightness). One could stipulate that the "possibility" in question just means that it is refutable that all nilsquares equal zero, i.e.,

$$\neg \forall \varepsilon \text{ in } \Delta (\varepsilon = 0),$$

which is true, whereas the direct statement of existence is not.[8] But this is a *consequence* of the combination of the KL Axiom and the restriction to intuitionistic logic, and cannot serve as a justification or grounding for either.

What of the KL Axiom itself? Here the problem is even more stark, for the axiom is an assertion of existence, viz. of the slope b of the "linelet" determined by the given function $f : \Delta \rightarrow R$. A constructive justification for this would consist in some explicit procedure for passing from (a presentation of) an arbitrary (smooth) function f to a b together with a proof that b uniquely satisfies

---

that the classical truth tables are applicable. In the case of pure mathematics, it is this assumption that intuitionism typically rejects (for various reasons that have been offered). *That* is the *locus* of the dispute with classicism (standard practice), not the correctness of a "law of logic." Sometimes this is understood but, in the interests of brevity, one tolerates talk of "giving up LEM," where this is taken in a purely formal sense, i.e., no axiom or rule of this syntactic shape is adopted in a particular system or context.

[7] See e.g., Kock [2003].

[8] Interestingly, we *cannot* construe the "possibility of non-zero nilsquares" as the claim that the double negation of existence is provable, i.e., that

$$\neg\neg \exists \varepsilon \text{ in } \Delta (\varepsilon \neq 0),$$

for then SIA would be inconsistent, as we do have

$$\forall \varepsilon \text{ in } \Delta \neg\neg (\varepsilon = 0),$$

whence

$$\neg \exists \varepsilon \text{ in } \Delta (\varepsilon \neq 0)$$

as already indicated.

$$f(\varepsilon) = f(0) + b\varepsilon,$$

for arbitrary ε in $\triangle$. But it is hard to see in what such a procedure could possibly consist that would not already be a method of finding the derivative of f at 0. But we cannot at this stage presuppose such a method, even if we take as given the restriction to smooth functions, for we do not yet know what "derivative" means. (Remember, we are supposed to be building up an independent concept of derivative based on the KL Axiom, and cannot presuppose standard notions based on the limit concept.) We conclude that, in contrast to principles of intuitionistic mathematics, the KL Axiom is no expression of constructive meanings and must be understood differently.[9]

Although a "meaning change" argument based on intuitionist/constructivist meanings is not available for SIA, that is not to say that no meaning change argument is possible. We shall return to this idea in connection with "identity" in Section 15.4, but let us now consider other strategies.

Indeed there is an alternative route to intuitionistic logic peculiar to SIA presented by its expositors. That route appeals to *smoothness requirements* governing "smooth worlds" that, in some sense, SIA is taken as describing. Bell presents two distinct arguments along these lines. The first seeks to establish the following.

(*) *If LEM holds, then discontinuous functions arise, contrary to the background assumption that (in a smooth world) all functions are continuous.*

The argument is almost trivial. Given LEM, we have for every x in R, either $x = 0$ or $x \neq 0$. Then we can introduce f defined as taking the value 1, say, at $x = 0$, and taking the value 0 everywhere else. (This is called the "blip" function.) f is defined everywhere but is obviously discontinuous, q.e.d.

But is it just the assumption of LEM and the restriction to smooth functions that leads to contradiction? Surely not, for the classicist certainly can and does reason consistently about the class of C^{∞} functions on $\mathbb{R}$. The further assumption, taken for granted in this argument, is that in a "smooth world," any functions that are *definable* from functions already given in such a world must also be smooth, in brief, such worlds are *closed under definability*. Only when this is combined with (*) does it follow that LEM "fails in smooth worlds."

Now of course, one is free to stipulate whatever one likes as to the sort of worlds one wants to describe or investigate mathematically, and the fact that classes of smooth functions classically conceived (as arising in the topos of discrete sets, as category theorists would put it) are not closed under

---

[9] A further consideration reinforcing our assessment (pointed out by an anonymous referee) is that topos theory, the general background theory in which SIA and SDG and models theoreof are fully developed, is itself not a constructive theory.

definability does not prevent one from exploring "alternative worlds". But the question that then arises, and that a skeptic would pose, is whether there *are*, mathematically speaking, any such alternatives! Normally, if a closure condition leads to a logical contradiction, we conclude that the condition cannot be fulfilled (despite our wishes); we do not normally weaken our logic to accommodate it. (Thus, I cannot insist that the domain of an unbounded linear Hermitian operator with domain included in infinite-dimensional Hilbert space be extended to the whole space by any kind of adjunction of new elements. It is a theorem that any linear Hermitian operator defined on the whole space is bounded.) The assumption that there *are* smooth worlds in the stipulated sense is precisely what the adherent of classical logic will challenge, and that assumption is nowhere discharged in the argument but rather plays an essential role. As it stands, the argument begs the question.

Let us consider the second argument from smoothness. This has to do with "subobject classification" in smooth worlds, that is, with the machinery for determining, given for example, a collection of objects in such a world, what subcollections exist. Classical set theory handles this in a familiar way susceptible to generalization in topos theory, viz. by identifying a subcollection with its characteristic function, a mapping from the original collection to a two-element set coding the truth values, "true" and "false," typically taken as the set, $2 =^{df} \{1, 0\}$. The classical power set of a set X is thus identified with the object 2^X, the set of functions from X into 2. Now above we saw that the smooth line R is indecomposable, the only detachable parts being the empty set and R itself. This illustrates the problem with identifying the "truth-values object" Ω with the discrete set 2. In a smooth world, the only functions that could represent subobjects of R with this choice of Ω are constant, hence the only subobjects are trivial. To get around this, one must allow Ω to "expand" (ultimately to a Heyting lattice of open sets of a topological space, when this is formalized in category theory). This forces us to give up bivalence and LEM.

Unlike the first argument, this one does not appeal to definitional closure. It does appear in this case that the mere demand that continua have non-trivial parts together with classical logic and the restriction to smooth functions lead to contradiction. This is true but only when the latter requirement, the "restriction to smooth functions," is understood in a particularly strong sense, namely as applying not just to the mathematical subject matter of, say, smooth manifolds, but also to *the internal representation of logic* afforded by suitably rich mathematical machinery, available in set theory and topos theory. That is, *if* it is required that subobject classification be part of a smooth world rather than handled by logical machinery external to such a world, then indeed LEM cannot be correct in the internal logic if there are to be non-trivial subobjects of continua. But, a critic might ask, just because classical set theory is able to represent internally what we take to be genuine logical operations systematized in external systems and governing our ordinary reasoning in presenting SIA, why should we expect internal representability – in putative "smooth worlds" – of "genuine logic" in this sense, as applied to the subject of smooth manifolds? Put another way, how do we assure ourselves that there *are* "smooth worlds" in the sense of meeting the strong requirement that logic applicable to reasoning about its objects, including of course non-trivial subobjects, be internally representable, i.e., by smooth logical valuation functions? Topos models do show that, mathematically, "smooth worlds" exist but in a weaker sense: An internal "smooth" logic arises with a rich "truth-values" object, Ω, obeying the laws of intuitionistic logic, and subobject classification is highly non-trivial. However, no reference to external or "genuine" logic is involved in this characterization of "smooth world"; there is no guarantee that the "truth values" occurring in Ω are really worthy of the name, as opposed to merely fulfilling a formal role.[10] Nor is there any guarantee that the logical connectives and quantifiers of the internal logic genuinely correspond to, or represent, the connectives and quantifiers with which we actually reason externally with the notions of SIA, which, as we have said, is thought to stand on its own prior to topos modelling. In particular, topos modelling does not tell us whether the consistency established for SIA is merely a formal artifact of suppressing inferences that, from an external point of view, are still perfectly justified by prior meanings associated with the symbols, for example, 'or', 'not', '=', etc. Thus, once again, the argument from smoothness is inconclusive regarding the sense in which LEM is "given up" and regarding how a genuine "revision of logic" might be justified if that is really what is involved.[11]

An elementary example of toposes with "extra truth values" leading to intuitionistic logic is instructive (see Mac Lane [1986], pp. 402–404). This involves "sets varying in time" (functions between pairs of indexed sets, e.g.,

---

[10] An analogy may be helpful here. In presenting semantics for paraconsistent logics, one has recourse to "truth-value gluts," assignments of "both T and F" to sentences, as well as gaps. But one can use such a semantics for formal logical purposes and even claim that this is necessary for a good systematization of certain informal reasoning patterns – i.e., one can "be a para-consistentist" – without becoming a full-fledged "dialethist" who believes that, e.g., paradoxical sentences are genuinely both true and false.

[11] It should be noted that there is significant variety among topos models, e.g., Archimedean and non-Archimedean, and, among the latter some admitting invertible infinitesimals over and above nilpotents (cf. Bell [1998], pp. 106, ff.). Especially relevant to questions of logic is the fact that some topos models actually respect LEM for (closed) sentences, i.e., the propositional law $p \vee \neg p$ is validated, although instances involving free variables are not, e.g.,

$$\forall x \forall y [x = y \vee x \neq y],$$

as they cannot be, the negation being provable. Such models reinforce our points above on the inapplicability of intuitionist meaning analysis to SIA, for intuitionistic meanings lead decisively to rejection of propositional LEM as well as instances with free variables.

X_0 and X_1, Y_0 and Y_1, as the objects, with certain pairs of functions running from the X_i and the Y_i, as the morphisms), and the "internal truth values" just correspond to more possibilities of "membership" than usually arise simply because a set may "change from one time to another." Those internal truth values, just three in this example, are even given temporal names, "now," "eventually," and "never," and it is crystal clear that there is no conflict whatsoever with an external description whose background logic is classical, with the familiar two truth values. Closer to home in the present context of SIA, consider the relation of the internal notion of "truth" to external of a well-adapted topos model **E** of SIA (described by Bell [1998], pp. 115–116). As Bell puts it, "If we examine the meaning of any statement of the internal language of **E** containing a variable over R, we find that it is true in **E** if and only if the corresponding statement in **Man** (the category of classical manifolds) is, in addition to being true for all points of $\mathbb{R}$, also *locally true* for all smooth maps to $\mathbb{R}$." For example, a smooth map $f: M \to \mathbb{R}$ (M a classical manifold) may have $f(a) = 0$ but not be constantly 0 on any neighborhood of a, so that neither $f = 0$ nor $f \neq 0$ is locally true, nor therefore their disjunction. Clearly, there is complete compatibility here between the "non-classical" internal logic's treatment of the formula and the external classical treatment. These examples are thus unproblematic. The problematic case is where the external reasoning involves the highly non-classical nilsquares object $\triangle$, i.e., reasoning of SIA prior to topos modelling. It cannot be classical on pain of contradiction, but topos modelling in itself does not tell us why we should not think that, for example, what Equation (1) negates is "really true," even though formally SIA is cleverly devised to avoid recognizing it. (The cartoon character who walks off the cliff with his eyes staring straight ahead never falls!)

Now the alert reader may well object to this objection, asking "Doesn't all this reference to 'genuine logic' conflict with the Carnapian pluralist stance espoused above?" Not if it is properly understood, I would say. Logical particles (connectives, quantifiers, etc.) in use have meanings (surely a good, Carnapian assumption, one that even Quine had trouble questioning (cf. Hellman [1986], reproduced as Chapter 11 in this volume), and, to a large extent, these determine which rules are sound. Tolerance allows one to use different particles with different meanings (perhaps determined by stipulated rules which, of course, may differ from the ones we are used to), and tells us then to evaluate the results in light of the relevant goals. But tolerance has its limits. We do not tolerate, for example, mistakes in subtraction by accountants on tax forms even if defended by appeal to some new rules governing '–' (perhaps even when they work out in our favor)! Either we say there is a contradiction with established usage, or we find a new meaning for the new usage which resolves the contradiction under a translation back to the old, in which case we fire the accountant for a confusing and misleading choice of symbols, something even worse than a simple error in subtraction (for which we know how to check)!

In the present case, the problem is to reconcile Equation (1) with what we normally understand by the symbols involved, especially '='. If we are not using the (other) logical symbols constructively, how can what Equation (1) negates fail to hold, whatever the objects in the range of the quantifier might be? Once we are speaking of *objects x, y, at all*, are we not committed to (a) its making sense to write '$x = y$' and '$x \neq y$' and to (b) the truth of their disjunction, that one or the other is in fact the case whether or not we have any method of ever determining which it is? An affirmative answer is captured in Quine's famous slogan: "No entity without identity!" Even if we are only entertaining objects as mere possibilities, we seem committed to (a) and to a modal form of (b), that the disjunction *would hold* in any situation in which the object(s) existed. Let us call the conjunction of (a) and (b) (or its modal version) the "Determinateness of Identity," DI. SIA accepts (a) but negates (b), which is Equation (1).[12] A substantive interpretation of SIA (transcending formalism) must make sense of this remarkable exception to DI. In the next section, we take up some strategies for doing this.

15.4   Reinterpretative Strategies: Does '=' Mean Identity?

Even without constructivist readings of logical vocabulary, there are other routes leading to truth-value gaps thereby limiting bivalence and the LEM. Predicates may, for example, have limited domains over which they are well defined so that, outside such a domain, atomic sentences and their negations formed with such predicates are taken as "meaningless" in the sense of simply lacking truth value (ordinary "true" and "false"). Ordinary usage is a source of many examples, for instance, "My cat is (or is not) > 17," "The number 7 is (or is not) pink," etc. Vagueness offers another common route: "This pebble [near the base of Mt. Kilimanjaro] is (or is not) part of the mountain" is naturally taken as indeterminate, simply because prior usage has never reflected the conventions that would be required for such refinements. In the first sort of cases – limited domains of definition – while there is indeed a sense of (total) exclusion negation which renders the negative cases of such examples true, we also make room for "choice negation" leading to gaps, especially in the colloquial formulations given. (To call attention to exclusion negation, we

---

[12] In the sketch of SIA above we called what Equation (1) negates the "decidability of identity," which is familiar terminology in constructive mathematics based on intuitionistic logic. But at this point, we do not want to identify the logical connectives of SIA with those of intuitionism or constructivism, at least semantically speaking. Thus, "determinateness of identity" seems a more appropriate term to use.

would naturally appeal to the formulation, "It is not the case that 7 is pink.") In the second sort of cases – vagueness – even appeal to exclusion negation is insufficient (for we simply do not think it is determinate e.g., whether or *not* the pebble is part of the mountain), and to restore bivalence we have recourse to "definitely" as in "That pebble is not definitely part of (or not part of) the mountain," but of course now we are dealing with a different predication. In any case, these familiar sources leading to relaxation of bivalence and LEM remind us that classical logic is an idealization. Bivalence is best viewed as a *presupposition* of applicability of classical systems, not as an assertion about any particular applications (or even an assertion that such applications exist!); and when the presupposition is met, LEM_{Cl} of course is satisfied. When it is not met, there is then the challenge to come up with a satisfactory logical system that works. In the case of vagueness, especially, this is a daunting one, full of complexity and plagued by paradox (the *sorites*), and many proposals and theories have been put forward. In the present context, however, we have a system, SIA, which is precisely formulated and which works beautifully (on its own terms). Our question here and in Section 15.5 is whether its logic – obtained from classical logic by simply omitting LEM and suitably weakening the order axioms – can be better understood in light of these familiar phenomena leading to truth-value gaps.

The key relations affected by the weakening of the logic are the ordering, $<$, and, of course, $=$. Concentrating first on $<$, note that nilsquares are not ordered with respect to 0 or with respect to one another. We cannot prove of any hypothetical nilsquare ε that $\varepsilon < 0$ or that $\varepsilon > 0$ (written as a notational variant of $0 < \varepsilon$), for in either case we would have $\varepsilon^2 > 0$, a contradiction. (An axiom insures that $<$ is irreflexive, so $<$ implies $\neq$.) Thus we *do* prove $\neg(\varepsilon < 0)$ and $\neg(\varepsilon > 0)$. Thus, these formulas do not present examples of truth-value gaps. Likewise with formulas such as $\varepsilon < \delta$, for arbitrary elements of Δ. (If $\varepsilon < \delta$, then it would be provable that $\varepsilon < 0 \vee 0 < \delta$, whence $0 < \varepsilon^2 \vee 0 < \delta^2$, contradicting the definition of Δ. Thus we actually prove $\neg(\varepsilon < \delta)$ for arbitrary ε, δ in Δ.) It follows that non-identity inside Δ is always *refutable*, i.e., SIA proves

$$\forall \varepsilon, \delta \in \Delta \ \neg(\varepsilon \neq \delta)$$

(where $\neq$ as usual abbreviates $\neg(\_ = \_)$) but, lacking the law of double negation, this does not collapse to $\varepsilon = \delta$. Thus, we have the curious situation that non-identity within Δ is determinately false, whereas identity remains indeterminate, as reflected in Equation (1). Still, the notion that predicates have limited domains of application may help answer the question *how it can be that parts of a linear continuum are not ordered*. The notion of linear ordering applies to points any two of which mark a *finite* separation from one another. It is no violation of entrenched *a priori* principle to withhold this ordering from infinitesimals which are conceived as *not* separated by any finite distance, no matter how small (a fact that can be expressed and derived in SIA with a natural numbers object, cf. Bell [1998], Exercise 8.1, p. 108).

This brings us back to the real culprit, =. How can it be that determinateness, DI, fails if we are speaking of objects at all and if '=' really means 'is identical with'? We cannot simply restrict the range of application of =. Even SIA proves '$\forall x(x = x)$,' for example, including '$\forall \varepsilon \in \Delta(\varepsilon = \varepsilon)$'. What we do do, however, is restrict the range of *true* application of $\neq$. In SIA, this can only obtain for objects which are finitely separated. This includes elements of Δ in relation to other elements of R, for example, we do have $\varepsilon \neq 10^{-40}$, $\varepsilon \neq 10^{-40} - \delta$, etc. (for arbitrary ε, δ in Δ). But as already noted, we never have $\varepsilon \neq \delta$ or $\varepsilon \neq 0$, and indeed have the negations of these, but without that implying that they are "one and the same object"! How can this be? Perhaps, although '=' really means 'is identical with,' '$\neq$' really means (or should be interpreted as) something other than 'is not identical with.' Bell in his presentation seems to suggest this, calling two "points," a, b (in an extended sense of "point" to allow for infinitesimals) "distinguishable" when they are "not identical" ($\neq$), and calling them "indistinguishable" in case $\neg(a \neq b)$, so that the latter does not imply that they are identical, $a = b$, in conformity with giving up the law of double negation. We do prove, for arbitrary ε in Δ, that $\neg\neg(\varepsilon = 0)$, so that nilsquares are, according to this usage, "indistinguishable from 0." This is an attractive reading, but, unfortunately, it is problematic. For, since the combination '$\neq$' is not a new primitive but just abbreviates '$\neg(\_ = \_)$', it suggests a special, non-classical meaning of '$\neg$', along constructivist lines. But we have already seen the problems with attempting to assign constructivist meanings systematically to the logical symbols of SIA, and it is unclear what anything short of a systematic constructivist interpretation would look like and whether it could be made coherent.

It is worth mentioning that, in intuitionist analysis, a new "distinguishability" relation, $\sharp$, is introduced (defined), read "apartness," applied to choice sequences eventually bounded away from each other (by a finite amount, of course). But this has no bearing on the present case of infinitesimals in SIA, which of course have nothing to do with sequences of rationals. Moreover, it turns out that, in intuitionist analysis, one proves that $\neg(x \sharp y) \rightarrow x = y$. (This follows since reals x, y are taken as (equivalence classes of) Cauchy sequences of rationals and the ordering of rationals is decidable, cf. Dummett [1977], pp. 40–41, Theorem 3.) Again, the ideas of intuitionist analysis are not helpful in interpreting SIA.

Perhaps '=' could be interpreted as some equivalence relation broader than strict identity. That would still leave the problem of making sense of the theorem (1), but perhaps that would be possible along the lines of "truth-value gaps," more so than in the case of strict identity in that there is nothing in the mere supposition that we are speaking of objects (in the plural) that requires them to fall within the range of application of "not $\approx$," i.e., the idea of choice negation might hold up more intuitively than in the case of strict identity. Or perhaps theorem (1) and non-classical theorems like it need not be upheld at all. Carrying out such an idea, however, can be expected to be a complicated affair, since elsewhere in SIA, especially in recovering classical theorems, presumably '=' should retain its role as true identity.[13] It is interesting therefore to note that a systematic treatment of SIA along these lines has actually been worked out in detail [Giordano, 2001]. This is, in fact, a *classical approximation* to SIA in which the set D of nilsquares (the surrogate for Δ) is introduced via the definition:

$$D = \{h \in \mathbb{R}^{\bullet} | h \approx 0\},$$

where $\mathbb{R}^{\bullet}$, the extended (classical) reals, is introduced as a quotient of sequences of continuous functions from $\mathbb{R} \rightarrow \mathbb{R}$ under an equivalence relation $\sim$ (viz. that for some m and all $n \geq m$, $(x_n(t) - y_n(t))/t \rightarrow 0$ as $t \rightarrow 0$)) and $\approx$ is a further equivalence relation on ($\sim$ – equivalence classes of) such sequences, viz.

$$\exists m \forall n \geq m : \lim_{t \to 0} \sup \left| \frac{x_n(t) - y_n(t)}{t} \right| < +\infty$$

read "x is close to y." $\mathbb{R}^{\bullet}$ is a ring and D is an ideal in it satisfying $h^2 = 0$. (Remember here = is in the quotient space.) Based on these constructions, a restricted version of the KL Axiom is derived as a theorem. There are, however, a number of differences from SIA.

1. The principal restriction in deriving the analogue of the KL Axiom is to a class of functions satisfying a Lipschitz condition (not a very severe restriction). In this case, the unique "slope" b of the microtangent to f at 0 given by the KL Axiom is proved to coincide with the derivative of f at 0 in the usual sense, and this generalizes to arbitrary arguments x of f in $\mathbb{R}$ (cf. Giordano [2001], Theorem 1.4, pp. 81–82).
2. Instead of the Microcancellation Principle there is a stronger cancellation principle: For x in $\mathbb{R}^{\bullet}$, $x \neq 0$, and r, s in $\mathbb{R} : x \cdot r = x \cdot s \Rightarrow r = s$.
3. Unlike SIA where for ε, δ in Δ we cannot prove $\varepsilon \cdot \delta = 0$, and indeed can refute the universal generalization, in Giordano's reconstruction, for h, k in D, we do have $h \cdot k = 0$.

---

[13] Indeed, if we naively try reading '=' as '= or only infinitesimally different', in accordance with the idea that '$\neq$' behaves like 'finitely separated,' we run into the problem that now, at least in the presence of the Archimedean axiom (with which SIA is compatible) double negation should hold, i.e., we should have $\neg\neg(\varepsilon = 0) \Rightarrow \varepsilon = 0$, and SIA would become inconsistent.

4. An ordering $\leq$ on the extended reals is introduced which is antisymmetric (unlike SIA, where $a \leq b$ means '$\neg\, b < a$', and where, for ε in Δ, we have both $\varepsilon \leq 0$ and $0 \leq \varepsilon$ but of course cannot prove $\varepsilon = 0$).
5. Most fundamentally, since the logic is classical, no theorems formally inconsistent with classical mathematics, such as (1), arise.

Note that this latter point holds regardless of what '=' means in (1). LEM is still available even where an equivalence relation is what matters in relating nilsquares to 0. That is just what one would expect from a classical reconstruction.

Because of these and related differences from SIA, we should not regard this as an "interpretation" of SIA but rather as a classical reconstruction or approximation thereof. Certainly it is, in any case, interesting in its own right, for it shows that the key idea of SIA of doing analysis and differential geometry with nilpotent infinitesimals is indeed coherent even from a wholly classical perspective. Although Giordano's reconstruction is abstract and somewhat elaborate, topos models are also abstract, built on certain function classes obeying ring axioms. Although the basic ideas of SIA are intuitive and straightforwardly presentable without such models (as Bell's [1998] *Primer* demonstrates), they are invoked to allay doubts about formal consistency that naturally arise due to the non-standard logic and axioms. Still, it seems clear that no classical reconstruction or approximation can itself answer our question of interpretation of the original, genuine article.

15.5   Vague Objects?

This brings us to consider some recent philosophical literature on vagueness, specifically the debate over whether it even makes sense to speak of inherently "vague objects," objects whose identity conditions are "objectively indeterminate," that is, indeterminate but in such a way that the source of indeterminateness cannot be located in our linguistic descriptions or other symbolic modes of representation.

A strong tradition going back at least to Russell [1923] has held that objective indeterminateness of identity really makes no sense, that all vagueness of identity statements must be traceable to vagueness in language. If there is indeterminateness of "Kilimanjaro = Kilimanjaro + that small pebble at the base of the mountain," it is because the precise boundaries of the designator 'Kilimanjaro' are not fixed by anything established in usage, a common form of linguistic vagueness.[14] We can also say that "part of a mountain," as well as "mountain" itself, are vague terms. Indeed, most would agree that "small,"

---

[14] For an interesting discussion of such examples and their ubiquity, see e.g., McGee and McLaughlin [2000].

"pebble," "base," i.e., all the terms in that sentence – with the exception of '='
itself – are vague, as are most terms in most ordinary sentences, although of
course the indeterminateness of any particular sentence may be due to vagueness of some terms occurring in it rather than others. (In this case, even if we
pretend that "small pebble" is absolutely precise, the indeterminateness is
unaffected.) But identity itself is thought to be different. What counts as an
object may well be indeterminate in certain ways, but once we are speaking of
objects at all, according to this "standard view," we are *ipso facto* committed to
recognizing objective facts as to non-identity of one from another. Ordinary use
of the plural commits us to "more than just one," and, although we may have no
method whatsoever of saying "*how* many" and indeed may not even make
sense of that question, still the idea of "many" carries with it the idea of cases of
$\neq$. Thus, we may not make sense of "How many mountains are there in the
Himalayan range," but we insist that the problem of counting is due to the
vagueness of the terms "mountain," and "Himalayan range"; once that is
(sufficiently) resolved, however arbitrarily, cases of $\neq$ will be determinately
true and enable us to answer the question (perhaps better, " a related question").
And even if, as in set theory, we recognize no cardinality at all of a mathematically precise notion, for example, of "all von Neumann ordinals," still we
recognize "many" cases of $\neq$ (indeed, as one says, "too many"!).

In this tradition, there is even a(n) (in)famous *proof*, due to Gareth Evans
[1978], that there cannot be "vague objects," i.e., that indeterminateness of =, if
treated objectively, independent of linguistic sources, leads to contradiction.
Using ∇ as a sentential operator meaning "it is indeterminate that" and
λ-abstract notation to designate properties expressed by formulas, the argument
assumes the following, given designators a, b.

1. $\nabla(a = b)$	Hypothesis
2. $\lambda x[\nabla(x = b)](a)$	1, λ-abstraction
3. $\neg\nabla(b = b)$	Law of self-identity
4. $\neg\lambda x[\nabla(x = b)](b)$	3, λ-abstraction
5. $\exists P[P(a)\wedge\neg P(b)]$	2, 4 Existential generalization
6. $\neg(a = b)$	Leibniz principle (non-identity of discernible)

6 contradicts 1, that "$a = b$" is indeterminate. In words, if the latter, then the
property "indeterminately = b" holds of a but not of b, so this property
discriminates a from b, so they are determinately non-identical after all.

Now if all vagueness is due to language, to the imprecision with which our
symbols attach to parts of reality, then this reasoning can be blocked by
pointing to a fallacy in passing from 1 to 2 associated with a kind of referential
opacity – failure of substitutivity of equals preserving truth value – of contexts

involving ∇, "it is indeterminate that ... " This prevents us from moving from a sentence of the form

$$\nabla \ (...a...)$$

to one of the form

a is such that $\lambda x[\nabla \ (...x...)]$ holds of it,

as symbolized in line 2. Lewis [1988] diagnoses this fallacy pointing to an analogy with "it is contingent that" in modal logic. We cannot infer from, for example, "It is contingent that the number of planets = 9" that "The number of planets is such that it is contingent whether it (that number) = 9." For another analogy, consider the operator, "It is uncertain whether ... " (U). Under a natural reading of this, we have (adapting a nice example of referential opacity of Quine's),

$$U[\text{Tegucigalpa} = \text{the capital of Nicaragua}]$$

but surely not

$$U[\text{The capital of Honduras} = \text{the capital of Nicaragua}]!$$

On the "linguistic doctrine" of vagueness, the Evans argument is blocked by noting the opacity of contexts created by ∇, arising from sensitivity to the particular information conveyed by the designating terms employed. On the other hand, if we are dealing with purported cases of entirely objective indeterminateness, there would seem to be no fallacy in the argument.

However, in making a persuasive case for the cogency of objective indeterminateness of =, Parsons and Woodruff [1996] find that the principle of λ-abstraction combined with EG as in step 5 in the Evans argument is essentially question begging. If, as they argue, objective indeterminateness of identity is coherent, then we simply cannot assume that λ-abstraction leads to designation of a genuine property open to existential quantification as in step 5.[15] They go on to develop a hypothetical example of how this could fail, and then they show how the linguistic doctrine surprisingly leads to the counterintuitive conclusion that virtually no ordinary designation can be definite. We need not examine their example and arguments here. Rather we want simply to suggest that SIA itself actually provides a striking example of the possibility they promote and nicely illustrates their criticism of the Evans argument.

---

[15] They also examine the original presentation by Evans which appears to bypass reference to properties, moving directly from lines 2 and 4 to 6, citing "Leibniz' Law," which Parsons and Woodruff take to be the perfectly correct "indiscernibility of identicals": from '$a = b$' and '$...a...$' to infer '$...b...$'. Here, as they point out, the problem is that it is really the *contrapositive* of this law that is needed, and this is not generally available when there are truth-value gaps, so that, on this presentation as well, the question is begged.

Consider that argument in the context of SIA. As presented, it cannot even get started for the simple reason that it presupposes individual designators a, b, whereas in the case of nilsquares, there are no such designators (with the sole exception of terms for 0)! As mentioned in our overview above, reasoning with nilsquares never invokes terms for particular items in Δ (other than 0) but uses variables ε, δ, etc., for *arbitrary elements of* Δ. Precisely because identity "inside Δ" is indeterminate, there can be no basis in the SIA framework for introducing individual designators for infinitesimals. That framework is a radical enough departure from ordinary usage so that Evans' starting point simply begs the question.

Now it might be suggested that the Evans argument be reformulated so as to use variables only. Here one must be careful. We cannot start by presupposing an existential claim such as,

$$\exists \varepsilon, \delta \text{ in } \Delta \ [\Phi(\varepsilon, \delta) \wedge \varepsilon \neq 0],$$

as this is refutable in SIA, as we have seen, as is

$$\exists \varepsilon, \delta \ \neg[\varepsilon = \delta \vee \varepsilon \neq \delta \ ].$$

The closest we come is the SIA theorem,

$$\neg \forall \varepsilon, \delta \ [\varepsilon = \delta \vee \varepsilon \neq \delta],$$

but this cannot lie behind step 1 in the Evans argument even if read with variables. We are not here requiring that SIA prove 1, which it cannot for lack of the operator ∇. We are allowing that the argument takes place formally outside SIA. Still its steps have to be grounded in the theory. Suppose then, that we simply allow step 1 as it stands but treat a, b, as variables (writing u, v), so that the whole argument would establish the universal conditional,

$$\forall u, v[\nabla(u = v) \rightarrow \neg(u = v)],$$

whence,

$$\forall u, v[\nabla(u = v) \rightarrow \neg(\nabla(u = v))],$$

i.e.,

$$\forall u, v[\neg(\nabla(u = v))].$$

Even on this formulation, however, we surely distort the theory rather badly at step 2 when we introduce '$\lambda x[\nabla(x = v)]$' and reason as though this picked out a property that can be predicated of a particular nilsquare (existentially generalized in step 5). Indeed, if λ-abstraction were not restricted, we would be committed to properties corresponding to terms such as $\lambda x(x = v)$, individual properties ("haecceities") which are just as problematic in SIA as individual designation (within Δ apart from '0'). λ-abstraction in this situation is highly suspect, for, as we have seen, in SIA we treat nilsquares entirely generically. Precisely because of indeterminate identity conditions, we refrain from recognizing properties that would imply non-identity and lead to contradiction. Thus, SIA actually provides a fairly clear example from mathematics itself of a logical possibility Parsons and Woodruff consider in the abstract.

To illustrate further SIA's radical departure from classical set-theoretic ideas, especially relevant to talk of "properties" of its objects, consider the question of *the cardinality* of Δ. Here our intuitive ideas and pictures can be misleading. Depicting or thinking of the microneighborhood Δ as a tiny stretch about 0 leads us to expect it to be uncountable, like an ordinary (finite) stretch. But the concept of "cardinality," of "how many," is intimately bound up with the concepts of '=' and '$\neq$', and as we know these are not treated standardly in SIA. In a topos setting, as in set theory, the cardinality of a collection C corresponds to the totality of maps from a terminal object 1 to C. (By definition, an object 1 is *terminal* in a category just in case, for any object O in the category, there is exactly one map from O to 1. In the topos **Set** any singleton is terminal.) Applying this to Δ, we have at least one map $f: 1 \to \Delta$ since 0 may be assigned. Are there any others? If $\exists g : 1 \to \Delta$ and $g \neq f$, then we infer $\exists \varepsilon$ in Δ such that $\varepsilon \neq 0$, which as we know leads to contradiction, i.e., we have proved

$$\neg[\exists g : 1 \to \Delta \ \wedge \ g \neq f],$$

whence

$$\forall g : 1 \to \Delta \ \Rightarrow \ \neg(g \neq f),$$

which we can take to express "f is *weakly unique*," correspondingly that "Δ is weakly of cardinality 1." But we cannot go on to infer "strong uniqueness," i.e.,

$$\forall g : 1 \to \Delta \ \Rightarrow \ g = f,$$

or that "Δ is strongly of cardinality 1." Thus, we can say that, in a weak sense, Δ has cardinality 1, but in a strong sense, it is indeterminate as to cardinality. In the strong ("un-double-negated") sense, the classical dichotomy, "Either a non-empty totality has exactly one member or it has more than one," is not available in SIA.

The situation may be summed up by saying that SIA replaces traditional, vague talk and imprecise treatments of "infinitesimal quantities" with a precise treatment of vague objects. (In this it is reminiscent of the old caricature of Impressionism as "a clear picture of a fog.") Limitations on the application of LEM arise from the fact that an entity, Δ, posited in the theory is conceived as consisting of entities indefinite in =-conditions, as in principle not susceptible of any particular, *individual* reference or property ascription whatsoever. Ordinary linguistic or other symbolic resources are simply *stipulated* to be inapplicable to "items in Δ," including even this oblique use of a plural construction were it to be taken literally in its ordinary sense. Yet, remarkably, SIA manages to permit *quantification* over Δ without ever implying any disallowed =-conditions, and the results are actually interesting and useful! Its surprisingly simple rules manage to allow it to say just enough – but not too much!

15.6   The Bearing of Structuralism: Peaceful Coexistence

Although we lack a clear resolution of the apparent conflicts between SIA and classical mathematics by appeal to constructivist meanings, our discussion suggests quite a different resolution along structuralist lines. *Despite the large measure of shared terminology, axioms, and applications, SIA and CA are really postulating and theorizing about distinct mathematical structures, each having good claims to model aspects of our experience of continuity and continuous phenomena.* Structural descriptions of classical continua are very familiar (carried out in set theory, or in a second-order (modal) logic, or in category theory). But structures for SIA (and SDG) can also be described, most readily in the terms of topos theory, and this in turn can be given a modal-structural description [Hellman, 2003]. Non-classical objects arise, viz. micro-neighborhoods, obeying special axioms and restricted logical rules, viz. the KL Axiom and the limitations on LEM reflecting indeterminateness of =, intrinsic vagueness of infinitesimals. The key points of this structuralist conception are the following.

1. The axioms of SIA, like those of CA and for other structures in mathematics generally, serve a dual function. They are of course the starting points of proofs, but also they are *defining conditions*, stipulating the sort of mathematical structure or structures the theory purports to investigate. As individual "statements," the axioms need not be actually true (as Hilbert insisted, against Frege's traditional pre-structuralist "direct description" view of axioms); it suffices if they are coherent. At most what must be true are claims of coherence, or the possibility of satisfaction, hence consistency, of the system as a whole. Furthermore, the axioms need not be "analytic" in the sense of "guaranteed true entirely by meanings of constituent symbols."

2. The framework of category theory is especially useful in describing "smooth structures" for SIA, as its "arrows only" formulations (without set-membership) characterize objects only "up to isomorphism," bringing out their functional-structural roles in abstraction from the intrinsic nature of objects. Indeed, questions about "the nature of nilpotent infinitesimals" can be completely bypassed in favor of an algebraic representation of the maps of $R^Δ$. And, as already explained, the appropriate, restricted logic

arises naturally through topos-theoretic generalization of classical subobject classification.

3. A modal-structural (MS) presentation of all this would distinguish *internal* satisfaction conditions, based on intuitionistic logic, from *external* logic, which may remain classical. Utilizing the algebraic representation alluded to in 2, one thereby avoids saying things like, "It is possible that there are nilsqures other than 0 itself." This would actually land us in contradiction. Rather one claims that a structure of the appropriate type, $\langle R, \Delta, <, 0, 1 \rangle$, obeying the axioms and rules of SIA, is possible. Thus, talk of nilpotent infinitesimals as "merely potential entities" is not to be understood in the way in which we understand talk of mathematical structures generally, for example, the classical continuum, as "merely possible."

The first of these points has been set out in detail elsewhere and need not be elaborated upon here.[16] It should be stressed, however, that, on this Hilbertian structuralist approach, *stipulations* which would be quite out of place in a direct description of actual phenomena, are perfectly in order. If, for instance, SIA were taken as a theory of the actual constitution of ordinary space, the burden of justification would be far heavier than on a structuralist approach, for one would confront such questions as, "Why should we believe there actually are such things as nilsquares, not identifiable with 0 but at the same time not $\neq 0$?", or "Why should we think space or time or space-time actually contains objects that are in principle indistinguishable from one another, not even not $=$ to one another, but still are somehow 'many'?" If all one need care about in pursuing the mathematics of SIA is the possibility or coherence of entertaining of such things, this being claimed sufficient for the theory's applicability, then such questions can be bypassed.

Concerning point 2, the construction of topos models for SIA begins with the category **Man** of smooth manifolds (whose arrows are smooth maps between smooth manifolds), and adds to this enough structure to incorporate the mathematical role of microobjects such as Δ (speaking here in the language of SIA prior to modelling). This is accomplished through a representation of the "coordinate ring" R^Δ of functions from Δ to the (classical) real line, $\mathbb{R}$. This arises as follows. The KL Axiom induces a correspondence between pairs of reals (a, b) and maps $\varphi_{a,b} : \Delta \rightarrow \mathbb{R}$ defined by $\varphi_{a,b}(\varepsilon) = a + b\varepsilon$. The Axiom implies that this correspondence, φ, sending (a, b) to $\varphi_{a,b}$ is a bijection between $\mathbb{R} \times \mathbb{R}$ and $\mathbb{R}^\Delta$. Now the latter is a ring with the operations $+, \cdot$ defined pointwise in the expected way $((f + g)(\varepsilon) = f(\varepsilon) + g(\varepsilon)$ and $(f \cdot g)(\varepsilon) = f(\varepsilon) \cdot g(\varepsilon))$. If we now introduce operations $\oplus$ and $\odot$ on $\mathbb{R} \times \mathbb{R}$ via $(a, b) \oplus (c, d) = (a + b, c + d)$ and $(a, b) \odot (c, d) = (ac, ad + bc)$, respectively, then we have that the

---

[16] See for example, Awodey [1996], Hellman [1989a, 2003], also Hellman [1996], reproduced here as Chapter 1, and Shapiro [1997].

correspondence φ is a ring isomorphism between $\mathbb{R}^\Delta$ and the ring $(\mathbb{R}\times\mathbb{R}, \oplus, \odot)$. (Note that in the definition of $\odot$, the expected algebraic "fourth term" for bd in the expansion of $f(\varepsilon) \cdot g(\varepsilon)$ is "missing" since this is a term with $\varepsilon^2 = 0$.) Thus, the latter ring provides all the relevant information of $\mathbb{R}^\Delta$ and it, along with similar structures for coordinate rings of microobjects for other manifolds, can be added to the coordinate rings of manifolds to obtain an algebraic structure that behaves "as if" microobjects were present. (This is not yet a topos, but the further constructions for that need not concern us here. See Bell [1998], Appendix, for a sketch, and Moerdijk and Reyes [1991] for full details.) In this way, topos modeling sidesteps metaphysical questions about "the nature" of nilsquare and nilpotent infinitesimals. All that matters for mathematics, whether it is SIA or CA or whatever, are the "structural roles" of its objects, and topos modeling realizes this insight.

Concerning point 3, it might then seem that the apparatus of modal-structuralism is not needed in telling this story. However, as we have argued elsewhere [Hellman, 2003], there is no generally agreed upon starting point or background framework – with an explicit logic along with comprehension and other mathematical existence axioms – for category theory and topos theory, conceived as autonomous from set theory. Modal-structuralism provides such a background. It too lacks set-membership as a primitive, although the essential mathematical content of $\in$ can be recovered through a combination of mereology and plural quantification (providing the strength of second-order logic, hence a full theory of relations); as a result, the notions of "category," "morphism," "functor," need not be taken as primitive. Modality enters to avoid commitments to actual infinities, and to avoid commitments to maximal, inextendable totalities (avoiding thereby the embarrasments of proper classes, the "absolute infinite," etc.).

When it comes to non-standard mathematics such as SIA, however, care must be taken in applying modal-structural ideas. The point is elementary. In normal cases of classical theories, there is no need to distinguish internal from external logic, as the MS framework is fully classical (although certain crucial restrictions on modal-comprehension axioms must be observed). Thus, in asserting, for example, the possibility of structures for classical number theory or analysis, one can simply write out suitable (second-order) axioms characterizing (standard) structures for those theories, relativized to a hypothetical domain X along with "functions and relations" over X realizing the conditions expressed in those axioms. (Officially, functions and relations are explained away in the mereological-plurals reduction, but this justifies the use of function and relation *variables*.) That is, one asserts the possibility of the relevant structures in direct, mathematical-logical terms, without the semantic ascent involved in model theory via the use of "satisfaction." Now clearly, in the case of SIA, this procedure will not work, due to the restrictions on LEM built into SIA. That is, if we attempt to assert the possibility of an SIA structure in the usual, direct way, we would be saying in effect that it is possible there is a $\langle R, \Delta, <, 0, 1 \rangle$ obeying the KL Axiom, Integration Principle, etc., but this implies the possibility that $\neg \forall x [x = 0 \vee x \neq 0]$ which directly contradicts the (MS-demonstrable necessitation of the) classical logical theorem, $\forall x [x = 0 \vee x \neq 0]$! To avoid this, either we must restrict LEM in the background logic of MS, or we must enforce a distinction between external and internal logics by introducing *satisfaction* as a relation between structures and formulas of SIA, framed so as to respect intuitionistic internal logic. Call this relation $\models_{Int}$. Then, we end up saying (in fewer words): "An SIA-structure bears $\models_{Int}$ to '$\neg \forall x [x = 0 \vee x \neq 0]$' (NB the quotes!), even though in our classical background, everything is either equal to or not equal to 0," and there is no inconsistency in this.[17] This latter route is the natural one to take, since we are attempting to apply a general MS framework to a wide variety of mathematical theories, and we have no reason to impose logical restrictions in this general setting. Rather we are trying to assert the possibility of structures which themselves allow "indeterminate objects," reflected in limitations on LEM internal to those structures. But with these adjustments, this can be accomplished as indicated, and the result is only somewhat more complicated than the MS treatment of classical analysis.[18]

This leaves us finally with the question of what sense can be made of the notion that nilsquare and nilpotent infinitesimals "exist only in a potential sense." It should be clear from the last paragraph that this cannot be understood simply as "possible existence in a model of SIA." It is manifestly *impossible* that

$$\exists x [x \neq 0 \wedge x \in \Delta],$$

in a model of the SIA axioms! "Merely potential existence" cannot mean "possible existence in a model." What then can it mean? The best I can think

---

[17] There is no general license for *disquotation,* that is, we cannot in general pass from truth of a sentence in a model to truth *simpliciter* or truth in our background framework. A helpful analogy might be the Skolem "paradox": In our classical set-theoretic background we recognize both that "The classical reals are uncountable" holds inside a countable model of the relevant axioms *and* that those reals are really countable as seen in our background framework.

[18] There are some delicate logical questions about just what background strength is required to insure a model of SIA. For a topos model, the "theory of large domains" sketched in Hellman [2003] would certainly suffice, but much of the topos machinery is "overkill" for a mere model of the basic system of SIA (analysis on the smooth line R), where a fragment of second-order number theory ought to suffice. Whether the background logic actually needs to be classical, or whether intuitionistic logic would suffice, is a further question, certainly relevant to a better understanding of the relation between SIA and CA, but one that need not be resolved for our purposes in this paper.

of is simply "not excluded," where this in turn is spelled out as *the possibility that*

$$\neg\forall x[x \in \Delta \rightarrow x = 0].$$

This as we know is a consequence of the KL Axiom but, without LEM, it does not yield the existential formula. Well, that is all consistent, but the phrase "exists only in a potential sense" now plays no explanatory role at all, but merely refers to this logical situation.

In summation: a structuralist approach to SIA and SDG, through topos theory in a modal-structural setting sustains these theories as consistent alternatives to classical analysis and differential geometry and, moreover, does so while subtly sidestepping metaphysical puzzles about "the nature of infinitesimals" that threaten to derail these theories. In this arena as in so many others, the utility and flexibility of the Hilbertian conception of mathematical axiomatics proves its worth, even in the absence of combinatorial consistency proofs. There is, however, this compromise with respect to "geometric intuition": While it can certainly play a motivating and even guiding role in coming up with axioms (e.g., the KL Axiom (of Microaffineness)), when it comes to questions as to the coherence and consistency of the system as a whole, intuitions run out of steam, and we appeal to model-theoretic constructions. At the same time, questions as to "the nature of the objects" fall by the wayside. As in the case of complex numbers and Hamilton's reduction to $\mathbb{R} \times \mathbb{R}$ with specially defined algebraic operations, we bypass such questions in favor of a structuralist treatment. (At a deeper level, much the same can be said with regard to $\mathbb{R}$, $\mathbb{Q}$, and $\mathbb{N}$ themselves, as we have maintained.) In the end, the algebraic structures of topos models seem as far removed from geometric intuition of continua as does the set-theoretic continuum at the hands of Dedekind. But precisely because we are dissatisfied with appeals to intuition as a response to questions of consistency, these models assume an unexpected, but it seems, unavoidable importance. Ironically, this lesson of modern logic and mathematics is, if anything, reinforced by "synthetic geometry."

Acknowledgement: Support of this work by the National Science Foudation, Award SES-0349804, is gratefully acknowledged. I am also grateful to Steve Awodey, John Bell, Colin McLarty, Stephen Read, and Dana Scott for helpful discussion and correspondence, and to an anonymous referee for useful comments.

References

Awodey, S. [1996] "Structure in mathematics and logic: a categorical perspective," *Philosophia Mathematica* **4**: 209–237.

Bell, J. L. [1998] *A Primer of Infinitesimal Analysis* (Cambridge: Cambridge University Press).

Carnap, R. [1937] *The Logical Syntax of Language* (London: Routledge & Kegan Paul).

Dummet, M. [1977] *Elements of Intuitionism* (Oxford: Oxford University Press).

Evans, G. [1978] "Can there be vague objects?," *Analysis* **38**: 208, reprinted in Keefe, R. and Smith, P., eds., *Vagueness: A Reader* (Cambridge, MA: MIT Press, 1996), p. 317.

Giordano, P. [2001] "Nilpotent infinitesimals and synthetic differential geometry in classical logic," in Berger, U., Osswald, H., and Schuster, P., eds., *Reuniting the Antipodes–Constructive and Nonstandard Views of the Continuum* (Dordrecht: Kluwer), pp. 75–92.

Hellman, G. [1986] "Logical truth by linguistic convention," in Hahn, L. E. and Schilpp, P. A., eds., *The Philosophy of W. V. Quine* (LaSalle, IL: Open Court), pp. 189–205.

Hellman, G. [1989a] *Mathematics without Numbers: Towards a Modal-Structural Interpretation* (Oxford: Oxford University Press).

Hellman, G. [1989b] "Never say 'never'! On the communication problem between intuitionism and classicism," *Philosophical Topics* **17**(2): 47–67.

Hellman, G. [1996] "Structuralism without structures," *Philosophia Mathematica* **4**(2): 100–123.

Hellman, G. [2003] "Does category theory provide a framework for mathematical structuralism?," *Philosophia Mathematica* **11**: 129–157.

Kock, A. [1981] *Synthetic Differential Geometry*, London Mathematical Society Lecture Note Series 51 (Cambridge: Cambridge University Press).

Kock, A. [2003] "Differential calculus and nilpotent real numbers," *Bulletin of Symbolic Logic* **9**: 225–230.

Lawvere, F. W. [1979] "Categorical dynamics," in *Topos Theoretic Methods in Geometry* (Aarhus: Aarhus University), pp. 1–28.

Lewis, D. [1988] "Vague identity: Evans misunderstood," *Analysis* **48**: 128–130, reprinted in Keefe, R. and Smith, P., eds., *Vagueness: A Reader* (Cambridge, MA: MIT Press, 1996), pp. 318–320.

Mac Lane, S. [1986] *Mathematics: Form and Function* (New York: Springer).

McGee, V. and McLaughlin, B. [2000] "The lessons of the many," *Philosophical Topics* **28**: 128–151.

Moerdijk, I. and Reyes, G. E. [1991] *Models for Smooth Infinitesimal Analysis* (Berlin: Springer).

Parsons, T. and Woodruff, P. [1996] "Worldly indeterminacy of identity," in Keefe, R. and Smith, P., eds., *Vagueness: A Reader* (Cambridge, MA: MIT Press), pp. 321–337.

Russell, B. [1923] "Vagueness," *Australasian Journal of Philosophy and Psychology* **1**: 84–92, reprinted in Keefe, R. and Smith, P., eds., *Vagueness: A Reader* (Cambridge, MA: MIT Press, 1996), pp. 61–68.

Shapiro, S. [1997] *Philosophy of Mathematics: Structure and Ontology* (New York: Oxford University Press).

Index

Lightning Source UK Ltd.
Milton Keynes UK
UKHW022105240321
380951UK00005B/141

9 781108 494182